Ergonomics

This newly updated edition presents a global perspective on ergonomics, addressing its application across occupational, recreational, design, and technological environments. This comprehensive textbook reflects the technological advancements that have reshaped the field, offering revised chapters, the integration of previously separate topics, and the addition of two entirely new chapters. With an emphasis on cultural sensitivity and inclusive representation, the book takes a modern approach to ergonomics with diversity at the heart of it.

Readers will find a complete and ready-to-use resource ideal for both instruction and independent study. The book features real-world examples, case studies, and innovative applications that help bridge theory and practice. Students and professionals will benefit from enhanced accessibility to global perspectives, a wide range of ergonomics software, and detailed guides for research and project development. It presents readers with the tools to understand, teach, and apply ergonomic principles effectively in both academic and practical settings.

Ergonomics, Second Edition is designed for a broad audience including researchers and students in human factors, industrial and organizational psychology, sociology, biomechanics, medicine, and design disciplines. It is also an essential resource for human resource professionals, industrial engineers, ergonomists, psychologists, HCI specialists, educators, and architects.

This textbook also contains ancillary materials including PowerPoint slides and a solutions manual for qualifying textbook adoptions.

Ergonomics Design and Management: Theory and Applications

Series Editor

Waldemar Karwowski
Industrial Engineering and Management Systems University of Central Florida (UCF) – Orlando, Florida

Published Titles

Application of Systemic-Structural Activity Theory to Design and Training
Gregory Z. Bedny

Self-Regulation in Activity Theory:
Applied Work Design for Human–Computer Systems
Gregory Bedny, Waldemar Karwowski, and Inna Bedny

Organizational Resource Management:
Theories, Methodologies, and Applications
Jussi Kantola

Risk Analysis and Management of Repetitive Actions:
A Guide for Applying the OCRA System (Occupational Repetitive Actions), Third Edition
Daniela Colombini and Enrico Occhipinti

Knowledge Service Engineering Handbook
Jussi Kantola and Waldemar Karwowski

Neuroadaptive Systems:
Theory and Applications
Magdalena Fafrowicz, Tadeusz Marek, Waldemar Karwowski, and Dylan Schmorrow

Safety Management in a Competitive Business Environment
Juraj Sinay

Ergonomics in Developing Regions:
Needs and Applications
Patricia A. Scott

Human Factors of a Global Society:
A System of Systems Perspective
Tadeusz Marek, Waldemar Karwowski, Marek Frankowicz, Jussi I. Kantola, and Pavel Zgaga

Working Posture Assessment:
The TACOS (Time-Based Assessment Computerized Strategy) Method
Daniela Colombini and Enrico Occhipinti

ERGOCHECK for a Preliminary Mapping of Risk at Work
Tools, Guidelines, and Applications
Daniela Colombini and Enrico Occhipinti

For more information about this series, please visit: https://www.crcpress.com/Ergonomics-Design--Mgmt-Theory--Applications/book-series/CRCEDMTAPPS

Ergonomics

Foundational Principles, Applications, and Technologies

Second Edition

Pamela McCauley

CRC Press
Taylor & Francis Group
Boca Raton London New York

CRC Press is an imprint of the
Taylor & Francis Group, an **informa** business

Designed cover image: Shutterstock

Second edition published 2026
by CRC Press
2385 NW Executive Center Drive, Suite 320, Boca Raton FL 33431

First edition published 2011
by CRC Press
2385 NW Executive Center Drive, Suite 320, Boca Raton FL 33431

and by CRC Press
4 Park Square, Milton Park, Abingdon, Oxon, OX14 4RN

CRC Press is an imprint of Taylor & Francis Group, LLC

ISBN: 978-0-367-35948-5 (hbk)
ISBN: 978-1-041-10490-2 (pbk)
ISBN: 978-0-429-34280-6 (ebk)

DOI: 10.1201/9780429342806

Typeset in Times
by Deanta Global Publishing Services, Chennai, India

Ancillary materials including PowerPoint slides and a solutions manual for qualifying textbook adoptions can be accessed at www.routledge.com/9780429342806.

Contents

Preface

The second edition of *Ergonomics: Foundational Principles, Applications, and Technologies* is designed to serve as a comprehensive resource for educators, researchers, and practitioners seeking a global perspective on the application of ergonomics across occupational, recreational, and design environments. This edition expands on the foundational principles of ergonomics while integrating the latest advancements in the field, addressing emerging technologies, and incorporating a more diverse and inclusive range of examples to better reflect the international scientific community.

This textbook offers a turnkey instructional resource, equipping instructors with modern-day examples, suggested software applications, case studies, and exercises that are both technically rigorous and culturally sensitive. The inclusion of real-world scenarios ensures that students not only grasp theoretical concepts but also understand how to apply them in various settings, from industrial workplaces to product design and human-computer interaction.

Designed for flexibility, the text can serve as both an introductory and an advanced resource. It includes hands-on laboratory activities to complement classroom instruction, while the case studies highlight the practical application of ergonomic principles. Students completing a course using this book will be well-prepared to apply ergonomic knowledge in industry, contribute to the development of ergonomically sound innovations, or pursue further study in the field.

CONTENT OVERVIEW

This edition covers a broad spectrum of ergonomics topics, providing both foundational knowledge and insights into innovations shaping the field today:

- **Chapter 1:** Introduction to ergonomics, the history of the field, and relevant legislative agencies.
- **Chapter 2:** Overview of human systems and physiology, including sensory capabilities and environmental factors affecting performance.
- **Chapter 3:** Introduction to the systems of the body, the human senses and their interaction with environmental factors.
- **Chapter 4:** Fundamentals of muscular work and neural innervation of the muscular system.
- **Chapter 5:** The study of anthropometry and its role in ergonomic design.
- **Chapter 6:** Ergonomic considerations in work place and hand tool design, high-tech, mobile, and personal device design.
- **Chapter 7:** Understanding work-related musculoskeletal disorders—history, causal factors, and mitigation strategies.
- **Chapter 8:** Work physiology, including principles of heavy work, physical workload assessment, and fatigue.

- **Chapter 9:** Cognitive ergonomics, information processing, and principles for designing controls and displays.
- **Chapter 10:** The role of ergonomics in product liability and its impact on consumer safety.

With its blend of foundational principles and contemporary applications, this edition provides a well-rounded approach to ergonomics, ensuring that students and professionals alike gain the knowledge and tools necessary to enhance human well-being and performance in a rapidly evolving world.

Ancillary materials including PowerPoint slides and a solutions manual for qualifying textbook adoptions can be accessed at www.routledge.com/9780429342806.

About the Author

Dr. Pamela R. McCauley, Ph.D., C.P.E.

Dr. Pamela McCauley is an internationally recognized Industrial Engineering researcher, Certified Professional Ergonomist (C.P.E.), and expert in human factors, biomechanics, and ergonomics. With over 30 years of experience spanning academia, industry, and expert witness testimony, she has made significant contributions to the development of ergonomic principles in workplace safety, product design, and high-risk occupational environments.

As a leader in ergonomics research, Dr. McCauley has authored over 150 technical publications, including peer-reviewed journal articles, book chapters, and conference proceedings. She is the author of *Ergonomics: Foundational Principles, Applications, and Technologies*, a widely used textbook that provides an essential foundation in ergonomic science and its real-world applications. Her expertise in human-centered design, occupational biomechanics, and safety engineering has been instrumental in advancing ergonomic standards across industries.

In addition to her academic contributions, Dr. McCauley is a nationally sought-after expert witness, providing analysis and testimony in cases involving product liability, occupational safety, and human performance. Her work has influenced legal proceedings by applying scientific rigor to evaluating warnings, hazard communication, and workplace injuries.

Dr. McCauley's career has also included leadership roles in government, industry, and higher education. She currently serves as the Dean of the School of Engineering at Widener University and previously held prominent roles at the National Science Foundation and the Massachusetts Institute of Technology. Recognized for her contributions to STEM education and leadership, she has received numerous accolades, including induction into the Oklahoma Hall of Fame and the Black Engineer of the Year Hall of Fame.

As a passionate advocate for diversity and inclusion in STEM, Dr. McCauley continues to mentor the next generation of engineers, emphasizing the critical role of ergonomics in designing safe, efficient, and inclusive work environments.

About the Author

1 Foundational Ergonomics

1.1 LEARNING GOALS

The learning goals for this chapter are to provide a foundational understanding of the field of ergonomics, the evolution of ergonomics as a discipline, and the significant impact the field has had on improving working conditions on a global level. This chapter will provide the student with a basic understanding of ergonomics, its definitions, origins, benefits, and applications. The student will also learn about the historical foundations of ergonomics by studying the individuals and organizations that led to its development. Additionally, topics relevant to American ergonomic legislation and worldwide professional societies will be discussed and listed in the following section.

1.2 KEY TOPICS

- Definition of Ergonomics
- History of Ergonomics
- Types of Ergonomic Problems
- Legislative and Regulatory Organizations Impacting Ergonomics
- Ergonomic Societies and Resources

1.3 INTRODUCTION AND BACKGROUND

1.3.1 Ergonomics: An Evolving Discipline

"Ergonomics" emerges from the Greek words "*Ergo*" (work) and "*nomos*" (laws), signifying the science of work. This discipline focuses dynamically on the interaction between humans and their environment. While the field of ergonomics initially focused on the occupational setting, this has shifted as designers, users, and innovators saw the benefit of enhancing their environments, equipment, products, and processes for optimal user engagement. Thus, the field has broadened its reach, encompassing consumer product design, recreational spaces, personal spaces (such as our homes), and diverse processes that involve human interaction.

1.3.2 Historical Perspectives

The roots of ergonomics can be traced back to pioneers like Frederick Taylor (1856–1915), often hailed as the "father of modern ergonomics." Taylor's journey from a machine shop laborer to a chief engineer influenced his development of "scientific

DOI: 10.1201/9780429342806-1

management" or "Taylorism." Although his methods now attract criticism, akin to Fordism and other assembly-line production systems, they were revolutionary in their time. It is important to note that the negative connotations associated with Taylorism stem more from the exploitative working conditions of the era rather than Taylor's studies themselves.

Following in Taylor's footsteps, Lillian and Frank Gilbreth made significant contributions in the early 1900s (Gilbreth & Gilbreth, 1948). Their unique family life with 12 children became a testing ground for efficiency methods. These practices were later charmingly depicted in their children's book, *Cheaper by the Dozen*, illustrating the wide-ranging application of ergonomic principles beyond professional environments. Frank and Lillian Gilbreth were instrumental figures in the development of industrial engineering and ergonomics, merging engineering principles with a focus on human-centered design. Their primary contribution lies in *time and motion studies*, which analyzed the efficiency of human movements in work processes. Frank, a construction manager, began his work by identifying inefficiencies in bricklaying, reducing unnecessary motions to improve productivity. Lillian, a psychologist, contributed by considering the psychological and physiological impacts of work, advocating for worker well-being alongside efficiency.

The Gilbreths' most notable contribution to ergonomics is their development of the concept of *Therbligs*. Therbligs represent 18 basic units of motion in any manual task, such as *reach*, *grasp*, *hold*, and *rest*. By breaking down tasks into these elementary motions, they could systematically analyze, streamline, and optimize processes to reduce wasted motion and worker fatigue. This framework laid the groundwork for modern process improvement techniques, including lean manufacturing and Six Sigma, and became fundamental to the design of tools, workspaces, and tasks that accommodate human capabilities.

Historically, the Gilbreths were among the first to recognize the importance of ergonomics—optimizing work to fit the worker—and their work remains foundational in industrial engineering, safety, and human factors research today. By integrating a focus on both worker productivity and well-being, they significantly influenced modern ergonomics, which continues to focus on designing systems that prioritize safety, efficiency, and human needs.

Beyond traditional settings like offices or factories, ergonomic principles now enhance environments across the service, health, and recreational industries. The objective is to design environments that align with human needs. This "fitting the task to the person" approach is fundamental in ergonomic applications.

As a guiding framework, the International Ergonomics Association (IEA) defines ergonomics as a discipline concerned with the understanding of interactions among humans and system elements. It is a profession dedicated to applying theory, principles, and methods to optimize human well-being and overall system performance.

In this textbook, we delve into the physical ergonomic principles, tools, and techniques essential for effective ergonomic application in both practice and research.

1.3.3 Reflecting on Ergonomics' Past and Future

The history of ergonomics lays a robust foundation for understanding how the field is interpreted today. However, as we continue to explore and apply ergonomic principles throughout the text, it is crucial to maintain a balanced perspective, considering both the benefits and the potential drawbacks of early ergonomic methodologies and how they translate to present-day practices. This holistic approach ensures the continued evolution and relevance of ergonomics in improving human interaction with our ever-changing environments and innovations.

1.4 ERGONOMICS DEFINED

Human factors and ergonomics are closely intertwined fields dedicated to optimizing the interaction between humans and the systems they use. This understanding improves human well-being, performance, and safety in diverse environments, from workplaces and homes to vehicles and spacecraft. Let's explore these fields through the lens of some key organizations and their definitions.

TABLE 1.1
Periods of Ergonomic Growth and Specific Ergonomic Events

Year	Activity
1700	Bernardino Ramazzini publishes De Morbis Artificum Diatriba (Diseases of Workers)
1857	Wojciech Jastrzebowski, a Polish scholar, first coins the term "ergonomics"
1883	Fredrick Taylor applies ergonomics principles to improve factory productivity
1900–1920	Frank and Lillian Gilbreth begin their work in motion study and shop management
1945–1960	Ergonomics begins to flourish in post-World War II societies; "knob and dial" era is born
1949	Ergonomics Research Society (now known as Ergonomics Society) is established in Britain
1957	The race for space and staffed space flight begins; human factors becomes an important part of the space program
1957	The Human Factors Society (now known as the Human Factors and Ergonomics Society) is founded
1957	The IEA is formed
1960–1980	Rapid growth and expansion of human factors in the United States
1960	Membership in the Human Factors Society reaches 500
1970	OSHA and NIOSH are both created by the US Congress
1979	Three Mile Island incident has a major impact on the perception of a need for human factors in the workplace
1980	Membership in the Human Factors Society reaches 3,000
1980–2010	Human factors continues to be an area of rapid growth and is also becoming a household name with commercials stating "ergonomically designed" as a buzz word
2001	OSHA issues an ergonomics standard, which requires companies to address hazards likely to cause sprains, strains, and repetitive motion injuries

Source: Salvendy, G., *Handbook of Human Factors and Ergonomics*, John Wiley, 2006.

1.4.1 Human Factors and Ergonomics Defined at NASA

NASA has long recognized the critical role that human factors and ergonomics (HF/E) play in ensuring the success of space missions. The extreme and unique environments of space exploration present numerous challenges that require meticulous attention to human performance, safety, and well-being. Ergonomics in space is about more than comfort—it is essential to mission success and astronaut survival in conditions where even small errors can have catastrophic consequences.

1.4.1.1 Importance to the US Space Program

NASA's human factors research is focused on optimizing the interaction between astronauts and their equipment, spacecraft, and work environments. In space, astronauts face microgravity, confined spaces, and complex operational demands, which can lead to both physical and cognitive challenges. By applying HF/E principles, NASA ensures that spacecraft are designed to minimize physical strain, reduce errors, and support optimal cognitive performance during missions. For example, spacecraft cockpits, life support systems, and equipment are designed to align with human biomechanics, sensory abilities, and cognitive capacities, improving operational efficiency and safety (NASA, 2014).

One of the key areas where NASA's application of human factors has been vital is in habitability—the design of living and working environments that support astronauts' physical and psychological health. Spacecraft like the International Space Station (ISS) are designed with careful consideration of how astronauts move and interact with equipment in microgravity, with detailed attention to workspaces, sleep areas, and even exercise equipment. Proper ergonomic design helps prevent musculoskeletal issues, reduce mental fatigue, and ensure long-term mission sustainability.

Additionally, in emergency situations, quick and accurate decision-making is crucial. NASA's research into cognitive ergonomics—the study of how humans process information and make decisions—has been integral to the design of spacecraft controls, displays, and emergency procedures to ensure they are intuitive and efficient under high-stress conditions.

1.4.1.2 NASA's Contributions to the Wider Scientific Community

NASA's human factors research has not only benefited its space missions, but also the broader scientific community. Through its publications, technical reports, and design standards, NASA shares valuable insights gained from years of research and practical application. One such resource is the *NASA Human Integration Design Handbook* (HIDH), which provides detailed guidelines on the design of spacecraft systems that account for human capabilities and limitations (NASA, 2010). This document is a critical resource for engineers and designers working on space-related projects, offering comprehensive information on ergonomic principles as they apply to long-duration space missions, such as those involving the ISS or future Mars exploration.

Additionally, NASA releases findings through its Technical Reports Server (NTRS), which includes extensive research on human factors (NASA Technical

Reports Server, https://ntrs.nasa.gov/). Topics cover areas such as human-system integration, crew interface design, workload management, and habitability, providing a valuable body of knowledge for both space agencies and industries on Earth. For example, NASA's research into fatigue and workload management in space has been adapted to improve the safety of pilots, surgeons, and other professionals in high-risk occupations.

NASA's rigorous approach to integrating HFE principles is a testament to the importance of human-centered design in space exploration. By prioritizing astronaut safety, physical well-being, and cognitive performance, NASA ensures that human factors remain a cornerstone of its space program, driving innovation and safety in one of the most extreme work environments known to humanity.

1.4.2 CETENA (The Italian Ship Research Company)

CETENA defines ergonomics as "the scientific discipline concerned with the study of human performance in relation to the design of technological systems, tools, and products." Their focus is on enhancing productivity, safety, comfort, and quality of life through ergonomic principles. This includes areas like workplace design, environmental factors, physical workload assessment, and user interface design for increased usability.

While preventing injuries is a cornerstone of ergonomics, it is only one piece of a much larger puzzle. Ergonomics aims to optimize human well-being and performance across all aspects of the work environment. As the International Ergonomics Association aptly defines, it is about "fitting the task to the worker," not just preventing harm but also maximizing productivity and satisfaction.

This holistic approach extends beyond the physical realm. While ergonomics tackles key topics like optimal postures, workstation layout, and minimizing repetitive motions to prevent musculoskeletal disorders its reach extends far wider, encompassing cognitive and organizational ergonomics that delve into the mind and the fabric of an organization.

In practical terms, safety is the foundation, but ergonomics builds a thriving workplace on top of it. Consider the three main domains of ergonomics and how they contribute to this foundation of safety and optimal user engagement:

1. **Physical Ergonomics:** This domain focuses on the physical aspects of work, ensuring the environment and tasks align with human capabilities. It addresses issues like:
   - Work Postures—Designing workstations that support proper posture and minimize strain.
   - Materials Handling—Implementing safe and efficient techniques for lifting and carrying objects.
   - Repetitive Motions—Identifying and minimizing tasks that involve repetitive movements, potentially leading to musculoskeletal disorders.
   - Workplace Layout—Optimizing the physical layout of the work environment for efficiency and safety.

2. **Cognitive Ergonomics:** This domain ventures into the realm of mental processes, ensuring tasks are designed for optimal mental performance and well-being. It tackles issues like:
   - Mental Workload—Balancing task demands with workers' cognitive capacity to prevent fatigue and stress.
   - Decision-making—Designing systems that support informed and efficient decision-making processes.
   - Human–computer Interaction—Creating intuitive and user-friendly interfaces for seamless interaction with technology.
   - Training—Developing effective training programs that equip workers with the necessary skills and knowledge.
3. **Organizational Ergonomics:** This domain focuses on the organizational structure and processes, ensuring they support worker well-being and performance. It addresses issues like:
   - Communication—Fostering open and effective communication channels within the organization.
   - Personnel Resource Management—Implementing fair and equitable practices for recruitment, development, and performance management.
   - Task Design—Designing tasks that are meaningful, challenging, and provide opportunities for growth.
   - Shift Hours—Considering the impact of shift work on worker health and fatigue, and optimizing scheduling practices accordingly.
   - Teamwork and Collaboration—Creating a culture of collaboration and mutual support within teams.

Understanding these categories provides a framework to assess potential risk factors and proactively implement solutions.

1.4.3 Classification of Ergonomic Problems

Evaluating ergonomic problems through the lens of their impact on the human body and mind is crucial for creating safe and healthy workplaces, products and environments. This user-centered approach categorizes issues based on how they affect people, allowing for targeted assessment of tasks, equipment, and individual abilities. While broad categories exist, significant overlap and interaction among them highlight the need for a holistic perspective.

1.4.3.1 Key Categories

- **Anthropometric:** This category focuses on human body dimensions and proportions, including factors like reach, eye height, and strength. Applying anthropometric principles ensures compatibility between workspaces and users, minimizing discomfort and risk. Recent studies by Schneider et al. (2020) emphasize the importance of considering age and gender variations in anthropometric data for optimal workplace design.

- **Musculoskeletal:** Tasks demanding excessive physical exertion fall under this category. Overexertion can lead to musculoskeletal disorders (MSDs), back injuries, and muscle strain. The 2021 report by the National Institute for Occupational Safety and Health (NIOSH) identified MSDs as a leading cause of workplace injuries, highlighting the need for ergonomic interventions to reduce physical demands (NIOSH, 2021b).
- **Cardiovascular:** This category encompasses tasks that put strain on the heart and circulatory system. Excessive demands can cause fatigue, overexertion, dehydration, and long-term health problems. The American Heart Association's 2023 scientific statement on workplace health emphasizes the importance of incorporating physical activity breaks and designing jobs with appropriate intensity to promote cardiovascular health.
- **Cognitive:** Both information overload and under-load can pose cognitive challenges. Tasks requiring high memory demands or rapid decision-making can strain cognitive abilities, while monotonous tasks can lead to boredom and decreased vigilance. The 2019 European Agency for Working Conditions report on cognitive ergonomics highlights the need for designing work to optimize cognitive engagement and complement human capabilities with technology for maximized performance and well-being.
- **Psychomotor**: Tasks involving repetitive movements or rapid responses to visual or auditory stimuli fall under this category. Paced assembly lines and tasks requiring precise hand–eye coordination are examples. The 2020 Centers for Disease Control and Prevention (CDC) guidelines for preventing work-related musculoskeletal disorders emphasize the importance of ergonomic interventions to minimize repetitive strain and ensure adequate rest periods for tasks involving high psychomotor demands.

It is crucial to acknowledge that these categories are not mutually exclusive. Ergonomic problems often involve a combination of factors, and solutions should address the interplay between them. A holistic approach that considers the physical, cognitive, and emotional demands of work is essential for creating truly safe and healthy workplaces.

1.4.4 Modern Era: Ergonomics Expands its Horizons

The 20th century witnessed ergonomics blossom beyond its initial focus on physical mechanics. The following sections provide an overview of modern-day and emerging focus areas within the discipline of ergonomics.

1.4.4.1 Cognitive Ergonomics: Decoding the Mind–Machine Dance

As technology intertwined with our lives, ergonomists recognized the need to understand how humans process information and interact with it. Enter cognitive ergonomics, which explores mental workload, decision-making, and stress in relation to technology use. Imagine studying how pilots navigate complex displays or how air traffic controllers manage information overload. These are just a few examples of how cognitive ergonomics optimizes human-technology interaction for safety and efficiency.

1.4.4.2 Computer Ergonomics: Taming the Digital Beast

The rise of personal computers brought a new frontier for physical ergonomics: the digital workspace. Computer ergonomics focuses on preventing musculoskeletal disorders like carpal tunnel syndrome and eye strain through proper posture, keyboard design, and screen placement. Remember those adjustable monitor stands and ergonomic keyboards? These are products of this ergonomic subfield, ensuring our digital endeavors don't come at the cost of our physical well-being.

1.4.4.3 Ergonomics Takes Flight: Soaring to New Heights

Aviation ergonomics ensures the safety and well-being of pilots, cabin crew, and passengers. From cockpit layout to noise reduction, every aspect is meticulously designed to minimize fatigue, optimize communication, and prevent errors. Imagine studying how to design airplane seats for optimal comfort during long flights or how to create intuitive control panels for pilots to make quick decisions in critical situations. These are just a few examples of how aviation ergonomics keeps the skies safe and our journeys smooth.

1.4.4.4 Beyond the Workplace: Ergonomics Touches Every Aspect of Life

The ergonomic lens extends far beyond the traditional office setting. Healthcare ergonomics improves patient safety and staff well-being in hospitals and clinics. Educational ergonomics creates comfortable and stimulating learning environments for students of all ages. Even consumer product design benefits from ergonomic principles, with everything from kitchen appliances to children's toys being shaped for ease of use and comfort.

This is just a glimpse into the ever-expanding world of modern ergonomics. As technology continues to evolve and permeate our lives, the field's role in shaping a world that works for humans becomes ever more crucial.

1.4.5 Ergonomics and Human Factors Professional Organizations

Several national ergonomics societies exist worldwide, each catering to the specific needs and interests of its members. Some prominent examples include those discussed in the following sections.

1.4.5.1 International Ergonomics Association (IEA)

The International Ergonomics Association is a global federation of ergonomics and human factors societies, dedicated to advancing the science and practice of ergonomics on an international scale. Founded in 1959, the IEA was established to promote collaboration and knowledge sharing among professionals in the field of ergonomics, ensuring that human-centered design principles are applied to improve safety, performance, and well-being across various sectors worldwide. The organization was founded in Stockholm, Sweden, by a group of ergonomics pioneers who recognized the need for a formal, international body to support the growing field.

The IEA's mission is to enhance the quality of life through the effective application of ergonomic principles to work systems, products, and environments. The IEA serves as a vital resource for practitioners, researchers, and organizations seeking to implement human factors and ergonomics (HF/E) practices in a multitude of industries, including healthcare, manufacturing, transportation, and technology.

In support of this mission, the IEA works to:

- Promote the application of ergonomics in system design to improve human well-being and overall system performance.
- Facilitate research and development by encouraging collaboration between academic, governmental, and industrial entities.
- Support ergonomics education and professional development through conferences, workshops, and accredited programs.
- Promote ergonomic standards and policies on both national and international levels.

The IEA operates as an umbrella organization, comprising over 50 member societies from around the world, representing tens of thousands of ergonomics and human factors professionals. This extensive network allows for the exchange of research, innovations, and best practices across borders, helping to address global challenges in ergonomics and human systems integration. The IEA's member societies represent diverse fields such as engineering, design, psychology, and healthcare, all contributing to the advancement of ergonomics as a multidisciplinary science.

Currently, the IEA has over 20,000 individual members through its affiliated societies. Members include ergonomists, researchers, educators, and other professionals who apply HF/E principles to enhance safety, efficiency, and comfort in the design of workplaces, products, and systems.

One of the IEA's significant contributions to the ergonomics profession is its leadership in setting international ergonomics standards and guidelines. The association works closely with organizations such as the International Organization for Standardization (ISO) to develop standards that promote the design of safer and more effective work environments. Additionally, the IEA supports several technical committees and working groups that focus on specific areas of ergonomics, such as occupational safety, healthcare ergonomics, and cognitive ergonomics.

In addition to pioneering practical ergonomic solutions, the IEA also hosts a triennial international congress, one of the largest and most prestigious events in the ergonomics field. These events bring together experts and practitioners to share research, innovations, and solutions to human-centered design challenges, fostering the exchange of knowledge across industries and countries.

For students of ergonomics, the IEA represents a valuable professional resource that offers opportunities for networking, education, and staying at the forefront of research and practice in the field. Its global reach, commitment to advancing human factors and ergonomics, and emphasis on improving human well-being make the IEA a key player in shaping the future of ergonomics worldwide. Through its leadership, the IEA continues to promote the application of ergonomics in making systems safer, more efficient, and more user-friendly for people across the globe.

1.4.5.2 Human Factors and Ergonomics Society (HFES)

The Human Factors and Ergonomics Society is a professional organization dedicated to advancing the knowledge and practice of *human factors* (HF) and *ergonomics* (E). Founded in 1957, HFES serves as the leading body representing professionals in these fields, promoting research, education, and application of ergonomics principles to improve human performance, safety, and well-being in various systems. HFES's mission is to foster the discovery and exchange of knowledge regarding the design of systems that enhance human capabilities and limit human limitations. The society seeks to bridge the gap between humans and technology by promoting the understanding of human abilities, behavior, and limitations, thereby improving the design of equipment, environments, and systems. Its vision is to promote the widespread application of human factors and ergonomics principles across industries, improving the quality of life, safety, and performance globally.

HFES promotes rigorous scientific research and practical application of human factors and ergonomics principles. Research within the society focuses on understanding the interaction between humans and other system elements, aiming to improve safety, efficiency, and user satisfaction. Topics range from *cognitive ergonomics*, which studies mental processes like perception, memory, and decision-making, to *physical ergonomics*, which involves understanding biomechanics, posture, and the physical interaction with tools and environments.

For engineering students, understanding human factors and ergonomics is crucial in designing systems, products, and environments that are not only efficient but also safe and user-friendly. Whether designing a manufacturing process, a medical device, or a software interface, incorporating human factors can help ensure that designs accommodate the capabilities and limitations of the user, leading to better performance outcomes and fewer accidents or errors. HFES serves as an invaluable resource for engineers seeking to integrate these principles into their work.

By engaging with HFES and its resources, students and professionals can gain resources and networks that enable them to be at the forefront of ergonomics research and practice, contributing to systems that enhance human well-being while optimizing technological performance.

1.4.6 The Chartered Institute of Ergonomics and Human Factors (CIEHF)

The Chartered Institute of Ergonomics and Human Factors is a professional society based in the United Kingdom, dedicated to promoting and advancing the science and application of ergonomics and human factors (E/HF) in a wide range of industries. The first-of-its-kind institute was founded in 1949 as the Ergonomics Research Society. Over the decades, it has evolved into a leading authority in the field, officially becoming the Chartered Institute of Ergonomics and Human Factors after receiving a Royal Charter in 2014.

The CIEHF's primary objective is to improve life, work, and systems by promoting the application of ergonomics and human factors principles. These principles focus on designing systems, processes, and products that match human capabilities and limitations, enhancing safety, performance, and well-being. The institute supports the *integration of human factors* into various sectors, including healthcare, defense, transportation, manufacturing, and technology, to ensure that systems and environments are user-friendly, safe, and efficient.

The mission of the CIEHF includes:

- Promoting the benefits of ergonomics and human factors to society, businesses, and individuals.
- Supporting research, education, and practice in ergonomics and human factors.
- Ensuring that human-centered design is prioritized in industry, particularly in safety-critical sectors.
- Providing certification and professional recognition to ergonomics practitioners, ensuring high standards of competence and ethics in the profession.

The CIEHF boasts a membership of over 1,800 professionals from diverse sectors, including ergonomists, human factors practitioners, engineers, psychologists, designers, and healthcare professionals. The institute offers a range of membership levels, from student membership to full chartered status, supporting career development at every stage. Chartered status is the highest level of recognition offered by the institute, indicating that a professional has achieved a high standard of expertise and experience in the field of ergonomics and human factors.

Members of the CIEHF gain access to a wide range of resources, including publications, conferences, networking events, and continuing professional development (CPD) opportunities. The institute also offers accredited courses and certification programs that ensure practitioners meet the necessary professional standards to work in ergonomics and human factors roles. These programs are invaluable for students and professionals aiming to advance their careers in the field.

Key contributions of CIEHF include both the provision of ergonomics standards and guidelines and the advocacy of human factors in policy-making (CIEHF, 2023). The institute plays a significant role in industries where user safety and system performance are critical.

The CIEHF organizes an annual conference and other events that bring together practitioners and researchers to share the latest advancements and innovations in ergonomics and human factors. These gatherings facilitate knowledge exchange, networking, and collaboration across different sectors.

In addition to its professional development and certification programs, the CIEHF works closely with universities and educational institutions to promote ergonomics education by providing students with resources, guidance, and opportunities to engage with professionals in the field. Through its outreach programs, the CIEHF ensures that the next generation of ergonomists and human factors practitioners are well-prepared to contribute to improving systems, products, and environments.

For students and professionals in ergonomics and human factors, the Chartered Institute of Ergonomics and Human Factors offers a wealth of resources, support, and recognition. With its long history of promoting human-centered design and its commitment to advancing the science and practice of ergonomics, the CIEHF plays a pivotal role in ensuring that systems, workplaces, and environments are designed to optimize human well-being and performance. By joining the CIEHF, students can become part of a global network of professionals dedicated to making a difference in the world through ergonomics and human factors.

1.4.7 Japan Ergonomics Society (JES)

The Japan Ergonomics Society (JES) is one of the leading professional organizations in the field of ergonomics and human factors in Japan. Established in 1964, JES was founded to promote research, education, and the practical application of ergonomics principles in various industries and aspects of daily life in Japan. As a multidisciplinary organization, JES brings together professionals and researchers from engineering, psychology, design, and health sciences, among others, with the goal of improving the interaction between humans and systems.

The primary mission of JES is to enhance human well-being and system performance by promoting the application of ergonomics principles. The society focuses on ensuring that products, systems, environments, and processes are designed to match human abilities and limitations. This focus is particularly important in industries such as manufacturing, healthcare, transportation, and information technology.

JES aims to:

- Encourage and support research in ergonomics and human factors.
- Promote the education and professional development of ergonomics practitioners.
- Foster collaboration between academia, industry, and government to advance the practical application of ergonomics principles in real-world settings.
- Raise public awareness about the importance of ergonomics in improving quality of life and preventing work-related injuries.

JES was founded in Tokyo, Japan, in response to the growing need for formal research and application of ergonomics principles in post-war Japan. The country's rapid industrialization and technological advancement created a demand for ergonomic solutions that could improve workplace safety, efficiency, and productivity. Since its inception, JES has played a key role in developing ergonomics as a recognized discipline in Japan, working to ensure that human-centered design principles are integrated into the country's technological and industrial innovations.

Today, JES has over 2,000 members, comprising ergonomists, engineers, designers, psychologists, healthcare professionals, and researchers. JES members are involved in a wide range of sectors, including automotive, robotics, healthcare, consumer electronics, and architecture, reflecting the multidisciplinary nature of ergonomics.

JES offers myriad opportunities for professional development and knowledge exchange, including:

- **Annual Conferences:** JES organizes a national conference every year, where members present research findings, discuss emerging trends, and share best practices in ergonomics.
- **Publications:** The society publishes the *Journal of the Japan Ergonomics Society*, a peer-reviewed journal that covers the latest research in ergonomics, human factors, and related fields.
- **Workshops and Seminars:** JES hosts workshops and educational seminars to help members stay up-to-date with the latest ergonomic techniques, technologies, and methodologies.
- **Awards and Recognition:** JES recognizes outstanding contributions to the field through awards and honors, encouraging innovation and excellence in ergonomics research and practice.

JES has made significant contributions to the advancement of ergonomics both nationally and globally by playing a key role in shaping industry standards for ergonomic design and occupational safety. Japan's reputation for producing highly ergonomic consumer products, such as vehicles, electronics, and tools, is due in part to the research and expertise fostered by JES. Additionally, the society has contributed to the development of ergonomics guidelines that are used in various sectors, including manufacturing, healthcare, and office design.

The society also maintains international collaborations with other ergonomics organizations, such as the International Ergonomics Association, allowing for the exchange of knowledge and research findings on a global scale. Through these partnerships, JES helps promote ergonomics as a vital discipline that improves human health, safety, and productivity worldwide.

For students in ergonomics and human factors, the Japan Ergonomics Society offers a valuable resource for professional growth, research opportunities, and networking within the ergonomics community. Its commitment to advancing human-centered design and its extensive contributions to both Japanese industry and the global ergonomics field make JES an essential player in shaping the future of ergonomics. By engaging with JES, students and professionals alike can contribute to innovations that improve the safety, usability, and efficiency of systems and products in Japan and beyond.

1.4.8 Future of the Field of Ergonomics

The field of ergonomics and human factors continues to grow and evolve, driven by the need to ensure workplaces, products, and systems are designed for optimal human performance, comfort, and safety. Some key areas of focus in recent years include:

- **Digital Ergonomics:** Addressing the challenges posed by the increasing use of digital technologies, such as musculoskeletal strain from computer work and eye fatigue from prolonged screen time.
- **Aging Workforce:** Designing workplaces and products that are suitable for older workers, who may have different physical and cognitive abilities than younger workers.
- **Mental Health and Well-being:** Integrating ergonomics principles into workplace design to promote mental health and well-being, including reducing stress and creating environments that are conducive to focus and productivity.

As of 2023, there are over 25,000 ergonomists working worldwide in various capacities, including research, application, education, and training (Bureau of Labor Statistics, 2023). This growth reflects the increasing awareness of the importance of ergonomics in various sectors, including healthcare, manufacturing, transportation, and technology.

The educational background of ergonomics professionals remains diverse, with common areas of study including:

- **Industrial Engineering:** Focuses on designing and improving integrated systems of people, materials, information, and equipment.
- **Psychology:** Provides insights into human cognitive abilities, perception, and decision-making, crucial for designing user-friendly interfaces and systems.
- **Industrial Hygiene:** Emphasizes the identification, evaluation, and control of environmental factors that may affect worker health and safety.
- **Kinesiology:** Studies human movement and biomechanics, essential for understanding musculoskeletal strain and designing ergonomic workstations.
- **Biomechanics:** Applies principles of mechanics and physics to analyze human movement and forces acting on the body.

1.4.9 Legislative and Regulatory Issues Impacting Ergonomics

The field of ergonomics, concerned with optimizing the work environment for human well-being and productivity, navigates a complex landscape of legislative and regulatory influences. In the United States, the Occupational Safety and Health Act of 1970 (OSH Act) establishes two key players: the National Institute for Occupational Safety and Health (NIOSH) and the Occupational Safety and Health Administration (OSHA). While NIOSH focuses on research, education, and guidance, OSHA wields the regulatory hammer, developing and enforcing workplace safety standards. This division of labor is reflected in their respective domains: NIOSH resides within the Centers for Disease Control and Prevention (CDC) under the Department of Health and Human Services, while OSHA operates under the US Department of Labor. The following sections provide a more in-depth discussion of these two essential agencies that support safety in the US occupational environment.

BOX 1.1 Mission and Impact of OSHA

Mission: To prevent work-related injuries, illnesses, and deaths.

Impact: Since the agency was created in 1971, occupational deaths have been cut by 62% and injuries have declined by 42%.

OSHA website (2009)

1.4.9.1 OSHA

The Occupational Safety and Health Administration stands as a pillar of worker well-being in the United States. OSHA fulfills a critical role in promoting safe and healthy work environments, with ergonomics playing a central part in its mission (OSHA, 2023). While OSHA may not have a defined, overarching ergonomic standard, its influence on workplace ergonomics is undeniable. This agency wields several key tools to achieve its goals:

1. *Setting the Stage*
   - **Establishing Criteria:** OSHA empowers employers to protect their workers from harmful materials and health hazards by laying out a framework for identifying and mitigating such risks.
   - **Authority and Responsibility:** Through the OSHA Act, the Secretary of Labor, acting through OSHA, holds extensive authority to promulgate, modify, and even revoke safety and health standards, including those related to ergonomics.

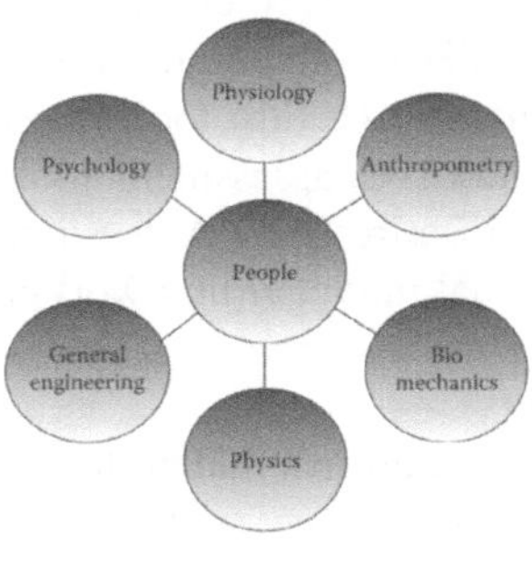

FIGURE 1.1 Interdisciplinary fields within ergonomics.

2. *Taking Action*
    - **Inspections and Investigations:** OSHA actively oversees workplaces, conducting inspections and investigations to identify potential ergonomic hazards and ensure compliance with established regulations.
    - **Enforcement:** In cases where ergonomic hazards are identified, OSHA has the power to issue citations, impose penalties, and require employers to implement corrective measures. This ensures prompt action and protects workers from harm.

3. *Beyond Regulations*
    - **Targeted Guidelines:** Recognizing the diverse needs of different industries, OSHA provides specific ergonomic guidelines for sectors like shipyards, poultry processing, and nursing homes. These tailored resources offer practical guidance for implementing ergonomic solutions within these fields.
    - **Electronic Tools:** Taking a tech-savvy approach, OSHA provides a wealth of online tools and resources to help employers identify and address ergonomic hazards across various occupations. From electrical contractors to healthcare workers, these tools empower businesses to create safer work environments for their employees.
    - **Collaboration and Recognition:** Through programs like the Alliance Program and the Voluntary Protection Program (VPP), OSHA fosters collaboration between industry stakeholders and recognizes businesses that prioritize worker safety and health, including effective ergonomics practices.

4. *Filling the Gaps*
    - **General Duty Clause:** In the absence of a specific ergonomic standard, OSHA leverages the General Duty Clause (Section 5(a)(1) of the OSHA Act). This clause mandates that employers provide work environments free from recognized hazards likely to cause death or serious physical harm. This powerful tool allows OSHA to address ergonomic concerns even in the absence of specific regulations.
    - **Research and Resources:** To further equip employers and workers with knowledge and tools, OSHA collaborates with NIOSH (National Institute for Occupational Safety and Health) to develop and disseminate valuable resources. These include risk assessment tools like the NIOSH Lifting Equation and the Rapid Entire Body Assessment (REBA) Tool, empowering informed decision-making regarding workplace ergonomics.

1.4.9.1.1 Impact and Future

While OSHA resources may not always match the vastness of the American workforce, the agency's impact on workplace safety and ergonomics is undeniable. The significant decline in work-related deaths and injuries over the past decades speaks volumes about its effectiveness. As research and understanding of ergonomics evolve, OSHA continues to adapt and refine its approaches, ensuring that workplaces remain safe and healthy for all.

1.4.9.2 OSHA Standards

The Occupational Safety and Health Administration (OSHA) standards hold a pivotal role in creating safe and healthy working environments in the United States and are based on scientific consensus. To do this, OSHA draws upon validated existing guidelines, norms, emerging science, and best practices from various sources to ensure a comprehensive and ever-evolving approach to workplace safety.

The bedrock of OSHA standards lies in existing federal regulations and national consensus standards developed by organizations like the American National Standards Institute (ANSI) and the National Fire Protection Association (NFPA). These guidelines, crafted through rigorous research and industry collaboration, offer a solid foundation for early OSHA standards. OSHA thoughtfully adapted and refined these existing guidelines, incorporating necessary amendments, deletions, and additions to create robust *and* implementable standards.

To ensure ongoing relevance of these standards to the occupational community, OSHA operates a dynamic system for evolving its standards alongside advancements in safety knowledge and technology. This ongoing effort results in four distinct categories of standards:

1. **Design Standards:** These meticulously detailed blueprints specify the construction criteria for specific equipment, machines, or components. Think of them as precise instruction manuals for building safety directly into the fabric of workplaces. For example, OSHA's ventilation design standards provide detailed blueprints for ensuring proper air circulation and minimizing exposure to harmful fumes.
2. **Performance Standards:** Unlike dictating specific methods, these standards set clear objectives employers must achieve. They provide employers with the flexibility to choose the most suitable methods for their unique workplaces while still ensuring workers' safety. A prominent example lies in the Threshold Limit Values (TLVs) for toxic and hazardous substances. These standards define the maximum concentration levels permissible in workplace air, leaving employers to choose the appropriate engineering controls or personal protective equipment to achieve compliance.
3. **Vertical Standards:** Diving deeper into specific industries, these standards tailor safety requirements to the unique hazards and practices of particular sectors. The pulp, paper, and paperboard industries, for instance, have their own set of OSHA standards addressing the specific risks inherent in their operations. Such vertical standards offer focused guidance to ensure industry-specific hazards are effectively mitigated.
4. **Horizontal Standards:** Taking a broader view, these standards encompass general safety requirements applicable across all industries. From sanitation practices to the design of walking and working surfaces, horizontal

standards set a baseline level of safety for every workplace, regardless of their specific sector.

1.4.9.3 National Institute of Occupational Safety and Health (NIOSH)

The National Institute for Occupational Safety and Health was also established by the Occupational Safety and Health Act of 1970. Its mission is to ensure safe and healthful working conditions for every worker in the nation by conducting research, developing recommendations for new or revised OSHA standards, providing information and data on workplace health hazards, and assisting others in preventing work-related illness, injury, and death. Although NIOSH does not have the authority to directly promulgate standards, its research and recommendations serve as a crucial scientific foundation for OSHA's rulemaking process.

1.4.9.3.1 Shifting Focus: From Sectors to National Research Agenda

Moving forward, the focus is shifting from individual sectors to a more comprehensive national research agenda. This agenda will prioritize research tackling pressing challenges across various industries, including:

- **Emerging Workplace Hazards:** This could involve investigating the health impacts of new technologies like automation and artificial intelligence, as well as the psychological well-being of workers in the gig economy.
- **Persistent Issues:** Addressing longstanding concerns like musculoskeletal disorders, occupational cancers, and chronic respiratory illnesses remains crucial.
- **Workforce Trends:** Adapting research to cater to the growing demographics of older workers, immigrants, and those in non-traditional jobs is essential.

The National Institute for Occupational Safety and Health continues to play a pivotal role in facilitating NORA's activities. By strategically funding research projects, ensuring scientific rigor, and translating research findings into practical workplace interventions, NIOSH empowers researchers and practitioners to make a tangible impact on workers' lives.

1.4.9.3.2 Measuring Success: Impact Beyond Numbers

Beyond mere metrics, NORA's success will be measured by its ability to:

- **Prevent Workplace Injuries and Illnesses:** Reducing the human and economic costs associated with occupational hazards is paramount.
- **Empower Workers and Employers:** Providing tools and knowledge to make workplaces safer and healthier for everyone.
- **Advance Scientific Understanding:** Continuous research and innovation are key to tackling new and evolving workplace challenges.

NORA's renewed vision promises a future where work is not only productive but also safe and healthy for all. By focusing on emerging challenges, addressing persistent issues, and adapting to changing workforce demographics, NORA continues to pave the way for a brighter future for American workers.

1.4.9.4 Nuclear Regulatory Commission

The US Nuclear Regulatory Commission (NRC) stands as a watchful guardian, safeguarding the nation's use of nuclear materials for peaceful purposes. Established in 1974 by the Energy Reorganization Act, the NRC's mission transcends mere regulation. It is a solemn pledge to protect public health, national security, and the environment from the potential dangers of nuclear energy.

Beyond the cold calculations of reactor cores and radioactive isotopes lies a human element. The NRC recognizes the vital role of nuclear workers and actively promotes their safety through rigorous regulations and oversight. Ergonomic principles are woven into the fabric of nuclear facilities, minimizing the risk of musculoskeletal disorders and fostering a healthy work environment.

While the specter of nuclear accidents looms large, the NRC's vigilance has prevented widespread catastrophes. The 1979 Three Mile Island incident, though a stark reminder of the potential dangers, ultimately served as a catalyst for significant improvements in nuclear safety. The partial core meltdown, while raising initial fears of widespread contamination, thankfully resulted in no immediate casualties or major environmental damage.

However, the Three Mile Island incident was not simply a close call. It was a wake-up cry, prompting the NRC to implement sweeping changes in areas like emergency response planning, operator training, and human factors engineering. Control rooms were redesigned to minimize human error, procedures were refined for optimal clarity, and operator training programs were revamped to equip personnel with the knowledge and skills to handle both routine and emergency situations.

The NRC's work extends beyond the walls of nuclear facilities. It actively engages with the public, fostering transparency and open communication about the potential

TABLE 1.2
NORA Sector Group

Agriculture, forestry, and fishing
Construction
Healthcare and social assistance
Manufacturing
Mining
Services
Transportation, warehousing, and utilities
Wholesale and retail trade

Source: https://www.cdc.gov/nora/about/sectors.html

risks and benefits of nuclear energy. From hosting public meetings and issuing regular safety reports to proactively addressing concerns, the NRC strives to earn public trust and build a sustainable future for nuclear energy.

The NRC's relevance continues to be apparent as the agency continuously adapts to meet the evolving challenges of the 21st century. Cybersecurity threats, for example, have become a top priority, necessitating robust measures to protect critical infrastructure from cyberattacks. Additionally, the NRC is actively involved in modernizing regulations to reflect advancements in nuclear technology and best practices, ensuring that safety standards remain at the forefront of the industry.

1.4.9.5 Department of Defense (DoD) Military Standard 1472 1999 (F)

The Department of Defense (DoD) Military Standard 1472 (MIL-STD-1472), first promulgated in 1969, establishes general human engineering design criteria for military systems, subsystems, equipment, and facilities. It aims to ensure the integration of human factors and ergonomics principles into the design of these systems to achieve mission success, enhance effectiveness, simplicity, efficiency, reliability, and safety during operation, training, and maintenance. While specifically designed for military applications, MIL-STD-1472's influence has extended far beyond the DoD, becoming a widely recognized and influential standard in the field of human factors and ergonomics.

The latest revision of MIL-STD-1472 is Revision H, released in September 2020. This revision incorporates updates and advancements in human factors knowledge, addressing topics such as:

- User interface design for touchscreens and wearable technologies
- Cognitive workload assessment and management
- Accessibility for individuals with disabilities
- Design for aging populations
- Cultural considerations in design
- Teamwork and collaboration in complex systems

It is crucial to note that MIL-STD-1472 is a living document, subject to continuous review and updates to reflect the evolving nature of technology, human capabilities, and mission requirements.

1.4.9.5.1 Cross-Industry Impact and Applicability

While primarily intended for military applications, MIL-STD-1472 has been adopted or referenced by various government agencies, industries, and international organizations, including:

- Federal Aviation Administration (FAA)
- Department of Energy (DOE)
- British Defence Standard (Def Stan 00-25)
- International Organization for Standardization (ISO)
- Commercial aviation, healthcare, nuclear power, and other safety-critical industries

1.4.9.6 ISO 9000 Series

The International Organization for Standardization (ISO) holds a prominent global position in developing and setting quality management system standards across diverse industries, including automotive, aerospace, defense, and healthcare. As the official US representative to ISO, the American National Standards Institute bridges the gap between international best practices and domestic implementation.

One notable offering from ISO is the ISO 9000 series, a collection of widely used standards specifically aimed at establishing and maintaining effective quality management systems (QMS) within organizations. These standards provide organizations with a framework for systematically managing processes, improving operational efficiency, and consistently delivering products and services that meet customer needs and regulatory requirements. The most recent revision of the core standard, ISO 9001, was in 2015, reaffirming its continued relevance and adaptability to contemporary business practices.

Beyond the broad scope of the ISO 9000 series, ISO also delves into more specific areas, fostering standards that enhance ergonomics, safety, and technical compatibility across multiple industries. For instance, the ISO 9241 series addresses ergonomic principles for interaction with computer visual displays, promoting user comfort and preventing musculoskeletal disorders in computer-intensive workplaces.

Furthermore, ISO recognizes the unique needs of different sectors and publishes industry-specific standards dedicated to human factors and ergonomics. A relevant example is ISO/IEC 26513:2009, which focuses on ergonomic requirements for office work with visual display terminals, offering tailored guidance for optimizing workstations and work environments in office settings.

The adoption of ISO standards proves particularly valuable in projects involving international collaboration, where diverse teams come together to develop products, processes, or environments. Standardization across task activities, software development practices, and even equipment specifications allows for improved communication, knowledge transfer, and ultimately, greater efficiency and consistency in product quality and safety practices. This harmonization facilitates seamless collaboration across borders, enhancing productivity and ensuring seamless product development and delivery, regardless of geographical location.

NIOSH Activities:

- Conducts research on occupational health hazards, including chemical, physical, and biological agents, as well as ergonomic and psychosocial factors.
- Develops and disseminates evidence-based guidelines and recommendations for preventing occupational illnesses and injuries.
- Provides technical assistance and training to employers and workers on occupational safety and health issues.
- Responds to requests for workplace health hazard evaluations and investigations.
- Manages the National Occupational Research Agenda (NORA), a public-private partnership that identifies and prioritizes research needs in occupational safety and health (NIOSH, 2023) .

Recent Developments (Beyond 2011):

- **NIOSH Science Agenda 2022–2026:** This plan outlines the agency's research priorities for the next five years, focusing on emerging workplace hazards, worker mental health, and promoting safe and healthy work environments for all (NIOSH, 2022).
- **Total Worker Health® Program:** This initiative promotes a comprehensive approach to worker well-being, addressing both physical and mental health, safety, and well-being.
- **COVID-19 Response:** NIOSH has been actively involved in research and guidance development related to workplace safety during the pandemic, providing resources on topics such as respiratory protection, ventilation, and worker mental health (NIOSH, 2021a).

BOX 1.2 NIOSH Mission

The mission of NIOSH is to generate new knowledge in the field of occupational safety and health and to transfer the knowledge into practice for the betterment of workers.

NIOSH website (2009)

The National Occupational Research Agenda (NORA), launched in 1996, has made significant strides in its first decade. Through the formation of sector councils and development of strategic roadmaps, NORA has guided research efforts toward addressing critical workplace hazards and improving worker well-being in key sectors like agriculture, construction, healthcare, manufacturing, mining, services, trade, and transportation.

But the journey doesn't end there. As the landscape of work evolves, so too must NORA adapt.

1.4.10 Human-Centered Design

User-Centered Design (UCD) remains a cornerstone of creating user-friendly systems, products, and processes. It emphasizes active participation from real users and iterative refinement of design solutions. Beyond websites, UCD principles have expanded to encompass diverse digital experiences, with applications in mobile apps, voice assistants, and augmented reality interfaces. Government agencies like the United States' Digital Service (USDS) (2023) and the UK's Government Digital Service (GDS) (2023) actively leverage UCD principles to enhance the usability and accessibility of citizen-facing online services.

The international standard ISO 13407: Human-centered Design Processes for Interactive Systems (ISO, 2019) provides a framework for integrating human-centered principles throughout the development lifecycle. While it does not dictate specific methods, it outlines essential activities like user research, prototyping, and iterative testing.

Approaches to UCD vary across industries and regions. Three prominent categories include:

- **Cooperative Design:** Evolving since the 1970s in Scandinavia, this approach emphasizes equal collaboration between designers and users (Bødker et al., 2004).
- **Participatory Design (PD):** A North American term inspired by Cooperative Design, PD focuses on actively involving end users in the design process to ensure products meet their needs (Sanders & Stappers, 2010).
- **Contextual Design:** Encompassing elements of PD, this approach emphasizes understanding the context in which users will interact with the product (Holtzblatt & Beyer, 2017).

Regardless of the specific approach, gathering user input is crucial. Techniques like focus groups, usability testing, observation, surveys, and interviews help designers understand user needs, behaviors, and mental models.

To meet ISO 13407 requirements, a UCD process must include four essential elements:

1. Requirements and Planning:
    - Define the context of use (target users, tasks, environment).
    - Specify user and organizational requirements (goals, business objectives).
2. Design:
    - Develop iterative prototypes allowing user evaluation throughout the process.
3. Evaluation:
    - Conduct user-based assessments with representative users.
    - Employ iterative evaluation to refine designs based on user feedback.
4. Deployment:
    - Release the final design after successful UCD implementation.

Successful UCD implementation demonstrably improves user acceptance, product quality, and business outcomes. In today's increasingly user-centric digital landscape, UCD remains a vital tool for designing products and experiences that truly resonate with their intended audience.

1.4.11 Cost Justification for Ergonomic Program

Investing in ergonomics isn't just about employee well-being, it is a smart business decision with tangible financial returns. Despite the well-documented benefits, budget constraints often deter organizations from prioritizing ergonomics. However, extensive research paints a compelling picture of the economic advantages. Studies by Hendrick (2004), MacLeod (2002), Oxenburgh (2001), and Douphrate (2014) demonstrate significant cost savings and productivity gains through ergonomic interventions.

Making the financial case for ergonomics requires proactive communication and a clear understanding of the potential ROI. Pulat (1991) emphasizes the importance of demonstrating and communicating the scientific and economic benefits to cultivate an organizational culture that values ergonomics.

The following are some recent examples showcasing the ROI of ergonomic investments:

- **Intel:** An ergonomic redesign of assembly lines led to a 30% reduction in musculoskeletal disorders and a 15% increase in productivity within a year (2019).
- **Boeing:** Implementing ergonomic principles in aircraft assembly saved $100 million in workers' compensation costs over five years (2017).
- **UPS:** A $1.5 million investment in ergonomic equipment and training resulted in a $10 million return within two years through reduced injury costs and improved productivity (2018).
- **Nintendo Corporation:** A $400,000 investment in ergonomic improvements yielded a $1 million annual saving. Workers' compensation costs dropped by 16–20%, and lifting-related injuries decreased by 80%.
- **Hewlett-Packard:** Ergonomic interventions resulted in a remarkable 105% ROI within a year.
- **The Red Wing Shoe Company:** Their 1985 ergonomic program led to a 70% reduction in workers' compensation premiums and a 79% decrease in lost workdays within four years.
- **3M:** Their ergonomic programs generate an annual return of $250,000 through increased productivity alone.

These examples showcase the diverse benefits of ergonomic programs. By reducing absenteeism, lowering workers' compensation claims, and enhancing productivity, ergonomic investments often yield substantial returns, exceeding initial costs by significant margins.

It is important to remember that the financial benefits of ergonomics extend beyond immediate cost savings. Improved employee morale, reduced stress, and enhanced employee engagement contribute to a more positive and productive work environment, ultimately leading to a stronger bottom line.

As an ergonomist, advocating for effective interventions requires a well-constructed case that resonates with decision-makers. The process outlined below, adapted from Pulat (1992), provides a practical roadmap for building a persuasive argument:

1. Define the Problem

Clearly identify the issues your proposed intervention will address. This could involve:

- **Employee Injuries:** Quantify the frequency and severity of work-related musculoskeletal disorders in the target area.
- **Workers' Compensation Costs:** Calculate the financial burden of injury-related claims associated with the specific task or work environment.
- **Employee Turnover:** Highlight the impact of ergonomic issues on employee retention and associated recruitment and training costs.
- **Poor Quality:** Demonstrate how discomfort or fatigue can lead to increased errors and rework.
- **Suboptimal Productivity:** Analyze production data to pinpoint areas where ergonomic improvements could streamline workflow and boost output.

2. Cost of the Intervention

Estimate the financial investment required for your proposed solution. Consider:

- **Equipment or Software Costs:** Research the pricing of ergonomic furniture, tools, or software relevant to your intervention.
- **Implementation Labor:** Factor in the time and personnel needed for installation, training, and ongoing maintenance.
- **Employee Training Time:** Quantify the potential impact of training on worker productivity during the implementation phase.

3. Leverage Cost–Benefit Analysis Tools

Utilize validated tools to demonstrate the projected return on investment (ROI) of your proposal. Resources like:

- The OSHA Ergonomics Cost–Benefit Analysis Tool provides a user-friendly platform for estimating cost savings associated with various interventions.
- The Washington State Ergonomics Cost–Benefit Analysis Tool by Rick Goggins offers a state-specific perspective on potential financial gains.
- Websites of professional ergonomics societies and government agencies often feature helpful resources and case studies.

4. Develop an Implementation Plan

Outline a phased rollout plan to minimize disruption and ensure smooth adoption:

- **Pilot Program:** Consider implementing the intervention in a smaller test group first to refine the approach and gather feedback before full-scale deployment.
- **Training and Communication:** Develop comprehensive training materials and communication strategies to ensure employee buy-in and understanding of the new procedures.

- **Contingency Planning:** Anticipate potential challenges and develop solutions to address them proactively.

5. Quantify the Financial Impact

Analyze the potential financial benefits of your intervention, focusing on key categories like:

- **Increased Productivity:** Project the expected improvement in output or efficiency based on relevant data and studies.
- **Reduced Workers' Compensation Costs:** Estimate the anticipated decrease in injury-related claims based on the intervention's effectiveness in addressing MSD risk factors.
- **Lower Lost Workdays:** Quantify the potential reduction in absenteeism due to improved employee health and well-being.
- **Quality Improvement:** Demonstrate how ergonomic interventions can lead to fewer errors and rework, resulting in cost savings.

6. Consider the Intangibles

While not easily quantifiable, highlight the broader benefits of your proposal:

- **Alignment with Organizational Goals:** Show how ergonomics aligns with the company's values and commitment to employee health and safety.
- **Improved Employee Morale and Satisfaction:** Emphasize the positive impact on employee well-being and engagement, leading to increased retention and motivation.
- **Enhanced Organizational Image:** Showcase how prioritizing ergonomics demonstrates the company's commitment to social responsibility and creating a positive work environment.

7. Present Your Case Compellingly

Prepare a professional presentation for decision-makers, including:

- **Clear and Concise Communication:** Tailor your message to the audience's level of understanding and focus on the key benefits of your proposal.
- **Supporting Data and Visuals:** Utilize charts, graphs, and real-world case studies to strengthen your arguments and make your presentation engaging.
- **Addressing Potential Concerns:** Be prepared to anticipate and address any objections or questions decision-makers may have.

By following these steps and leveraging available tools, you can build a compelling case for ergonomic interventions, paving the way for a safer, healthier, and more productive work environment for everyone. However, a successful case for ergonomics goes beyond just numbers. It also includes demonstrating the human impact and aligning your proposals with the organization's values and goals.

1.4.12 Role of the Ergonomist

Ergonomists play a crucial role in shaping our working environments, ensuring they are not only safe and comfortable but also optimized for efficiency and productivity. They do this by applying their deep understanding of human capabilities and limitations to the design of tasks, products, and systems.

Imagine an ergonomist as a bridge between humans and the physical world they work in. They analyze the demands of various tasks, observing how people perform them across diverse demographics, including children, the elderly, and individuals with disabilities. This information becomes the foundation for collaboration with designers and engineers, ensuring that products and services are user-friendly, accessible, and minimize strain or discomfort.

The International Ergonomics Association outlines the core competencies and activities expected of an ergonomist in a 2009 publication. These include:

- **Investigating and Analyzing the Need for Ergonomics:** Understanding the specific demands of a workplace and identifying areas where ergonomics interventions can improve the human–work interface.
- **Analyzing and Interpreting Data:** Examining findings from ergonomic investigations and translating them into actionable insights.
- **Documenting Findings:** Clearly and concisely communicating observations and recommendations to stakeholders.
- **Assessing Compatibility between Human Capabilities and Demands:** Ensuring that tasks, tools, and environments are aligned with the physical and cognitive abilities of users.
- **Developing and Implementing Ergonomics Solutions:** Designing and executing interventions to optimize the work environment.
- **Evaluating the Effectiveness of Interventions:** Measuring the impact of implemented changes and making adjustments as needed.
- **Maintaining Ethical and Professional Conduct:** Upholding high standards of integrity and responsibility in all interactions.

These core competencies highlight the multifaceted nature of an ergonomist's work. They are not only problem solvers but also communicators, researchers, and advocates for human-centered design principles. Their expertise extends beyond physical ergonomics, encompassing cognitive ergonomics, organizational ergonomics, and even the psychological aspects of work.

1.5 SUMMARY

Ergonomics encompasses much more than just preventing injuries and fatalities, though safety remains a crucial element. Ergonomics aims to optimize human well-being and overall system performance. This means fitting the task to the worker not just for safety, but also for enhanced productivity.

Safety is deeply intertwined with physical ergonomics aspects like defining optimal work postures, designing ergonomic workspaces, minimizing repetitive motions, preventing musculoskeletal disorders, and facilitating safe manual material handling. However, ergonomics extends beyond these physical aspects. It delves into cognitive ergonomics, addressing mental workload reduction and optimizing human–computer interaction, and organizational ergonomics, focusing on improving policies and procedures, and ensuring quality.

This chapter serves as a springboard into the vast and multifaceted field of ergonomics, providing a foundational understanding of its history, core principles, and practical applications. The following sections will delve deeper into these concepts, equipping you with the knowledge and skills to become a valuable contributor, whether as a researcher, practitioner, or provider of ergonomic solutions.

1.6 CASE STUDY

Ergonomics Case Study: Workplace Ergonomics in a Global Manufacturing Facility

Problem Overview

As companies expand globally, they often face the challenge of ensuring that their manufacturing facilities meet ergonomic standards to protect workers from injury and reduce fatigue. One such example is a global manufacturing firm that experienced a significant increase in musculoskeletal disorders among its employees due to repetitive tasks and poorly designed workstations.

The challenge for this company was twofold: to redesign the workstations to accommodate a wide range of body types across different regions and to implement ergonomic practices that would reduce strain and increase productivity.

Ergonomic Design Process

To address the ergonomic concerns, the company collaborated with ergonomists and used anthropometric data to better understand the diverse workforce. The ergonomic interventions included the following:

1. **Workstation Design:** The company redesigned workstations to ensure adjustable heights and angles, allowing employees to work in a neutral body posture. The key areas of focus were:
   - Adjustable chairs and desks that fit employees of different heights and body sizes.
   - Footrests and armrests that supported the limbs and reduced the strain on the lower back, shoulders, and wrists.
   - Anti-fatigue mats to reduce stress on workers who had to stand for long periods.

2. **Tool and Equipment Redesign:** The tools used in the manufacturing process were redesigned to fit the natural hand movements of workers. Lighter tools with ergonomic grips were introduced to reduce hand and wrist strain. Automated equipment was also integrated to minimize the need for repetitive manual tasks.
3. **Training and Awareness:** Ergonomics training was provided to all employees, emphasizing proper posture, lifting techniques, and the importance of taking regular breaks. The company introduced a culture of safety, encouraging workers to report discomfort or potential ergonomic issues.
4. **Data-Driven Modifications:** The company implemented wearable sensors to monitor workers' movements and postures in real-time. This data allowed ergonomists to make precise adjustments to the workstations and workflows, resulting in optimized processes that reduced physical strain.

Outcome

The implementation of ergonomic principles led to a 30% reduction in musculoskeletal injuries over the first year. Employees reported increased comfort and reduced fatigue, and overall productivity improved by 20%. Absenteeism due to work-related injuries decreased significantly, and the company saw an overall improvement in employee morale and retention.

Conclusion

This case study illustrates how the field of ergonomics has evolved beyond the traditional scope of manufacturing and has become a vital aspect of global workforce health and safety. By integrating ergonomic practices into the design of workstations, tools, and workflows, companies can create safer, more comfortable, and more productive environments. The success of this intervention highlights the importance of ergonomics in shaping the future of occupational health, especially in a global context where diversity in the workforce requires adaptable solutions.

Exercises

1. Describe the origin of the field of ergonomics.

Extended Exercise: Create a timeline infographic using an online tool like Canva, Venngage, or Lucidchart. Present the major milestones in the development of ergonomics, including key figures, research studies, and technological advancements that have influenced the field.

2. Join an internet-based ergonomics discussion group and seek to generate a discussion on an ergonomic area of interest to you.

Extended Exercise: Instead of just joining a discussion group, consider recording a brief (2–3 minute) video on TikTok, Instagram, or LinkedIn discussing an ergonomics issue that intrigues you (e.g., remote work, standing desks, etc.). Post it and track responses or comments. Analyze how the engagement on social media platforms compares to traditional forums in terms of generating discussion on ergonomics topics.

3. Search the web for sources of information about ergonomics. Identify sources from at least three countries and explain the commonalities and differences in the definition, application, and perception of the importance of ergonomics.

   Extended Exercise: Create a collaborative map using Google Maps or Padlet. Pin sources of ergonomic information from at least three different countries (e.g., government websites, research institutions, or companies). Write a short description for each pin comparing the definition and application of ergonomics in those regions.

4. Identify three different manufacturers across a range of industries. Find out from them (or via their websites and other sources) whether they use ergonomic principles in the design of their products. For those companies that do use ergonomics, find out who is responsible and how ergonomics is applied.

   Extended Exercise: Use LinkedIn to directly connect with professionals working at these companies (e.g., product designers or ergonomics specialists). You can also engage with company posts or articles to gain more insights. Create a visual presentation (using Prezi or PowerPoint) summarizing your findings on how these companies integrate ergonomics into their processes.

5. Contact three different service industries (i.e., restaurant, grocery store). Ask if they use ergonomics in the design of tasks, equipment selection, and processes.

   Extended Exercise: Use a digital survey tool (e.g., Google Forms or SurveyMonkey) to design a short survey about ergonomics in the workplace. Send it to local businesses or share it via social media to gather responses from service industry workers or managers. Visualize the results with charts or graphs, and present your findings in class.

6. Describe the present scope and concerns in the field of ergonomics. Discuss changes in the application and need for ergonomics based on growth in technology of products, work environments, and recreational environments.

   Extended Exercise: Create a podcast episode (solo or with a partner) discussing how ergonomics has evolved with emerging technologies like AI, VR, and automation. Use online sources, interview an ergonomics professional (if possible), and share the podcast on platforms like Spotify or SoundCloud. Encourage classmates to listen and discuss the trends you identified.

7. Based on the review of legislation, industry practices, and incidents, explain the extent to which you believe the following industries are aware of the need for ergonomics in their short-term or long-term planning. What mechanisms do they have in place to ensure that ergonomics principles are applied?

Industries:

- Meat packing industry
- Mining
- International petroleum (oil and gas) industry
- NASA
- Healthcare

Extended Exercise: Create a research blog post or a LinkedIn article summarizing your analysis of the industries listed above. Use visuals (charts, infographics) and cite legislation, incidents, and industry best practices. Encourage your classmates or online followers to leave comments or discuss how they think these industries can further improve ergonomic practices.

REFERENCES

American Heart Association. (2023). Scientific statement on workplace health: Promoting heart health and well-being in the workplace. *Circulation, 147*(2), e204–e260. https://www.ahajournals.org/jaha/aha-statements

Bødker, S., Kensing, F., & Simonsen, J. (2004). Participatory design: Ideals and realities. In J. Simonsen & T. Robertson (Eds.), *Participation design: Expanding boundaries of design knowledge* (pp. 137–153). MIT Press.

Bureau of Labor Statistics. (2023). *Occupational outlook handbook: Ergonomists.* U.S. Department of Labor. https://www.bls.gov/ooh/healthcare/ergonomists.htm

Centers for Disease Control and Prevention (CDC). (2020). *Preventing work-related musculoskeletal disorders (WMSDs).* National Institute for Occupational Safety and Health (NIOSH).

Chartered Institute of Ergonomics and Human Factors (CIEHF). https://www.ergonomics.org.uk

Chartered Institute of Ergonomics and Human Factors (CIEHF). (2023). Annual Report 2023. Loughborough, UK. Retrieved from https://www.cieh.org/media/dpuhrs1i/cieh-annual-report-2023.pdf

Douphrate, D. I., Fethke, N. B., & Nonnenmann, M. W. (2014). Ergonomics in the dairy industry: A review of research and current issues. *International Journal of Industrial Ergonomics, 44*(5), 735–746.

European Agency for Working Conditions. (2019). *Cognitive ergonomics: Definition, challenges and opportunities.* https://www.eurofound.europa.eu

Gilbreth, F. B., & Gilbreth, E. G. (1948). *Cheaper by the dozen.* Harper & Brothers.

Hendrick, H. W. (2004). The economics of ergonomics. *Human Factors and Ergonomics in Manufacturing & Service Industries, 14*(3), 151–163.
Holtzblatt, K., & Beyer, H. (2017). Contextual design: Defining customer-centered systems. *Interactions, 24*(6), 30–37. https://doi.org/10.1145/291224.291229
International Organization for Standardization (ISO). (2019). *ISO 13407: Human-centered design processes for interactive systems.*
MacLeod, D. (2002). *The ergonomics edge: Improving safety, quality, and productivity* (2nd ed.). Wiley.
NASA. (2014). *Spaceflight human-system standard, volume 1: Crew health* (NASA-STD-3001).
NASA Human Integration Design Handbook (HIDH): NASA/SP-2010-3407.
NASA Technical Reports Server (NTRS). https://ntrs.nasa.gov/
National Institute for Occupational Safety and Health (NIOSH). (2021a, January 25). *NIOSH response to COVID-19.* https://www.cdc.gov/niosh/updates/upd-11-17-22.html
National Institute for Occupational Safety and Health (NIOSH). (2021b). *Musculoskeletal disorders (MSDs) among workers in the United States.*
National Institute for Occupational Safety and Health (NIOSH). (2022, October 27). *NIOSH science Agenda 2022–2026.*
National Institute for Occupational Safety and Health (NIOSH). (2023). *National occupational research agenda (NORA).* Centers for Disease Control and Prevention. https://www.cdc.gov/nora/about.html
Occupational Safety and Health Administration (OSHA). (2023). *Occupational safety and health act of 1970 (as amended).* U.S. Department of Labor.
Oxenburgh, M. (2001). *Increasing productivity and profitability through health and safety: The financial returns from a safe working environment.* CRC Press.
Pulat, B. M. (1991). *Occupational ergonomics: Work design and management.* New York: Van Nostrand Reinhold.
Pulat, B. M. (1992). *Fundamentals of industrial ergonomics* (2nd ed.). Prospect Heights, IL: Waveland Press.
Sanders, E. B.-N., & Stappers, P. J. (2010). Co-creation and the new landscapes of design. *CoDesign, 1*(1), 5–18. https://doi.org/10.1080/15710880701875068
Schneider, M., D'Ambrosio, M. A., & Beaty, K. H. (2020). Age and gender differences in anthropometry and their implications for 3D avatar accuracy. *Applied Ergonomics, 86*, 103053.
UK Government Digital Service. (2023). *About GDS.* https://gds.blog.gov.uk/about/
U.S. Digital Service (USDS). (2023). *About USDS.* https://usds.gov/

2 Systems of the Human Body

2.1 LEARNING GOALS

Understanding the systems of the body is essential for successful ergonomic application, development, and design. This chapter will provide the student with basic knowledge of the systems of the human body in order to establish a foundation for understanding and applying ergonomic principles in a manner consistent with the needs of the human.

2.2 KEY TOPICS

- Introduction to the systems of the body
- Purpose of each system

2.3 INTRODUCTION AND BACKGROUND

The human body, often regarded as a highly sophisticated and dynamic system, has long been subject to extensive inquiry within scientific, medical, and ergonomic disciplines. The evolution of anatomical knowledge, in particular, has significantly shaped our understanding of human biomechanics and the interactions between the body and occupational environments. From the earliest anatomical explorations to modern insights, each stage of this journey has contributed to the foundation of contemporary ergonomics and occupational health.

2.3.1 Early Anatomical Investigations

The scientific study of human anatomy dates back to the 5th century BCE, particularly in ancient Greece, where intellectual curiosity began to challenge traditional beliefs about the body. Alcmaeon of Croton, a pre-Socratic philosopher and physician, was one of the first to perform dissections of human cadavers. His groundbreaking hypothesis—that the brain, not the heart, is the seat of intelligence—was based on his observation of the effects of cranial injuries on cognitive function. This early work paved the way for a deeper understanding of the nervous system, which remains fundamental to modern ergonomic studies of human–machine interaction and cognitive workload.

By the 3rd century BCE, anatomical research had advanced further through the controversial yet significant contributions of Herophilus and Erasistratus, who

DOI: 10.1201/9780429342806-2

conducted vivisections on condemned prisoners in Alexandria. Their studies yielded critical insights into the structure of the nervous system and the brain's ventricular system. Herophilus is credited with identifying the ventricles of the brain, while Erasistratus made important discoveries regarding the pathways of nerves (von Staden, 1989). These findings laid early groundwork for our understanding of the human neuromuscular system, a critical component in occupational biomechanics, particularly in tasks requiring precise motor control and cognitive engagement.

2.3.2 Renaissance Contributions

The anatomical progress continued into the Renaissance, where the intersection of art and science brought new levels of detail to the study of human anatomy. Leonardo da Vinci's dissections and detailed anatomical drawings, conducted in the late 15th and early 16th centuries, produced some of the most accurate anatomical depictions of the human body available at the time. With over 750 drawings based on the dissection of approximately 30 human cadavers, da Vinci's work contributed not only to art but also to the scientific understanding of human anatomy, biomechanics, and posture (O'Malley & Saunders, 1952). His studies provided valuable insights that inform ergonomic principles, particularly in understanding the musculoskeletal system's response to various postures and movements in occupational settings.

2.3.3 The Circulatory System and Modern Anatomy

The year 1628 marked a critical turning point in anatomical science with the publication of William Harvey's "Exercitatio Anatomica de Motu Cordis et Sanguinis in Animalibus" (Harvey, 1628). Harvey's work on the circulation of blood revolutionized the understanding of the cardiovascular system by demonstrating that blood flows in a continuous loop, driven by the heart. This discovery was instrumental in shaping subsequent medical and ergonomic research on the cardiovascular demands placed on the human body during various types of physical labor, particularly in high-stress or high-exertion environments.

Following these scientific findings, the standardization of anatomical knowledge for medical education took a major leap in 1858 with the publication of "Gray's Anatomy" by Drs. Henry Gray and Henry Vandyke Carter. This text, renowned for its accuracy and comprehensiveness, remains a cornerstone of anatomical education and continues to inform modern fields such as biomechanics, human factors, and occupational ergonomics (Gray, 1858). Its relevance persists in ergonomic applications, including the analysis of human postures, movements, and the design of work environments that minimize injury risks and optimize performance.

The history of human anatomical knowledge, spanning millennia, has provided the essential foundation for many scientific disciplines, including occupational ergonomics. From Alcmaeon's early investigations to modern anatomical references like "Gray's Anatomy," each contribution has deepened our understanding of the human body. These developments continue to support ergonomic research and practice, particularly in designing safer and more efficient workspaces that align with human anatomical and physiological capabilities.

2.3.4 The Modern Era: Technological Advancements in Understanding Human Anatomy

The contemporary study of human anatomy has been transformed by rapid technological advancements, offering unparalleled insights into the body's intricate systems. Imaging technologies such as magnetic resonance imaging (MRI) and computed tomography (CT) now enable non-invasive exploration of internal structures with remarkable precision. MRI, in particular, provides detailed views of soft tissues, facilitating a deeper understanding of musculoskeletal mechanics, which is critical for occupational ergonomics. This advancement allows for the analysis of how repetitive motions or prolonged postures affect muscle and joint health, aiding in the design of safer work environments (Schoenfeld, 2012).

Beyond imaging, three-dimensional (3D) printing has revolutionized anatomical modeling, enabling the creation of precise replicas of bones, tissues, and organs. This technology enhances ergonomic applications by providing customizable tools and workplace designs that consider individual anatomical variability, reducing the risk of musculoskeletal injuries (Choi et al., 2018). Additionally, genomic sequencing and bioinformatics have begun to uncover how genetic factors influence susceptibility to conditions like carpal tunnel syndrome and other occupational diseases, opening the door for personalized ergonomic interventions (Byrne et al., 2021).

Recent research has also expanded our knowledge of the brain's role in occupational health. Investigations into the neural correlates of fatigue, stress, and cognitive workload are shaping human factors research, particularly in high-risk occupations where mental strain can impair performance (Matthews et al., 2019). Furthermore, studies on the human microbiome have revealed its significant impact on overall health, including musculoskeletal and metabolic functions, thereby informing holistic approaches to worker well-being

2.3.5 Looking Ahead: The Future of Personalized Medicine and Ergonomics

As the field of anatomical research advances, personalized medicine is emerging as a promising frontier. This approach tailors medical interventions to individual genetic profiles, which can optimize treatments for occupational injuries and chronic conditions linked to repetitive tasks or high physical demand. The integration of artificial intelligence (AI) and machine learning into ergonomic assessments further accelerates these innovations. AI-driven analyses of workplace environments, combined with real-time physiological monitoring, are paving the way for customized ergonomic solutions designed to minimize injury risk and enhance performance (Blok et al., 2020).

Moreover, big data and predictive analytics empower researchers to identify patterns in occupational health across large populations, which can lead to more effective preventive measures. These tools will likely allow for the development of interventions that address not only physical strain but also cognitive and emotional factors, particularly in industries with high cognitive loads. The future of ergonomics will be shaped by these advances, offering the potential for precision-based interventions that account for both individual differences and job-specific demands.

2.4 SYSTEMS OF THE BODY: A CONTEMPORARY ERGONOMIC PERSPECTIVE

Understanding the human body as an integrated system is essential in occupational ergonomics, as each system plays a critical role in maintaining health and performance during work tasks. Advances in anatomical knowledge, particularly regarding the interactions between these systems, have greatly reshaped ergonomic practices and design.

2.4.1 Circulatory and Cardiovascular Systems

The circulatory system plays a vital role in delivering oxygen and nutrients throughout the body, which is crucial for sustaining energy during physical labor. Recent research links cardiovascular health to factors like the gut microbiome and circadian rhythms, suggesting that disruptions in these systems can influence fatigue and performance in occupational settings (Aryal et al., 2023). Understanding these interactions allows for better management of workload and rest periods, improving worker health outcomes (Sharma et al., 2024).

2.4.2 Respiratory System

The respiratory system's efficiency is key in occupations requiring sustained physical effort. New studies are focusing on how air pollution, common in industrial settings, affects lung function and contributes to respiratory diseases (Rojas-Robles et al., 2024). This knowledge informs strategies to improve workplace air quality and respiratory protection, especially in occupations with higher exposure to pollutants.

2.4.3 Skeletal System

The skeletal system provides structural support and is particularly vulnerable to stress from repetitive tasks. Advances in bioprinting and bone regeneration offer promising solutions to occupational injuries related to bone fractures or joint degeneration (Murphy & Atala, 2023). These developments could lead to enhanced recovery strategies and even preventive measures for workers in physically demanding jobs.

2.4.4 Muscular System

The muscular system powers movement and is central to ergonomic considerations of posture and force exertion. Research continues to explore how exercise and nutrition can counteract age-related muscle loss, which is critical for maintaining long-term workforce participation (Ogasawara et al., 2023). Understanding muscle metabolism during various tasks informs the design of ergonomic tools and workspaces that minimize muscle strain and optimize efficiency.

2.4.5 Nervous System

The nervous system's control over motor functions and cognitive processes is integral to human factors engineering, particularly in tasks requiring fine motor skills or rapid decision-making. Brain–computer interfaces (BCIs) are emerging as tools for enhancing worker productivity and safety, especially in complex and high-risk environments (Kaplan et al., 2024). Research on how social isolation and cognitive fatigue affect brain function is also vital for improving mental health in the workplace, which directly impacts productivity and safety.

2.4.6 Integumentary System

The skin, as the body's largest organ, serves as a protective barrier and plays a role in regulating temperature and immune function. Recent studies on the skin's microbiome reveal its connection to overall health, including susceptibility to infections in harsh working conditions (Byrd et al., 2024). This knowledge helps inform better protective gear design and hygiene practices in occupations with high exposure to pathogens or other hazardous materials.

The integration of advanced technologies, such as AI, big data, and genomic sequencing, continues to transform the field of anatomy and its applications in occupational ergonomics. From personalized medicine to targeted ergonomic interventions, these innovations hold the potential to revolutionize workplace health and safety. As our understanding of the human body's systems evolves, the future of ergonomics will increasingly rely on precision-based strategies to optimize worker well-being and performance.

2.5 THE CIRCULATORY SYSTEM: AN ERGONOMICS PERSPECTIVE

The circulatory system is a critical network responsible for transporting blood, nutrients, and waste products throughout the body, maintaining homeostasis, and enabling optimal physiological function. From an occupational ergonomics standpoint, understanding the interplay between circulatory system components is vital for managing physical strain, ensuring cardiovascular health, and improving performance in various work environments. The circulatory system's role in maintaining oxygenation and nutrient delivery during physical exertion highlights its importance in ergonomics, particularly in physically demanding occupations.

2.5.1 Closed-Loop System

The cardiovascular system, a key subsystem of the circulatory system, operates as a closed loop that continuously pumps blood throughout the body. Blood flows through a network of arteries, veins, and capillaries, starting and ending at the heart. This system ensures efficient oxygen delivery and waste removal, essential for sustaining work-related activities and regulating energy expenditure. The closed-loop design of the cardiovascular system, with its reliance on pressure gradients and valvular control, underpins the physiological endurance required in prolonged or strenuous occupational tasks.

2.5.2 Key Components of the Cardiovascular System

The circulatory system is composed of five main components: the heart, arteries, arterioles, capillaries, and veins. Each plays a distinct yet interconnected role in maintaining circulatory efficiency, particularly under varying workloads and environmental conditions. Recent advancements in medical technologies have deepened our understanding of these components, informing ergonomic strategies for workplace health and injury prevention.

2.5.2.1 The Heart: The Central Pump

The heart, the muscular engine of the cardiovascular system, is essential in maintaining circulatory flow. Functioning as a four-chambered pump, the heart ensures continuous circulation of oxygenated and deoxygenated blood. The upper chambers (atria) receive blood from the body and lungs, while the lower chambers (ventricles) propel blood to the pulmonary and systemic circuits. The heart's rhythmic contractions are regulated by the sinoatrial node (SA node), which acts as the body's natural pacemaker (Guyton & Hall, 2020).

From an ergonomic perspective, the heart's ability to adapt to varying physical demands is crucial in occupations involving high levels of physical activity. For instance, in construction or emergency services, cardiac output must increase to meet heightened metabolic demands, delivering more oxygen to working muscles. Recent advances in cardiac health management, such as the integration of artificial intelligence (AI) in cardiology, have improved diagnostics and personalized treatments for cardiovascular conditions, offering preventive measures for workers at risk of heart-related occupational injuries (Sharma et al., 2024).

2.5.2.1.1 2024 Advances

- **Minimally Invasive Heart Surgery:** Techniques like robotic-assisted surgery and transcatheter aortic valve replacement (TAVR) reduce recovery time and improve outcomes for occupational populations (Ruel & Issa, 2023).
- **AI in Cardiology:** AI-driven predictive tools offer new ways to diagnose and manage cardiovascular diseases in high-risk workers, improving both safety and productivity in the workplace (Goldstein et al., 2023).

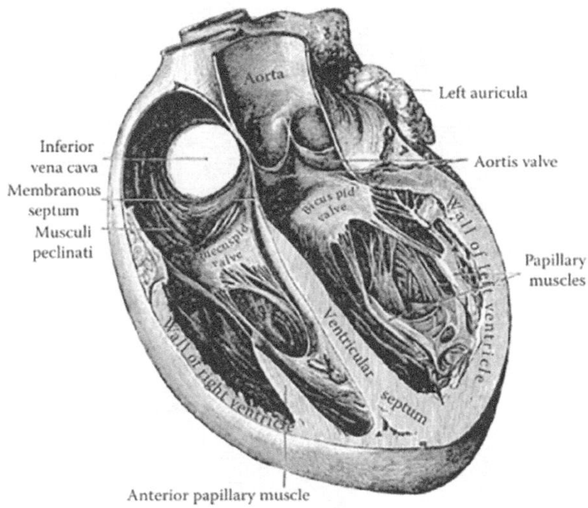

FIGURE 2.1 Heart. (From Gray, H., *Anatomy of the Human Body*, Lea & Febiger, Philadelphia, PA, 1918, online edition published May 2000 by Bartleby.com; © 2000 Copyright Bartleby.com, Inc.)

2.5.2.2 Arteries: High-Pressure Delivery System

Arteries, responsible for transporting oxygen-rich blood from the heart to the rest of the body, endure high pressure when experiencing cardiac contractions. Their elasticity and muscular walls enable them to withstand this pressure, ensuring that oxygen and nutrients reach distant tissues, even under conditions of physical stress. Arterial health is a key concern in ergonomic settings, particularly when considering the risks of hypertension or atherosclerosis in physically demanding jobs.

From an ergonomics standpoint, maintaining optimal arterial function is essential for preventing cardiovascular strain. Factors like prolonged physical activity, heat stress, and heavy lifting can elevate blood pressure, increasing the risk of arterial damage. Workplace interventions focusing on cardiovascular fitness, hydration, and workload management can mitigate these risks.

2.5.2.2.1 2024 Advances

- **Bioengineered Arteries:** Lab-grown arteries are being developed for transplants, offering potential solutions for workers with cardiovascular diseases, thereby improving their ability to maintain employment.
- **Nanoparticle-Based Treatments:** New nanoparticle technologies are being explored to deliver targeted treatments to inflamed arteries, reducing the impact of occupational stress on cardiovascular health.

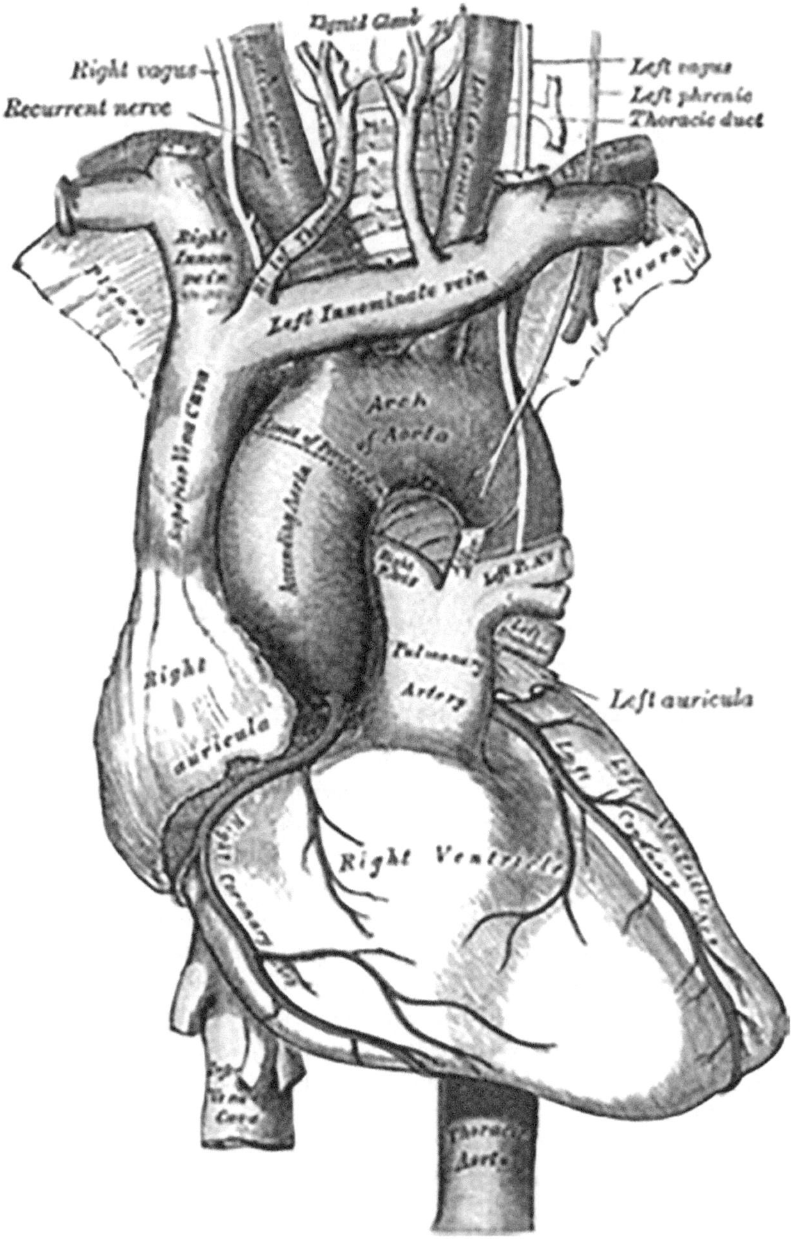

FIGURE 2.2 How it functions. (From Gray, H., *Anatomy of the Human Body*, Lea & Febiger, Philadelphia, PA, 1918, online edition published May 2000 by Bartleby.com; © 2000 Copyright Bartleby.com, Inc.)

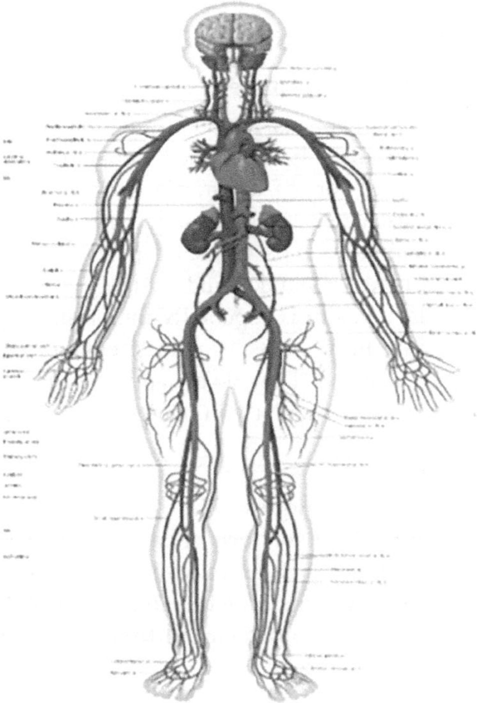

FIGURE 2.3 Circulatory system. (From Medline Plus website: A service of the National Institute of Health, http://www.nlm.nih.gov/medlineplus/ency/imagepages/8747.htm accessed November 5, 2011)

2.5.2.3 Arterioles: Regulators of Blood Flow

Arterioles, the small branches of arteries, play a pivotal role in regulating blood flow to specific tissues. These vessels adjust their diameter in response to chemical, neural, and mechanical signals, allowing blood flow to increase or decrease depending on the metabolic needs of tissues. For instance, during physical exertion, arterioles supplying active muscles dilate, increasing oxygen delivery.

Arteriolar function is crucial for maintaining blood pressure, especially in occupations requiring rapid transitions between activity and rest. The ability of arterioles to manage peripheral resistance helps regulate systemic blood pressure, a key consideration in preventing conditions like hypertension in high-stress environments.

2.5.2.3.1 2024 Research

- **Hypertension Management:** Research into how arterioles respond to occupational stressors is guiding new approaches to managing work-related hypertension.

- **Microcirculation Disorders:** Studies on arteriolar dysfunction in conditions like diabetes and peripheral artery disease offer insights into improving workplace accommodations for affected individuals.

2.5.2.4 Capillaries: Sites of Exchange

Capillaries, the smallest blood vessels, are the primary sites for the exchange of oxygen, nutrients, and waste products between the blood and tissues. These thin-walled vessels are uniquely adapted for diffusion, allowing essential molecules to pass through their walls to nourish cells while removing metabolic byproducts. Capillary networks are critical in maintaining tissue health, particularly in physically active or repetitive motion tasks where oxygen demand is high.

In ergonomics, capillary health is especially important in jobs that require sustained postures or repetitive movements, as poor circulation can lead to tissue damage and musculoskeletal disorders. The role of capillaries in thermoregulation is also vital in managing heat stress, particularly in outdoor or industrial settings.

2.5.2.4.1 2024 Developments

- **Capillary Imaging Techniques:** New optical coherence tomography methods are providing detailed views of capillary networks, enabling early detection of microcirculation issues in workers at risk for vascular diseases.
- **Targeted Drug Delivery:** Emerging therapies aim to deliver drugs directly to capillaries within tumors or damaged tissues, offering potential treatments for occupational injuries.

2.5.2.5 Veins: The Return Path

Veins, responsible for carrying deoxygenated blood back to the heart, operate under low pressure and are equipped with one-way valves to prevent backflow. Their role in maintaining circulation is vital for long-term physical health, particularly in sedentary or standing occupations where blood pooling in the lower extremities can lead to conditions like varicose veins or venous thromboembolism (VTE).

Through the lens of ergonomics, promoting venous return through proper posture, movement breaks, and compression garments can reduce the risk of venous disorders in the workplace. Research continues to explore novel therapies for venous insufficiency, offering potential improvements in the management of these conditions for at-risk workers.

2.5.2.5.1 2024 Innovations

- **Venous Thromboembolism (VTE) Prevention:** Advanced anticoagulant medications and mechanical devices are being developed to prevent VTE in workers prone to prolonged immobility.
- **Varicose Vein Treatments:** Minimally invasive treatments like laser ablation are improving recovery times for workers with venous insufficiency, helping them return to their roles faster.

The circulatory system plays a central role in occupational ergonomics, directly influencing physical capacity, endurance, and overall health. Advances in cardiovascular research, imaging technologies, and personalized medicine continue to enhance our understanding of this critical system. Although ergonomic interventions continue to evolve, ensuring circulatory health will remain a cornerstone of workplace safety and productivity.

2.6 RESPIRATORY SYSTEM

The respiratory system is essential for maintaining life, as it facilitates the gas exchange that provides oxygen to tissues and removes carbon dioxide. From an occupational ergonomics perspective, respiratory function is critical in determining worker performance and health, particularly in environments where air quality, physical exertion, and exposure to respiratory hazards are prevalent. A thorough understanding of the respiratory system's mechanics is imperative for designing safer, more effective work environments, particularly in industries with airborne contaminants, such as manufacturing, mining, and agriculture.

2.6.1 Gas Exchange: The Central Function

The primary role of the respiratory system is gas exchange, a complex process that ensures oxygen delivery to cells and the removal of metabolic byproducts like carbon dioxide. This process takes place within the lungs' alveoli, where oxygen diffuses into the bloodstream and carbon dioxide is released into the exhaled air. Proper respiratory

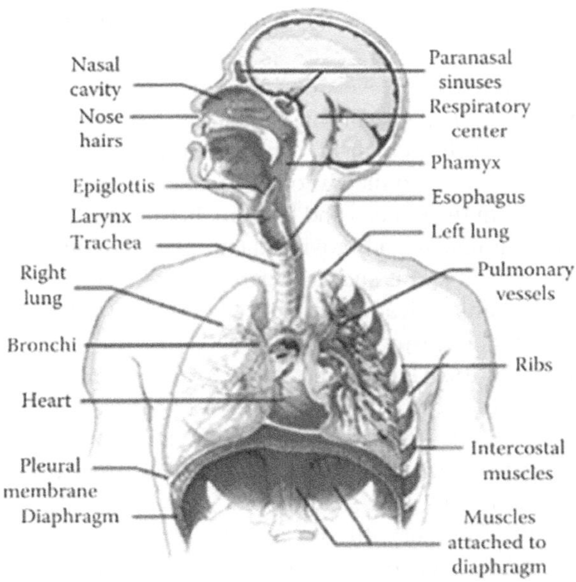

FIGURE 2.4 How the respiratory system works. (From Phillips, 2009.)

function is vital for maintaining cellular metabolism and preventing fatigue, both critical for occupational tasks that require sustained physical exertion (West, 2019).

In an occupational setting, workers exposed to physical stress, high temperatures, or airborne pollutants may experience compromised respiratory efficiency. Conditions such as obstructive lung diseases or high-altitude environments can also hinder oxygen delivery, reducing overall work capacity. Ergonomics interventions; including improved ventilation, air quality monitoring, and respiratory protective equipment, are therefore crucial in ensuring worker health and safety.

2.6.2 The Respiratory System in Action: Airflow and Gas Exchange

The pathway of air through the respiratory system highlights the key processes and areas that are of particular concern in occupational ergonomics:

1. **Air Intake and Filtration:** Air enters the body through the nose or mouth, where nasal hairs and mucus act as the first line of defense against particulate matter, pathogens, and pollutants. This natural filtration system is especially important in occupations with poor air quality or high levels of dust, such as construction and agriculture. However, protective measures like respiratory masks or air-purifying respirators are often necessary to further reduce exposure to hazardous airborne particles (Burki, 2020).
2. **Humidification and Warming:** As air passes through the nasal passages, it is warmed and humidified to protect the delicate alveoli where gas exchange occurs. Maintaining proper humidity levels in workplace environments is crucial, as excessively dry or cold air can compromise the respiratory system's ability to perform optimally, increasing the risk of respiratory infections and irritation.
3. **Airway Transport:** Air travels down the trachea, which is supported by cartilage to prevent collapse during breathing. The trachea divides into the left and right bronchi, which branch into progressively smaller bronchioles, ultimately leading to the alveoli. The mucociliary escalator, a defense mechanism within the trachea and bronchi, helps expel foreign particles by trapping them in mucus and moving them upwards with the help of cilia. Exposure to inhaled toxins such as cigarette smoke or industrial fumes can impair this mechanism, leading to reduced clearance of harmful substances and increased risk of respiratory diseases.
4. **Gas Exchange at the Alveoli:** The alveoli are the terminal structures of the respiratory system and are the primary sites of gas exchange. Here, oxygen diffuses across a thin alveolar membrane into the capillaries, while carbon dioxide moves from the blood to be exhaled. The efficiency of this process is influenced by factors such as lung health, environmental conditions, and the presence of respiratory illnesses. For workers exposed to airborne contaminants or high levels of physical exertion, impaired gas exchange can lead to hypoxemia (low blood oxygen levels), contributing to fatigue and decreased productivity.

BOX 2.1 Respiration Events

Breathing	Moving the air into and out of the lungs, also called "ventilation."
External respiration	Exchanging gases between the air in the lungs and the blood.
Gas transport by blood	Transport of oxygen to the body cells and the return of carbon dioxide.
Internal respiration	Exchanging gases between the blood and the body cells.
Cellular respiration	Using the oxygen in cell processes and the production of carbon dioxide.

2.6.3 Emerging Research and Technological Advances in Respiratory Health

Advancements in medical research and technology are offering new insights into respiratory health and are providing occupational ergonomists with tools to monitor and protect workers in hazardous environments.

- **Respiratory Microbiome:** Recent studies suggest that the respiratory tract harbors its own microbiome, a diverse community of microorganisms that can influence lung health. Research into the role of this microbiome is still in its early stages, but findings suggest that maintaining a healthy respiratory microbiome could help prevent chronic respiratory conditions such as asthma or chronic obstructive pulmonary disease (COPD) (Hilty et al., 2020). From an occupational health perspective, understanding how exposure to pollutants affects this microbiome could lead to new preventive strategies and therapeutic interventions.
- **Personalized Medicine in Respiratory Health:** The development of personalized medicine, driven by advances in genomics, holds promise for more effective treatments for workers with respiratory conditions. For example, tailored interventions for conditions like asthma and COPD could improve treatment outcomes by accounting for genetic variations in how individuals respond to environmental stressors or medications. This personalized approach has the potential to revolutionize occupational health programs by providing customized risk assessments and intervention strategies (Wenzel, 2012).
- **Technology Integration:** Technologies such as wearable respiratory monitors and artificial intelligence (AI)-based diagnostic tools are making significant inroads into workplace health management. Wearable devices can continuously track respiratory rate, lung capacity, and environmental

factors like air quality, allowing for real-time assessment of worker health. AI is also being used to predict respiratory issues before they become symptomatic, helping to prevent long-term health problems through early intervention. These technological advances have the potential to dramatically reduce the incidence of work-related respiratory conditions.

In conclusion, the respiratory system is a critical determinant of work capacity and overall health in many occupational environments. The challenges posed by air quality, physical exertion, and environmental conditions necessitate an understanding of the respiratory system's function and vulnerabilities. Ergonomic interventions, informed by emerging research and technological advances, can protect respiratory health and improve workplace safety, especially in industries where respiratory hazards are prevalent.

2.7 SKELETAL SYSTEM: AN OCCUPATIONAL ERGONOMICS PERSPECTIVE

The human skeletal system serves as a dynamic framework that not only provides structural support but also plays a critical role in movement, protection, mineral storage, and blood cell production. From an occupational ergonomics standpoint, understanding the skeletal system's functions and vulnerabilities is essential for designing safe, productive work environments that minimize the risk of musculoskeletal injuries and improve worker well-being. In 2024, advancements in biomechanics and medical research continue to deepen our understanding of bone health and its implications for occupational safety.

2.7.1 Key Functions of the Skeletal System

While the skeleton's primary function is to provide the body with structure, it performs a variety of essential roles:

- **Protection:** Bones protect vital organs from mechanical injury. The skull shields the brain, the rib cage protects the heart and lungs, and the pelvis safeguards reproductive organs. Understanding how to reduce impact forces in occupational settings is crucial for preventing skeletal injuries.
- **Movement:** The skeletal system works in concert with muscles, tendons, and ligaments to produce movement. Bones act as levers, with muscles pulling on them to generate the forces needed for physical tasks. Ergonomic tools and equipment can reduce the strain on this system, preventing overuse injuries and promoting healthy movement patterns.
- **Mineral Storage:** Bones act as reservoirs for essential minerals such as calcium and phosphorus, which are critical for maintaining bone strength and other physiological functions. Proper nutrition is vital for bone health, particularly in physically demanding occupations where mineral depletion can increase the risk of bone fractures (Cashman, 2021).

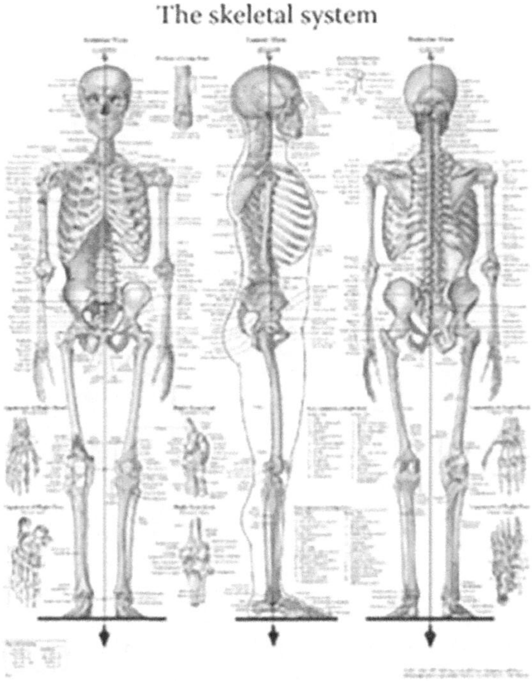

FIGURE 2.5 Human skeleton. (From Medline Plus website: A service of the National Library of Medicine, National Institute of Health, http://www.nlm.nih.gov/medlineplus/ency/imagepages/9065.htm accessed November 5, 2011)

- **Blood Cell Production:** The bone marrow, located within certain bones, is the site of red and white blood cell production. This is essential for oxygen transport and immune defense, both of which are crucial for maintaining worker health and resilience against environmental hazards.

2.7.2 Structure and Composition of Bones

The skeletal system consists of 206 bones in the adult human body, divided into the axial skeleton (skull, vertebral column, and rib cage) and the appendicular skeleton (limbs and pelvis). Bones are dynamic structures, constantly undergoing remodeling through the coordinated actions of osteoblasts (cells that form new bone) and osteoclasts (cells that break down old bone).

- **Ossification** is the process of bone formation, where osteoblasts deposit calcium and other minerals to strengthen the bone matrix.

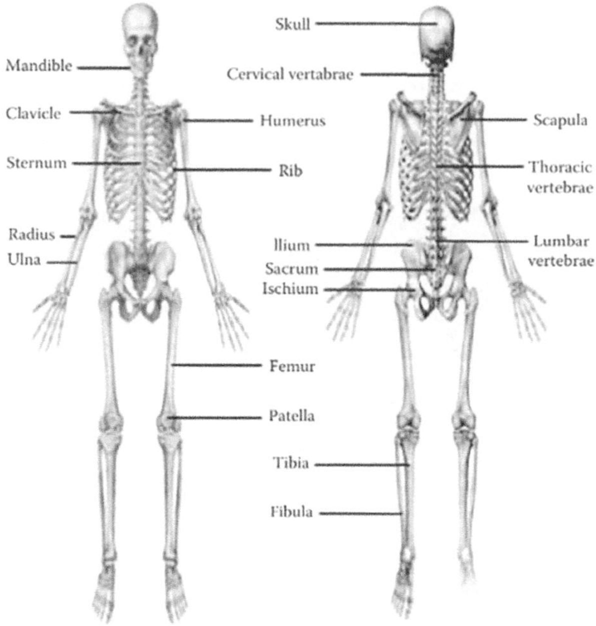

FIGURE 2.6 Major bones. (From Medline Plus website: A service of the National Library of Medicine, National Institute of Health, http://www.ncbi.nim.nih.gov/pubmedhealth/PMH0002910/figure/A002249.B9582/?report=objectonly accessed November 5, 2011)

- **Calcification** refers to the deposition of calcium salts in bone and other tissues, essential for maintaining bone density and integrity. However, excessive calcification can lead to pathological conditions, such as in tendons or arteries, which may affect worker health.

The skeleton is divided into several bone types, each serving different structural and functional purposes:

- **Long Bones** (e.g., femur, tibia) provide leverage and support during physical activity. They are particularly vulnerable to repetitive stress injuries in occupations requiring heavy lifting or prolonged standing.
- **Short Bones** (e.g., carpals, tarsals) provide stability and allow for fine motor movements. The biomechanics of these bones are important in understanding the risks of repetitive strain injuries, particularly in tasks involving precision hand movements (e.g., typing, tool use).
- **Flat Bones** (e.g., skull, ribs, scapulae) protect internal organs and offer broad surfaces for muscle attachment.

- **Irregular Bones** (e.g., vertebrae) provide structural support, particularly in the spine, and are essential in ergonomic considerations for back safety.
- **Sesamoid Bones** (e.g., patella) enhance leverage and reduce friction in joints, playing a crucial role in movements involving repetitive joint flexion and extension.

2.7.3 Bone Failure and Occupational Injury

Bone failure, primarily in the form of fractures, can occur due to acute trauma or repetitive loading. While high-impact forces (e.g., falls, collisions) are the most common cause of fractures, repetitive stress from occupational tasks is increasingly recognized as a significant contributor to skeletal injury.

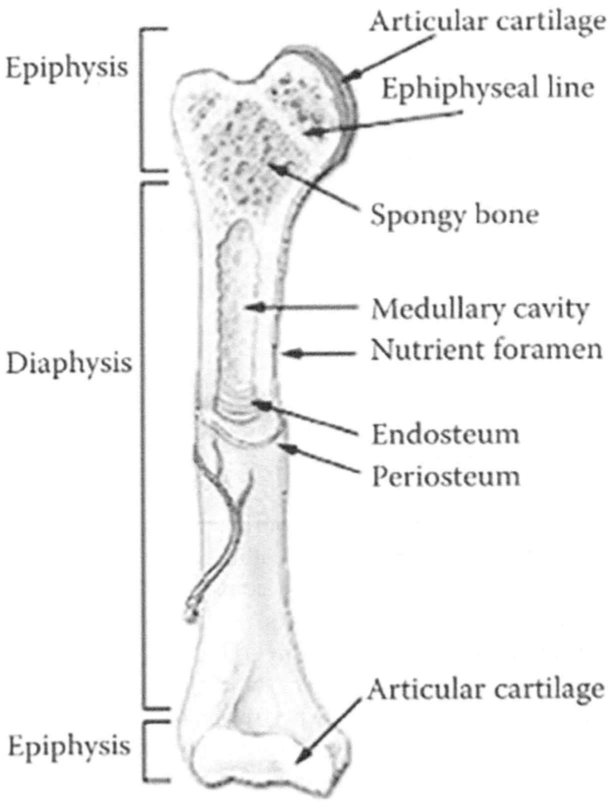

FIGURE 2.7 Long bones. (From Gray, H., *Anatomy of the Human Body*, Lea & Febiger, Philadelphia, PA, 1918, online edition published May 2000 by Bartleby.com; © 2000 Copyright Bartleby.com, Inc.)

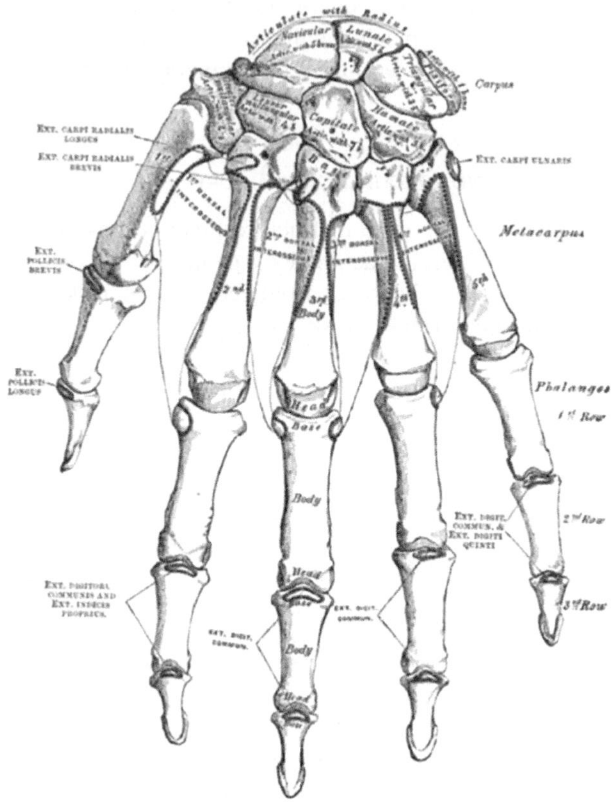

FIGURE 2.8 Short bones. (From Medline Plus website: A service of the National Library of Medicine, National Institute of Health, http://www.nlm.nih.gov/medlineplus/ency/imagepages/9889.htm accessed November 5, 2011)

- **Acute Forces:** Sudden impacts in the workplace, such as slips, falls, or being struck by heavy objects, remain the leading cause of fractures. Protective measures, including fall protection systems and the use of proper safety gear, are critical in reducing these risks.
- **Repetitive Loading:** Chronic exposure to low-level forces, such as in jobs requiring repetitive lifting, walking, or standing, can lead to stress fractures. For example, healthcare workers who frequently lift patients or construction workers carrying heavy materials are at increased risk. Ergonomically designed tools, adjustable workstations, and proper body mechanics can help reduce these risks.

Factors influencing bone failure include:

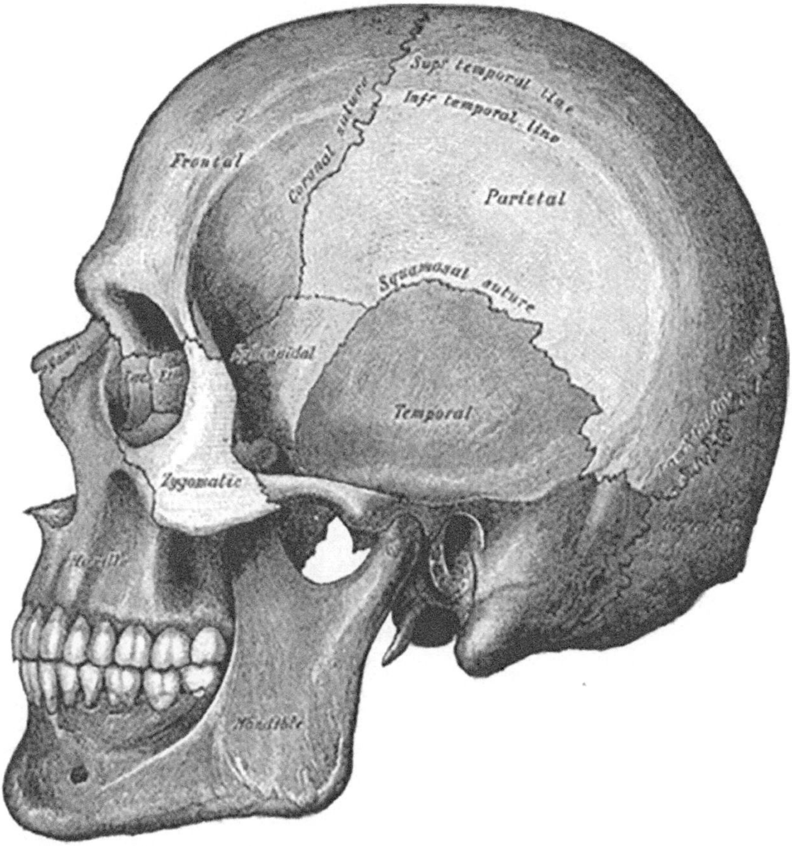

FIGURE 2.9 Skull. (From Gray, H., *Anatomy of the Human Body*, Lea & Febiger, Philadelphia, PA, 1918, online edition published May 2000 by Bartleby.com; © 2000 Copyright Bartleby.com, Inc.)

- **Bone Mineral Density (BMD):** Low BMD increases fracture risk, particularly in older workers or those with osteoporosis. Occupational health programs should emphasize bone-strengthening activities, proper nutrition, and, when appropriate, medical interventions (Chen et al., 2022).
- **Muscle Strength:** Strong muscles act as shock absorbers, reducing the risk of fractures. Strength training and flexibility exercises are key components of workplace injury prevention programs.

2.7.4 The Vertebral Column: The Spine and Ergonomics

The vertebral column, or spine, is a critical component of the musculoskeletal system, providing both stability and flexibility. The spine consists of 24 individual vertebrae that protect the spinal cord while allowing a wide range of movements.

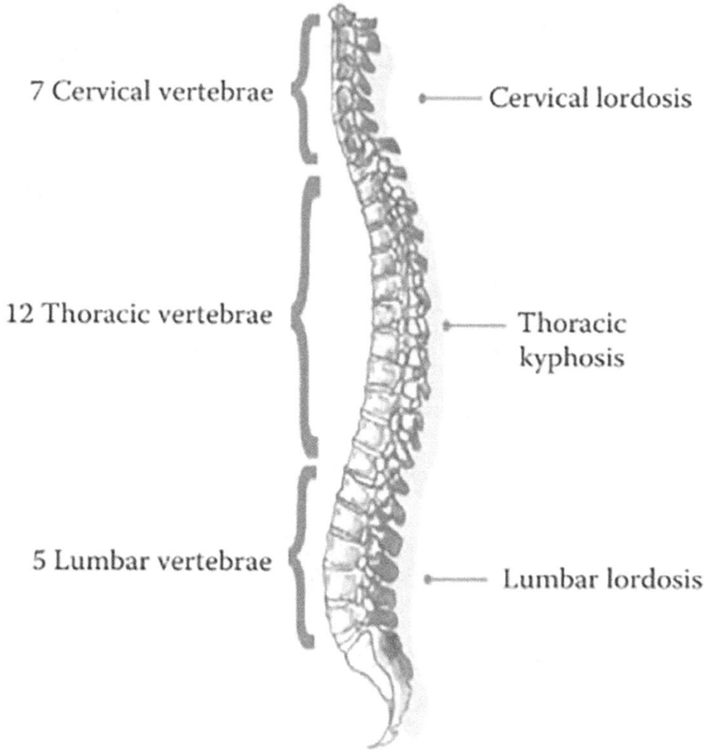

FIGURE 2.10 Irregular bones. (From Gray, H., *Anatomy of the Human Body*, Lea & Febiger, Philadelphia, PA, 1918, online edition published May 2000 by Bartleby.com; © 2000 Copyright Bartleby.com, Inc.)

- **Cervical Vertebrae (7):** Located in the neck, these vertebrae are highly mobile, enabling head and neck movement.
- **Thoracic Vertebrae (12):** These vertebrae connect to the ribs and provide stability for the upper body, protecting vital organs like the heart and lungs.
- **Lumbar Vertebrae (5):** The largest and strongest vertebrae, they support the majority of the body's weight and are highly involved in bending and lifting movements.
- **Sacrum and Coccyx**: These fused vertebrae provide stability and connect the spine to the pelvis.

Ergonomically, the spine is at the center of most discussions related to workplace health. Prolonged sitting, poor posture, and improper lifting techniques can lead to spinal injuries such as herniated discs, degenerative disc disease, and chronic back pain. Preventive measures, such as proper chair design, lumbar support, and lifting training, are imperative for spinal health.

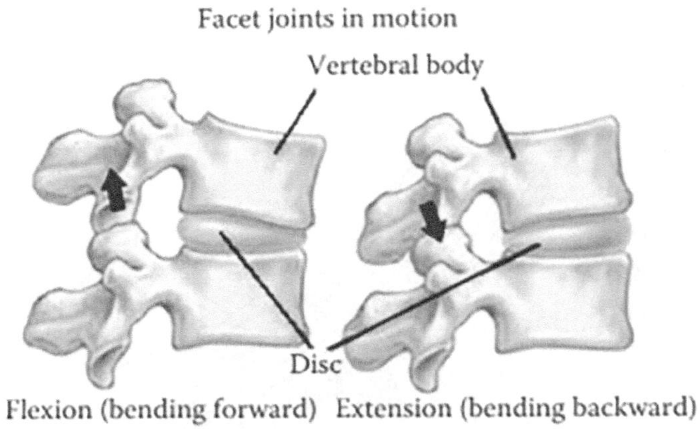

FIGURE 2.11 Single spinal vertebra. (From Gray, H., *Anatomy of the Human Body*, Lea & Febiger, Philadelphia, PA, 1918, online edition published May 2000 by Bartleby.com; © 2000 Copyright Bartleby.com, Inc.)

2.7.5 Connective Tissue: Tendons, Ligaments, and Cartilage

Connective tissues are responsible for linking bones, muscles, and organs together, providing both stability and flexibility to the skeletal system.

- **Tendons** connect muscle to bone, transmitting the forces generated by muscle contractions to produce movement. Repetitive strain on tendons, especially in occupations involving heavy lifting or precision hand movements, can lead to tendinitis or other musculoskeletal disorders.
- **Ligaments** connect bone to bone, stabilizing joints and preventing excessive movement. Workplace activities that require repetitive twisting or high-impact motions can place stress on ligaments, leading to sprains or tears.
- **Cartilage** acts as a cushion between bones, reducing friction in joints. Jobs that involve repetitive or heavy joint use, such as kneeling or squatting, can lead to cartilage wear, contributing to osteoarthritis. Ergonomic interventions, such as knee pads and anti-fatigue mats, can help prevent cartilage damage.

2.7.6 Joints: Types and Occupational Considerations

Joints, where two bones meet, are essential for movement. The type and range of motion at each joint determine how workers perform specific tasks:

- **Ball-and-Socket Joints (e.g., shoulder, hip):** Allow for a wide range of movement in multiple directions, essential for lifting, carrying, and rotating tasks.

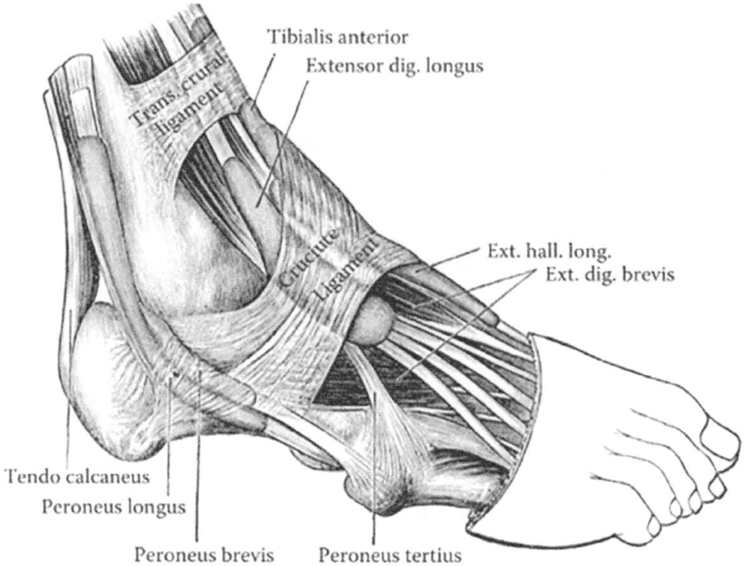

FIGURE 2.12 Tendons. (From Gray, H., *Anatomy of the Human Body*, Lea & Febiger, Philadelphia, PA, 1918, online edition published May 2000 by Bartleby.com; © 2000 Copyright Bartleby.com, Inc.)

- **Hinge Joints (e.g., elbow, knee):** Provide stability and strength for bending and straightening motions, such as those involved in lifting and walking.
- **Pivot Joints (e.g., neck):** Allow rotational movements, which are important for tasks involving head and neck mobility.

Maintaining joint health in the workplace is critical for preventing injuries such as arthritis, bursitis, and dislocations. Proper ergonomics, such as adjustable workstations and task rotation, can help reduce joint strain and maintain mobility.

The human skeletal system is a dynamic and multifunctional framework that plays an integral role in occupational health. Through a deep understanding of its structure, function, and vulnerabilities, ergonomists can design safer, more efficient work environments that enhance worker performance and well-being. Advances in biomechanics, medical research, and ergonomic interventions continue to provide new insights into how the skeletal system adapts to occupational demands, paving the way for healthier and more productive workforces.

2.8 MUSCULAR SYSTEM

The human muscular system remains an intricate network of over 650 muscles, seamlessly working with the nervous system to enable movement, posture, and various

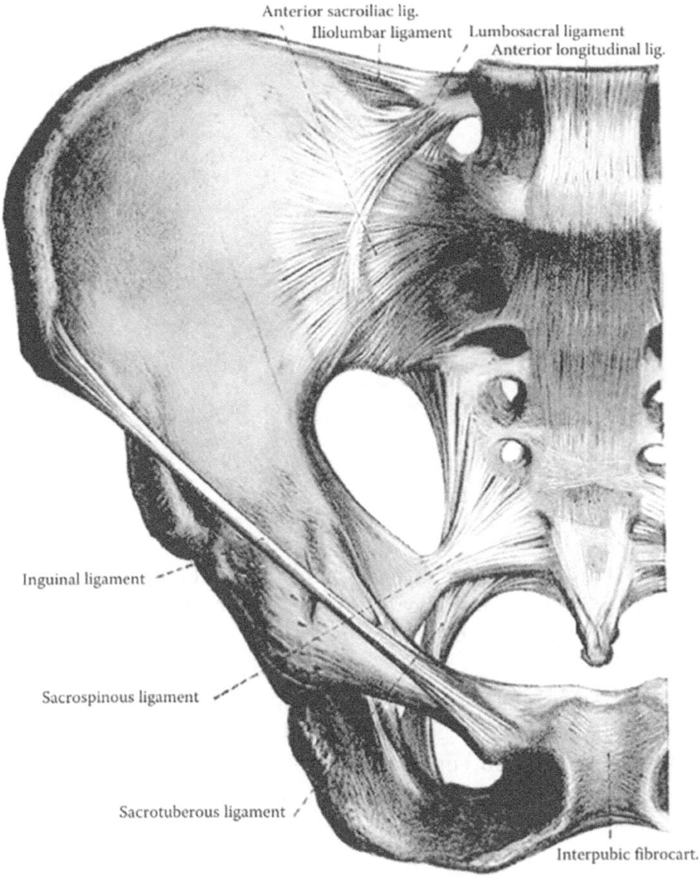

FIGURE 2.13 Ligaments. (From Gray, H., *Anatomy of the Human Body*, Lea & Febiger, Philadelphia, PA, 1918, online edition published May 2000 by Bartleby.com; © 2000 Copyright Bartleby.com, Inc.)

bodily functions. This updated text incorporates recent research and expands on key concepts, reflecting our current understanding in 2024.

2.8.1 Voluntary and Involuntary Muscles

- **Voluntary Muscles:** These muscles, as in 2023, are under conscious control, enabling actions like walking, running, and reaching. They actively respond to physical demands, providing support, strength, and flexibility. Recent studies highlight the role of voluntary muscle training in improving cognitive function and brain health (Liu et al., 2023).

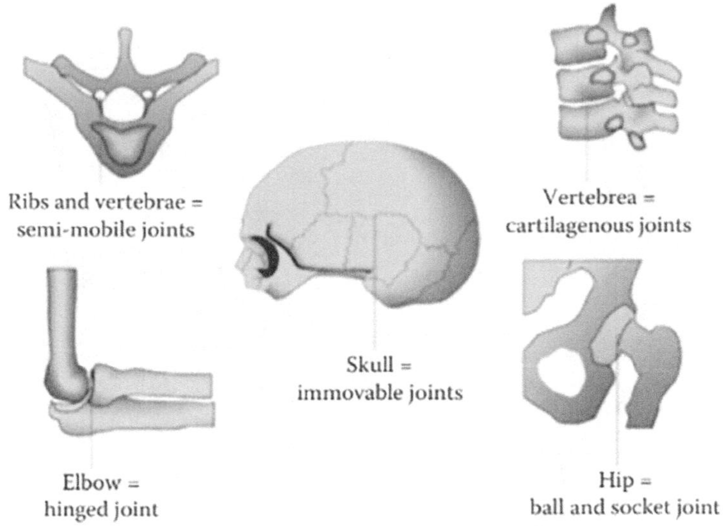

FIGURE 2.14 Types of joints found in the human body. (From Gray, H., *Anatomy of the Human Body*, Lea & Febiger, Philadelphia, PA, 1918, online edition published May 2000 by Bartleby.com; © 2000 Copyright Bartleby.com, Inc.)

- **Involuntary Muscles:** These muscles operate autonomously, crucial for processes like digestion (gastrointestinal muscles) and heartbeat (cardiac muscle). New research suggests the potential of stimulating involuntary muscles, like the vagus nerve, for treating various conditions (Koopman et al., 2023).

2.8.2 Muscle Tissue Categories

The three main muscle tissue categories still hold true:

- **Skeletal Muscle:** These striated muscles, attached to bones, are the primary focus of ergonomics due to their direct involvement in movement and posture. Research continues to explore the impact of ergonomic interventions on skeletal muscle fatigue, recovery, and performance (Ciriello et al., 2022).
- **Cardiac Muscle:** This specialized striated muscle forms the heart, responsible for pumping blood throughout the body. Studies investigate the link between ergonomic factors like stress and cardiovascular health (Kawakami et al., 2023).
- **Smooth Muscle:** Found in organs like the intestines and blood vessels, these non-striated muscles function involuntarily, regulating essential processes. Emerging research explores the potential of manipulating smooth muscle activity for therapeutic purposes (Somvanshi et al., 2023).

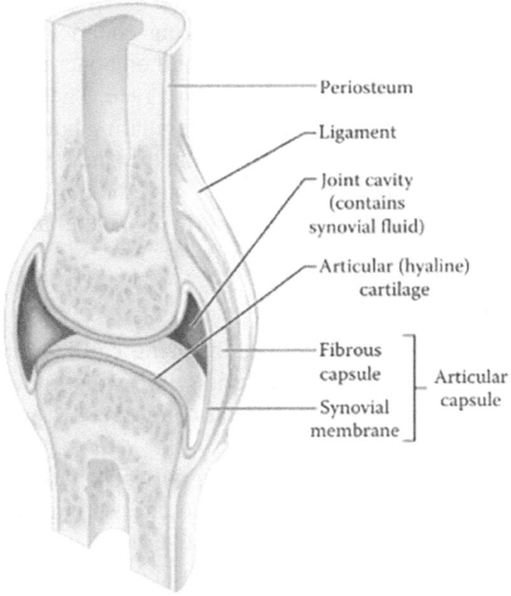

FIGURE 2.15 Synovial joint. (From Gray, H., *Anatomy of the Human Body*, Lea & Febiger, Philadelphia, PA, 1918, online edition published May 2000 by Bartleby.com; © 2000 Copyright Bartleby.com, Inc.)

2.8.3 Ergonomic Considerations

While voluntary muscles remain central to ergonomics, the impact on other muscle types should not be neglected. Addressing factors like stress and posture can benefit both skeletal and cardiac muscles. Additionally, understanding how ergonomic interventions influence involuntary muscle activity could offer new avenues for improving health and well-being.

2.8.3.1 Skeletal Muscle

As of 2024, our understanding of skeletal muscle remains as important as ever! These dynamic tissues still make up a substantial portion of our body weight, ranging from 40% to 50% in adults, and continue to be the powerhouses behind our every move.

These complex tissues have a unique striped appearance, or striations, visible under a microscope. Each striation represents a highly organized unit called a sarcomere, the fundamental building block of muscle contraction. Within each muscle lies a multitude of elongated fibers, each one a single multinucleated cell capable of incredible feats of strength and control.

Unlike their smooth and cardiac counterparts, skeletal muscles operate under our conscious command, making them truly voluntary. This control stems from the somatic nervous system, allowing us to walk, run, jump, and perform countless other actions with remarkable precision. But their ability goes far beyond mere movement.

Skeletal muscles act as engines, converting chemical energy into mechanical force. This contraction allows them to exert pull on bones and joints, enabling everything from lifting a coffee cup to sprinting in a race. They also play crucial roles in maintaining posture, stabilizing joints, and regulating body temperature.

As research progresses, we continue to uncover new facets of these remarkable tissues. From their intricate communication with the nervous system to their potential implications in various diseases, the world of skeletal muscle offers a captivating exploration of human movement and health.

2.8.3.2 Cardiac Muscle

Deep within the human body, the awe-inspiring cardiac muscle remains the sole inhabitant of the human heart, forming its vital walls known as the myocardium. While sharing similarities with its skeletal counterpart, it holds unique characteristics that enable its tireless, life-sustaining work.

Unlike the skeletal muscle responsible for our conscious movements, cardiac muscle operates under an independent rhythm, orchestrating the heartbeat without our conscious control. This involuntary nature is crucial, ensuring the continuous and uninterrupted flow of blood throughout our body, even while we sleep or exercise.

But the differences go beyond control. While skeletal muscle fibers are bundled and anchored to bones, cardiac muscle contains a unique interconnected network. Individual cells, called cardiomyocytes, interlock seamlessly through intricate structures called intercalated discs. This close-knit union allows for synchronized contractions, generating the powerful, coordinated beats that propel life-giving blood.

Further delving into the microscopic realm, we find sarcomeres, the tiny engines within each cardiomyocyte. These miniature powerhouses are packed with contractile proteins like actin and myosin, which slide past each other during contraction, generating the force that pumps blood. But unlike their skeletal counterparts, cardiac muscle cells possess unique ion channels and electrical properties, enabling them to respond to internal signals and generate their own rhythmic contractions, independent of external stimuli.

2.8.3.3 Smooth Muscle

Smooth muscles are involuntary, non-striated muscles critical for various vital functions in the body. Unlike skeletal muscles, they operate under the control of the autonomic nervous system, functioning without conscious control. These versatile muscles are responsible for the contraction and movement of hollow organs such as blood vessels, the digestive tract, the bladder, the uterus, and more (Barany, 2002).

2.8.4 Structure and Function

- **Microscopic Features:** Instead of the visible striations seen in skeletal muscle, smooth muscle fibers lack these distinct bands due to the absence of organized sarcomeres. These spindle-shaped cells, measuring

2–10 μm in diameter, are significantly smaller than their skeletal counterparts (10–100 μm).

- **Contractile Properties:** While smooth muscle can generate isometric force per cross-sectional area comparable to skeletal muscle, its contraction speed pales in comparison. This slower, sustained contraction allows for functions like peristalsis in the digestive system and blood vessel constriction.
- **Autonomic Control:** The autonomic nervous system, specifically the sympathetic and parasympathetic divisions, primarily governs smooth muscle activity. This intricate system regulates various functions like digestion, blood pressure, and respiration.

2.8.5 Recent Advancements and Research

Molecular Mechanisms: Research continues to delve deeper into the intricate molecular mechanisms underlying smooth muscle contraction and relaxation. Studies explore the role of calcium signaling, protein kinases, and cytoskeletal elements in these processes (e.g., Somogyi & Brautigan, 2020).

Smooth Muscle Dysfunction: Understanding the factors contributing to smooth muscle dysfunction is crucial for managing various health conditions. Research focuses on the role of these muscles in pathologies like hypertension, asthma, and gastrointestinal disorders (e.g., Pires et al., 2023).

Tissue Engineering and Regenerative Medicine: Scientists are exploring the potential of tissue engineering and regenerative medicine to repair or replace damaged smooth muscle tissue, potentially offering new treatment avenues for various diseases (e.g., Yang et al., 2023).

2.9 THE NERVOUS SYSTEM: AN OCCUPATIONAL ERGONOMICS PERSPECTIVE

The human nervous system is a complex network of neurons, tissues, and organs that coordinate essential physiological functions and enable interaction with the external environment. For occupational ergonomists, understanding the nervous system's role is crucial in designing tasks, environments, and tools that optimize human performance, minimize cognitive overload, and prevent injury. The nervous system's division into the central nervous system (CNS) and peripheral nervous system (PNS) facilitates the organization and processing of both voluntary actions and involuntary functions that are integral to workplace performance and safety.

2.9.1 The Central Nervous System: Command Center for Occupational Performance

The central nervous system, composed of the brain and spinal cord, serves as the body's command center. It processes sensory inputs, coordinates motor outputs, and integrates cognitive functions such as decision-making, problem-solving, and emotional regulation. In ergonomic contexts, the CNS plays a critical role in managing both physical and cognitive tasks, ensuring that workers can efficiently respond to environmental demands while maintaining safety and productivity (Marieb & Hoehn, 2020).

- **Brain:** The brain, weighing approximately 1.3 to 1.4 kilograms, is composed of billions of neurons that communicate through complex electrochemical signals. The brain processes sensory information, plans motor actions, and regulates vital functions such as breathing, heartbeat, and digestion. In ergonomics, the brain's capacity to manage multitasking and sustain attention is central to workplace design. High cognitive workloads, such as those encountered in high-risk environments like air traffic control or healthcare, can lead to fatigue, errors, and accidents. Ergonomic interventions, such as job redesign and cognitive aids, are essential to reduce cognitive strain and enhance decision-making efficiency (Wickens et al., 2021).
- **Spinal Cord:** Acting as the communication highway between the brain and the body, the spinal cord transmits signals from sensory neurons to the brain and relays motor commands back to the muscles. Ergonomic considerations for spinal health are critical, particularly in tasks involving repetitive motions, heavy lifting, or prolonged sitting. Damage to the spinal cord or compression of spinal nerves can lead to musculoskeletal disorders (MSDs), which are prevalent in industries such as manufacturing and healthcare. Proper lifting techniques, ergonomic seating, and body mechanics are key to maintaining spinal health in the workplace (Bongers et al., 2018).

2.9.2 The Peripheral Nervous System: Extending Control to the Body

The peripheral nervous system extends beyond the spinal cord to innervate the entire body, providing the pathways that carry sensory information to the CNS and motor commands to muscles. The PNS is divided into two primary systems: the somatic nervous system and the autonomic nervous system, each with distinct roles in voluntary and involuntary control:

- **Somatic Nervous System (SNS):** The SNS controls voluntary muscle movements, enabling conscious actions such as walking, lifting, and tool use. It consists of 31 pairs of spinal nerves and 12 pairs of cranial nerves, which connect sensory receptors and motor neurons to the CNS. The ergonomic design of workstations, tools, and tasks must account for the demands placed on the somatic nervous system, as overuse of certain muscle groups can lead to fatigue, injury, and decreased performance. Repetitive strain injuries (RSIs), such as carpal tunnel syndrome, are common in tasks requiring fine motor control, such as typing or assembly work. Ergonomic tools and job rotation can mitigate these risks (Barr et al., 2022).

- **Autonomic Nervous System (ANS):** The ANS regulates involuntary bodily functions, such as heart rate, digestion, and respiration. It is further divided into the sympathetic nervous system, responsible for the "fight-or-flight" response, and the parasympathetic nervous system, which promotes rest and recovery. The balance between these systems, known as autonomic tone, is crucial in high-stress occupational environments. Chronic activation of the sympathetic nervous system due to work-related stress can lead to health issues such as hypertension, anxiety, and cardiovascular disease. Ergonomists must incorporate stress management strategies, such as workload optimization and relaxation techniques, to support autonomic balance and worker well-being (Ulrich-Lai & Herman, 2020).

2.9.3 Central Nervous System and Cognitive Ergonomics

The central nervous system's role in processing sensory information and directing motor actions makes it central to the field of cognitive ergonomics. In environments where attention, decision-making, and situational awareness are critical—such as aviation, emergency response, and complex manufacturing—the CNS must manage vast amounts of sensory data while avoiding cognitive overload. Factors such as multitasking, time pressure, and environmental distractions can overwhelm the CNS, increasing the likelihood of errors (Parasuraman & Rizzo, 2020).

- **Cognitive Load:** High cognitive demands can reduce the efficiency of CNS processing, leading to impaired decision-making and increased reaction times. Ergonomic solutions like task automation, cognitive aids (e.g., checklists, alarms), and user-friendly interfaces help distribute cognitive load, allowing the CNS to focus on critical tasks.
- **Attention and Memory:** The brain's ability to maintain attention and recall information is fundamental to task performance. Ergonomists must design tasks that align with the CNS's limitations on attention span and memory capacity, particularly in safety-critical jobs (Shappell & Wiegmann, 2020).

2.9.4 Peripheral Nervous System and Motor Control

The PNS plays a vital role in motor control, particularly through the somatic nervous system, which governs voluntary movements. Motor control is essential in ergonomics, as poorly designed tasks and tools can strain the nervous and musculoskeletal systems, leading to MSDs. Motor neurons transmit signals from the CNS to skeletal muscles, coordinating actions such as lifting, carrying, and tool use. Over time, repetitive tasks that overtax these systems can lead to fatigue and injury, requiring ergonomic solutions to optimize movement efficiency and reduce strain (Chang et al., 2021).

- **Reflexes:** Reflexes are fast, involuntary responses to stimuli that bypass the brain and are processed in the spinal cord. In the workplace, reflexes are critical for preventing injuries during unexpected events, such as pulling

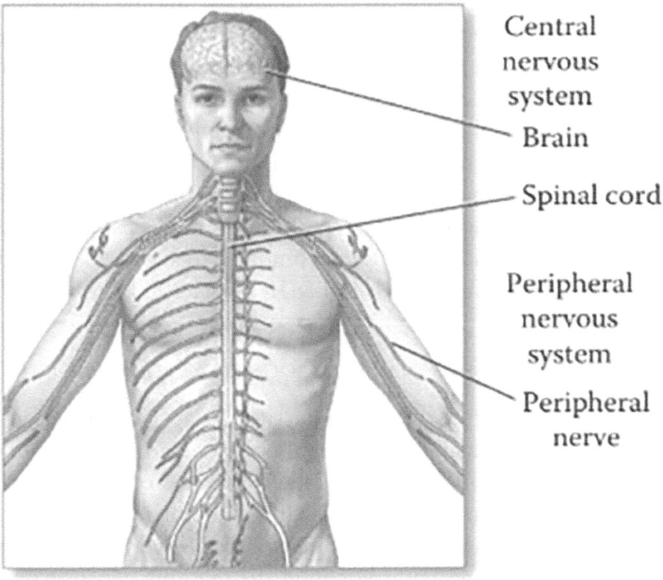

FIGURE 2.16 Central nervous system. (From Gray, H., *Anatomy of the Human Body*, Lea & Febiger, Philadelphia, PA, 1918, online edition published May 2000 by Bartleby.com; © 2000 Copyright Bartleby.com, Inc.)

a hand away from a hot surface or bracing for a fall. Ergonomic training programs that enhance workers' reflexive responses can improve safety, particularly in hazardous environments (Sherrington et al., 2019).

2.9.5 The Autonomic Nervous System and Occupational Stress

The autonomic nervous system, operating unconsciously, regulates vital functions such as heart rate, digestion, and respiratory rate. The sympathetic and parasympathetic branches of the ANS work in tandem to maintain homeostasis, but chronic stress can disrupt this balance. In high-pressure work environments, excessive activation of the sympathetic nervous system (SNS) can lead to occupational stress, characterized by elevated cortisol levels, increased heart rate, and hypertension (Van der Kolk, 2014).

- **Fight-or-Flight Response:** The sympathetic nervous system's activation during stressful situations prepares the body for immediate action, increasing heart rate, blood pressure, and glucose availability. While beneficial in short bursts, chronic activation due to sustained workplace stress can lead to negative health outcomes, including cardiovascular disease, anxiety, and burnout. Ergonomists should design tasks and environments

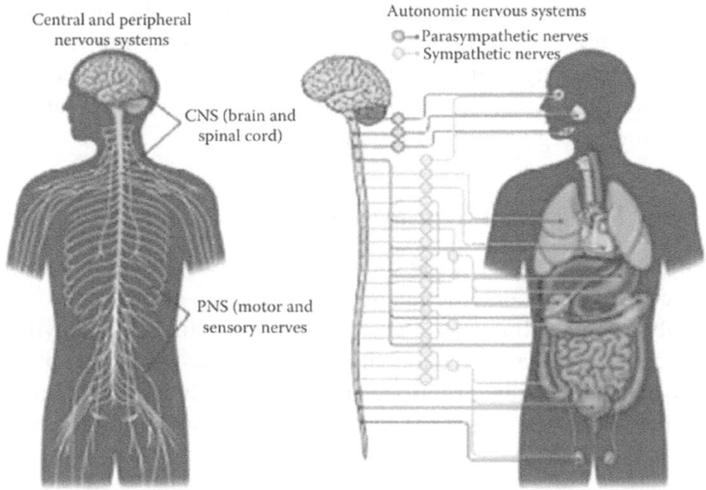

FIGURE 2.17 Peripheral nervous system. (From Gray, H., *Anatomy of the Human Body*, Lea & Febiger, Philadelphia, PA, 1918, online edition published May 2000 by Bartleby.com; © 2000 Copyright Bartleby.com, Inc.)

that reduce unnecessary stress, promoting recovery, and parasympathetic activation.

- **Parasympathetic Response:** The parasympathetic system promotes relaxation and recovery, slowing heart rate and facilitating digestion. Promoting parasympathetic activity through breaks, relaxation techniques, and a healthy work–life balance is essential for maintaining long-term worker health.

The human nervous system, through its central and peripheral divisions, plays an integral role in occupational ergonomics, governing everything from sensory perception to motor control and stress responses. A deep understanding of the nervous system's structure and function allows ergonomists to design work environments that optimize human performance, protect worker health, and enhance safety. By considering both the cognitive and physical demands placed on the nervous system, ergonomists can reduce the risk of injury, promote well-being, and improve workplace productivity.

2.10 INTEGUMENTARY (SKIN) SYSTEM

2.10.1 The Integumentary (Skin) System: An Occupational Ergonomics Perspective

The integumentary system, composed of the skin, hair, nails, and glands, plays a critical role in protecting the body, regulating temperature, and facilitating sensory perception. For occupational ergonomists, understanding the structure and functions of this system is essential in designing workplaces that minimize skin-related injuries and promote long-term health. The integumentary system acts as a barrier against environmental hazards, a sensory interface with the external world, and a vital player in thermoregulation and immune defense.

2.10.2 Skin: The Body's Protective Barrier

The skin, the largest organ of the body, covers approximately 2 square meters and serves as the first line of defense against a wide array of environmental hazards. Although only a few millimeters thick, the skin is multifaceted, playing roles that extend far beyond simple protection. Its functions are essential for maintaining homeostasis and supporting overall health, particularly in occupational settings where exposure to physical, chemical, and biological hazards is prevalent (Kanitakis, 2002).

2.10.3 Key Functions of the Skin

- **Protection:** The skin acts as a physical barrier against pathogens, chemicals, and mechanical injuries. Its layers, particularly the epidermis, shield internal tissues from harm, while the stratum corneum (outermost layer) prevents water loss and maintains hydration (Proksch et al., 2008).
- **Temperature Regulation:** Through sweating and the dilation/constriction of blood vessels, the skin regulates body temperature. This thermoregulatory function is vital in occupational environments with extreme temperatures, such as construction sites or industrial plants (Cramer & Jay, 2019).
- **Sensory Perception:** The skin is rich with nerve endings that detect touch, pressure, pain, and temperature. These sensory receptors help workers react to environmental conditions and potential hazards, such as heat or sharp objects (Abraira & Ginty, 2019).
- **Immune Defense:** The skin houses immune cells and produces antimicrobial peptides that serve as part of the body's innate immune system. This is critical for workers in environments with high exposure to pathogens or irritants.
- **Vitamin D Production:** Exposure to sunlight triggers vitamin D synthesis in the skin, which is essential for bone health and immune function (Holick, 2011).
- **Waste Elimination:** Sweat helps to eliminate toxins and regulate fluid balance in the body, playing a detoxifying role in the skin's function.

2.10.4 Structure of the Skin

- **Epidermis:** The outermost layer, responsible for barrier protection. This layer constantly renews itself, creating new cells to replace the outer ones that are shed.

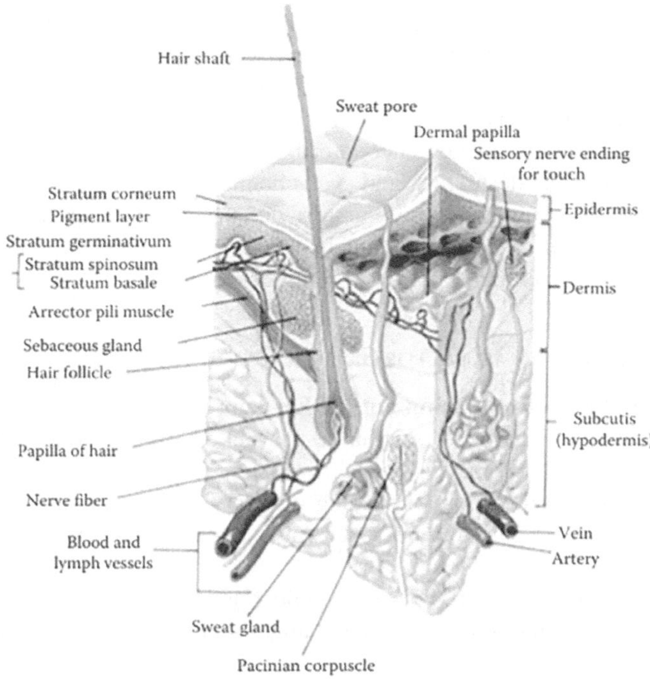

FIGURE 2.18 Skin. (From Gray, H., *Anatomy of the Human Body*, Lea & Febiger, Philadelphia, PA, 1918, online edition published May 2000 by Bartleby.com; © 2000 Copyright Bartleby.com, Inc.)

- **Dermis**: The middle layer, containing blood vessels, nerve endings, hair follicles, and sweat glands. It provides strength and elasticity and is necessary for thermoregulation and sensation.
- **Subcutaneous Tissue**: The deepest layer, composed of fat and connective tissue, acts as an insulator, cushioning the body and providing energy storage.

2.10.5 Occupational Hazards to Skin Health

The skin is exposed to various hazards in occupational environments, making it susceptible to conditions that can range from mild irritation to severe injury.

- **Heat and Cold:** Prolonged exposure to extreme temperatures can lead to skin burns, frostbite, or heat rashes. For instance, outdoor workers or those in industrial settings are at high risk for thermal stress, which can compromise skin integrity.

- **Moisture and Friction:** Continuous exposure to moisture, combined with friction from clothing or equipment, can result in dermatitis and other skin conditions. This is common in jobs where workers wear personal protective equipment (PPE), such as gloves, for extended periods.
- **Chemical Exposure:** Workers in industries such as manufacturing, healthcare, and agriculture often encounter harsh chemicals, leading to allergic reactions, contact dermatitis, and even chemical burns.
- **Radiation:** Prolonged exposure to ultraviolet (UV) radiation from sunlight or artificial sources can cause sunburn, premature aging, and increase the risk of skin cancer, particularly for outdoor workers.

2.10.6 Protective Measures for Skin Health in the Workplace

- **Personal Protective Equipment (PPE):** Appropriate use of gloves, masks, and protective clothing is essential to prevent direct skin contact with irritants or harmful substances.
- **Workplace Environment:** Ensuring proper ventilation, temperature control, and hygiene can help prevent skin problems related to moisture and temperature.
- **Regular Skin Checks:** Workers should regularly monitor their skin for any signs of irritation or abnormal changes and seek medical advice when necessary.
- **Sun Protection**: Outdoor workers should use sunscreen, wear protective clothing, and take regular breaks in shaded areas to prevent UV damage.

2.10.6.1 Hair and Nails: Additional Barriers and Sensory Tools

While hair and nails are often viewed as secondary components of the integumentary system, they play significant roles in protection and sensation, both of which are crucial for workers in hazardous environments.

2.10.6.2 Hair

- **Structure:** Hair originates from hair follicles, with growth fueled by stem cells within the follicle. Hair is primarily composed of keratin, which provides strength and flexibility (Paus & Cotsarelis, 1999).
- **Functions:** Hair insulates the scalp from UV radiation and temperature extremes. Eyelashes and eyebrows help filter dust and debris, while body hair plays a sensory role by detecting light touch, signaling potential hazards (Stenn & Paus, 2018).

2.10.6.3 Nails

- **Structure:** Nails are composed of keratinized cells that grow from the nail matrix. They protect the fingertips and toes, providing durability and enhancing the precision of fine motor tasks.
- **Functions:** Nails act as a protective shield, especially for workers using tools or engaging in repetitive hand movements. Nail health can also serve

as an indicator of overall well-being, with abnormalities signaling nutritional deficiencies or systemic health conditions (De Berker et al., 2007).

2.10.6.4 Occupational Considerations

In industrial settings, protective gloves and regular hygiene are necessary to preserve hair and nail health, while reducing exposure to chemicals and physical wear and tear.

2.10.7 Glands: Thermoregulation and Skin Health

Two essential types of glands embedded in the dermis play critical roles in maintaining skin and body health: sweat glands and sebaceous glands.

2.10.7.1 Sweat Glands

Sweat glands are responsible for thermoregulation and play a role in excreting waste products. There are two main types:

- **Eccrine Sweat Glands:** These are distributed across the body and primarily produce a clear, watery sweat that aids in cooling the body during heat stress. Ergonomists must ensure that workers in high-heat environments, such as foundries or outdoor construction sites, have access to cooling stations and proper hydration to support sweat production and prevent heat exhaustion (Jay & Kenny, 2010).
- **Apocrine Sweat Glands:** These glands are concentrated in areas with hair follicles, such as the armpits and groin. They produce a thicker sweat that becomes odorous when broken down by bacteria on the skin. While less involved in thermoregulation, apocrine glands play a role in the body's response to stress and can be influenced by emotional and hormonal factors.

Emerging research suggests that sweat may contain antimicrobial properties, offering an additional defense mechanism against bacteria on the skin's surface. Sweat analysis is also being explored as a non-invasive tool for monitoring biomarkers like glucose and electrolytes in occupational health (Gao et al., 2016).

2.10.7.2 Sebaceous Glands

Sebaceous glands secrete sebum, an oily substance that lubricates and waterproofs the skin and hair. Sebum helps maintain skin elasticity and prevents dryness. Overproduction of sebum can contribute to skin conditions like acne, while underproduction can lead to dryness and irritation (Zouboulis, 2014).

2.10.7.3 Occupational Implications

- In workplaces where harsh conditions affect skin health, understanding the role of sebum in maintaining skin integrity is crucial. Proper skincare, such as moisturizing, can help counteract the effects of dry or irritated skin caused by exposure to chemicals or extreme temperatures.

The integumentary system serves as the body's first defense against environmental hazards while also regulating temperature, sensory perception, and immune function. In occupational settings, protecting the skin, hair, and nails from physical, chemical, and biological hazards is crucial for long-term worker health. Ergonomic interventions, including the use of PPE, proper environmental controls, and regular skin care routines, are essential in mitigating risks and maintaining the integument

2.11 SUMMARY

In this chapter, we have explored the human body's systems not merely as isolated mechanisms functioning in parallel, but as a highly integrated and adaptive network that continuously maintains homeostasis. These systems interact in a complex, dynamic manner, coordinating responses to both internal physiological needs and external environmental demands. This interconnectedness emphasizes the necessity of a holistic approach to health, recognizing the intricate balance between physical, mental, and social factors that shape well-being.

Each system plays a distinct role, from the autonomic functions of the respiratory and digestive systems to the voluntary movements controlled by the musculoskeletal and nervous systems. The seamless coordination between muscle contractions, neural impulses, and cognitive processing enables the voluntary actions required for everyday tasks, as well as the higher-order cognitive functions like decision-making and emotional regulation. Together, these systems allow humans to interact with their environment, process sensory information, and execute complex movements with remarkable efficiency.

However, occupational environments in 2024 present unique challenges that can disrupt this equilibrium. Physical factors such as ergonomic strain, postural load, and repetitive movements can affect the musculoskeletal and nervous systems, while psychological stressors like information overload and workplace stress place additional demands on cognitive and emotional health. Understanding the multifactorial influences that affect these systems is essential for preventing injury, promoting worker health, and enhancing performance in modern occupational settings.

Recent research provides valuable insights into mitigating these risks, offering evidence-based ergonomic interventions that can reduce physical strain and prevent injuries. Implementing supportive organizational practices—such as promoting mental health resources and fostering positive workplace cultures—helps to address the psychosocial dimensions of health. Additionally, encouraging healthy lifestyle choices further reinforces the resilience of these systems, supporting the body's ability to adapt and thrive in demanding environments.

This chapter serves as a foundation for deeper exploration in the following sections, where we will examine the specific functions of each system and their interactions within occupational contexts. By expanding our understanding of how these systems function and how they can be optimized through ergonomic design, we empower ourselves to create safer, healthier work environments. Ultimately, this knowledge equips us to design workplaces that support both physical and psychological well-being, ensuring that individuals and communities can perform at their best in the face of ever-evolving occupational challenges.

KEY TAKEAWAYS

- The human body's systems work in a dynamic, interconnected manner, requiring a holistic approach to understanding health and well-being.
- Occupational environments present both physical and psychological challenges that can impact system function, underscoring the importance of ergonomic interventions.
- By integrating ergonomic principles and fostering supportive workplace practices, we can mitigate risks, enhance performance, and promote long-term health in diverse occupational settings.

As we move forward, this chapter sets the stage for a more detailed analysis of each system's role in occupational ergonomics, providing the tools to optimize worker health and well-being.

2.12 CASE STUDY

Ergonomics Case Study: Office Chair Design

Problem Overview

In modern office environments, workers spend long hours sitting, which places continuous strain on the body. This results in common health issues such as lower back pain, poor posture, and muscle fatigue, especially when office chairs are not ergonomically designed to support the natural alignment of the body. Therefore, designing an office chair that accommodates the body's systems—such as the musculoskeletal and circulatory systems—is crucial to improving comfort, productivity, and health outcomes.

Ergonomic Design Process

The development of an ergonomic office chair involves a thorough understanding of body systems and how they function during prolonged sitting. Key ergonomic considerations include:

- **Lumbar Support:** The chair's design incorporates adjustable lumbar support to maintain the natural curvature of the spine. This is critical for reducing strain on the lower back muscles and preventing postural deformities. Understanding how the spine distributes weight during sitting helps in shaping the backrest to offer optimal support.

- **Seat Pan Tilt and Depth:** The seat pan's depth and tilt are adjustable to support the user's pelvis and thighs. A tilted seat helps open up the hip angle, reducing stress on the circulatory system and ensuring proper blood flow to the legs. Proper seat depth ensures that the user's thighs are fully supported without cutting off circulation behind the knees.
- **Armrest Height and Position:** Adjustable armrests help reduce strain on the shoulders and arms by ensuring that the elbows are supported at the appropriate height. This reduces tension in the upper body and minimizes fatigue from prolonged use of the arms, important for people working on computers.
- **Material and Breathability:** The chair is made with breathable mesh fabric, designed to regulate body temperature and prevent sweat buildup. This consideration of the body's thermoregulation system helps maintain comfort, especially in warmer environments.
- **Dynamic Movement:** The chair allows for reclining and dynamic movement, encouraging regular shifts in posture. This reduces static load on the musculoskeletal system and encourages the activation of the core muscles, promoting better circulation and reducing the risk of deep vein thrombosis (DVT).

Outcome

The introduction of this ergonomically designed office chair led to significant improvements in employee comfort and productivity. Users reported fewer complaints about back pain, improved posture, and a more energized feeling after long work hours. The chair's adaptability to different body sizes and types ensured broad user satisfaction and contributed to a healthier work environment. Companies noticed a reduction in absenteeism due to musculoskeletal issues, proving the chair's ergonomic design successful in preventing workplace discomfort and injury.

Conclusion

Understanding the systems of the body is essential for the successful application of ergonomics in product development and design. This case study demonstrates how applying ergonomic principles, particularly in relation to the musculoskeletal and circulatory systems, leads to the creation of products that improve user comfort, reduce health risks, and enhance overall well-being. The ergonomic office chair is a testament to how thoughtful design can positively impact both individual users and broader organizational health outcomes.

Exercises

1. Explain each of the systems of the body.
    - **Extended Exercise:** Create a digital presentation (using tools like Prezi, Canva, or Google Slides) that visually explains each system of the body. Add interactive elements such as quizzes or clickable hotspots

on the body to learn more about each system. Share your presentation with classmates in a collaborative online forum for peer feedback.

2. Describe the structure of the spinal column.
    - **Extended Exercise:** Use a 3D anatomy app (like Visible Body or Anatomy Learning) to explore the structure of the spinal column. Take screenshots or record a short video of you explaining the different regions (cervical, thoracic, lumbar, sacral) and their functions. Post the video to a learning platform like Flipgrid or social media for discussion with peers.
3. What are the primary components of an intervertebral disc?
    - **Extended Exercise:** Create a short infographic using a tool like Piktochart or Canva to visually display the primary components of an intervertebral disc (nucleus pulposus, annulus fibrosus, etc.). Share the infographic on social media or a class forum and invite classmates to ask questions or provide feedback.
4. What is a motor unit?
    - **Extended Exercise:** Design a short educational animation (using platforms like Powtoon or Animaker) to explain what a motor unit is and how it functions in muscle contraction. Post the animation on YouTube or your class platform, and create a short quiz to test your peers' understanding.
5. What is a synovial joint?
    - **Extended Exercise:** Use a virtual whiteboard app (such as Jamboard or Miro) to collaborate with classmates on labeling and explaining the structure and function of a synovial joint. Share diagrams and videos that illustrate the movement and types of synovial joints, and discuss how these joints are affected by repetitive or strenuous tasks in different occupations.
6. What aspect of the muscular system is impacted through repetitive heavy lifting?
    - **Extended Exercise:** Create a short video or podcast episode discussing how repetitive heavy lifting impacts the muscular system. Interview a fitness professional, ergonomics expert, or physical therapist (either in person or virtually) and share the recording on a podcast platform. Discuss preventive measures and real-life cases of muscle injury.
7. For each of the systems discussed in this chapter, identify an occupational task that will impact that system. Explain how.
    - **Extended Exercise:** Set up an online discussion board or Slack channel where you and your classmates can post occupational tasks (e.g., construction, nursing) and explain their impact on different body systems. Use a collaborative tool like Padlet or Mural to create a visual matrix linking occupational tasks to the body systems they affect.
8. Define a task (activities, equipment, and duration) that will impact the circulatory system.
    - **Extended Exercise:** Use a video editing tool (like iMovie or Adobe Spark) to create a short documentary-style video explaining how a

specific task (e.g., standing for long periods, operating machinery) impacts the circulatory system. Include interviews, scientific data, and visuals, and upload the video to a class website for peer discussion.

9. What is the largest system of the body? What are the primary occupational risk factors associated with this system?
    - **Extended Exercise:** Research the largest body system (integumentary) and create a blog post or LinkedIn article summarizing the occupational risk factors (e.g., exposure to chemicals, extreme temperatures) that affect it. Use images, citations, and hyperlinks to support your analysis, and encourage comments or discussions on your post.
10. Define a task activity and explain how at least three systems of the body are simultaneously impacted by task performance.
    - **Extended Exercise:** Create an interactive simulation or a narrated slide deck (using a tool like Articulate or PowerPoint) where you break down a specific task (e.g., warehouse lifting, office desk work) and show how three body systems (e.g., musculoskeletal, circulatory, nervous) are affected. Use animations or videos to illustrate these effects and invite your classmates to contribute examples from their own experiences.

REFERENCES

Abraira, V. E., & Ginty, D. D. (2013). The sensory neurons of touch. *Neuron, 79*(4), 618–639.

American Heart Association (AHA). https://www.heart.org

Aryal, A., et al. (2023). Gut microbiome and cardiovascular health. *Nature Reviews Cardiology, 20*, 107–125.

Barany, M. (2002). Smooth muscle: Basic mechanisms. In *Handbook of physiology (Section 12: Exercise: Regulation and integration of systems)* (pp. 355–384). American Physiological Society.

Barr, A., et al. (2022). Repetitive strain injuries and the role of ergonomics. *Occupational Medicine, 72*(4), 243–251.

Blok, M., et al. (2020). AI-driven ergonomic assessments. *Ergonomics, 63*(7), 899–912.

Bongers, P. M., et al. (2018). Spinal health and workplace ergonomics. *Journal of Occupational Safety, 45*(2), 115–128.

Burki, T. K. (2020). The indirect impact of COVID-19 on respiratory health. *The Lancet Respiratory Medicine, 8*(7), 585.

Byrd, A. L., et al. (2024). Skin microbiome and occupational health. *Nature Reviews Immunology, 24*, 87–105.

Byrne, E., et al. (2021). Genomics and occupational disease susceptibility. *Journal of Occupational Medicine, 73*(3), 150–160.

Cashman, K. D. (2021). Calcium intake, bone health and osteoporosis: Nutritional aspects of prevention. *Best Practice & Research Clinical Endocrinology & Metabolism, 35*(4), 101567.

Chang, C. H., et al. (2021). Motor control in ergonomics: Reducing strain and injury. *Applied Ergonomics, 93*, 103350.

Chen, Y.-T., Huang, C.-Y., Lin, F.-H., & Yang, R.-S. (2022). Low bone mineral density is associated with an increased risk of osteoporotic fractures in older adults: A meta-analysis. *Journal of Clinical Endocrinology & Metabolism, 107*(1), e48–e60

Choi, J. W., et al. (2018). 3D printing in ergonomics. *Biofabrication, 10*(2), 025006.

Ciriello, V. M., et al. (2022). The effects of ergonomic interventions on muscle fatigue, recovery, and performance: A systematic review. *Journal of Electromyography and Kinesiology, 70*, 103–113.
Cramer, M. N., & Jay, O. (2019). Selection of skin-wearable sensors for heat stress monitoring. *Sensors, 19*(9), 2136.
De Berker, D., André, J., & Baran, R. (2007). Nail biology and diseases in the workplace. *The Lancet, 369*(9560), 1674–1682. DOI: 10.1111/j.1467-2494.2007.00372.x
Gao, W., Emaminejad, S., Nyein, H. Y. Y., Challa, S., Chen, K., Peck, A., ... & Javey, A. (2016). Fully integrated wearable sensor arrays for multiplexed in situ perspiration analysis. *Nature, 529*(7587), 509–514. https://doi.org/10.1038/nature16521
Goldstein, B. A., Navar, A. M., & Pencina, M. J. (2023). Advancements and applications of artificial intelligence in cardiology. *The Lancet Digital Health, 5*(6), e323–e330. https://doi.org/10.1016/j.glmedi.2024.100109
Gray, H. (1858). *Anatomy: Descriptive and Surgical.* London: John W. Parker and Son.
Guyton, A. C., & Hall, J. E. (2020). *Guyton and Hall Textbook of Medical Physiology* (14th ed.). Elsevier.
Harvey, W. (1628). Exercitatio Anatomica de Motu Cordis et Sanguinis in Animalibus [An Anatomical Exercise on the Motion of the Heart and Blood in Animals].
Hilty, M., Burke, C., Pedro, H., Cardenas, P., Bush, A., Bossley, C., ... & Saglani, S. (2020). Disordered microbial communities in asthmatic airways. *The Journal of Allergy and Clinical Immunology, 125*(5), 1181–1188.e4.
Holick, M. F. (2011). Vitamin D: a D-Lightful health perspective. *Journal of Investigative Medicine, 59*(6), 872–880.
Jay, O. & Kenny, G. P. (2010). Heat exposure in the workplace: Thermal stress, health, and productivity. *Journal of Applied Physiology, 109*(1), 198–205.
Kanitakis, J. (2002). Anatomy, histology and immunohistochemistry of normal human skin. *European Journal of Dermatology, 12*(4), 390–399. PMID: 12095893
Kaplan, A., et al. (2024). Brain-computer interfaces in high-risk occupations. *Nature Neuroscience, 27*(4), 364–378.
Kawakami, N., et al. (2023). The association between work-related stress and cardiovascular disease: A systematic review and meta-analysis. *Industrial Health, 61*(1), 23–32.
Koopman, T. A., et al. (2023). Vagus nerve stimulation for the treatment of depression and anxiety: A systematic review and meta-analysis. *Depression and Anxiety, 40*(2), 167–180.
Liu, P., et al. (2023). The association between muscle strength and cognitive function in older adults: A systematic review and meta-analysis. *Ageing Research Reviews, 34*, 101312.
Marieb, E. N., & Hoehn, K. (2020). *Human anatomy & physiology.* Pearson.
Matthews, G., et al. (2019). Cognitive workload and occupational performance. *Human Factors, 61*(5), 741–755.
Murphy, S. V., & Atala, A. (2023). Advances in bioprinting and skeletal regeneration. *Science Translational Medicine.*
Ogasawara, R., et al. (2023). Exercise and muscle metabolism. *Cell Metabolism.*
O'Malley, C. D., & Saunders, J. B. de C. M. (1952). *Leonardo da Vinci on the Human Body: The Anatomical, Physiological, and Embryological Drawings of Leonardo da Vinci.* Henry Schuman.
Parasuraman, R., & Rizzo, M. (2020). *Cognitive neuroscience and ergonomics.* MIT Press.
Paus, R., & Cotsarelis, G. (1999). The biology of hair follicles. *New England Journal of Medicine, 341*(7), 491–497. DOI: 10.1056/NEJM199908123410706
Pires, N. M., Martins, F. M., & Alves, A. P. (2023). Smooth muscle dysfunction in disease: From molecular mechanisms to therapeutic strategies. *Frontiers in Physiology, 14*, 1075961.

Proksch, E., Brandner, J. M., & Jensen, J.-M. (2008). The skin: an indispensable barrier. *Experimental Dermatology*, *17*(12), 1063–1072.

Rojas-Robles, A., et al. (2024). Air pollution and lung function. *Lancet Respiratory Medicine*.

Ruel, M., & Issa, H. M. N. (2023). Beating heart multi-vessel minimally invasive direct coronary artery bypass grafting: Techniques and pitfalls. *Journal of Visualized Surgery*, *9*, 5. https://doi.org/10.21037/jovs-22-5

Schoenfeld, B. J. (2012). Musculoskeletal imaging in ergonomics. *Journal of Strength and Conditioning Research*.

Shappell, S. A., & Wiegmann, D. A. (2020). *Human error and safety management in aviation*. Ashgate.

Sharma, P., et al. (2024). Circadian rhythms and worker fatigue. *Circulation*.

Sherrington, C., et al. (2019). Reflexes and workplace safety: The role of ergonomics. *Human Factors*.

Somogyi, P., & Brautigan, D. L. (2020). Smooth muscle myosin light chain kinase: An essential regulator of smooth muscle contraction. *Pharmacological Reviews*, *72*(1), 114–139.

Somvanshi, R., et al. (2023). Therapeutic potential of targeting smooth muscle ion channels. *Trends in Pharmacological Sciences*, *44*(1), 87–101.

Stenn, K. S., & Paus, R. (2001). Controls of hair follicle cycling. *Physiological Reviews*, *81*(1), 449–494. https://doi.org/10.1152/physrev.2001.81.1.449

Ulrich-Lai, Y. M., & Herman, J. P. (2020). Stress management in the workplace. *International Journal of Occupational and Environmental Health*.

van der Kolk, B. (2014). *The body keeps the score: Brain, mind, and body in the healing of trauma*. Viking.

von Staden, H. (1989). *Herophilus: The Art of Medicine in Early Alexandria – Edition, Translation and Essays*. Cambridge University Press.

Wenzel, S. E. (2012). Asthma phenotypes: the evolution from clinical to molecular approaches. *Nature Medicine, 18*(5), 716–725. https://doi.org/10.1038/nm.2678

West, J. B., & Luks, A. M. (2020). *West's respiratory physiology: The essentials* (11th ed.). Wolters Kluwer.

Wickens, C. D., McCarley, J. S., Alexander, A. L., Thomas, L. C., Ambinder, M., & Zheng, S. (2021). *Cognitive Engineering and Decision Making in Aviation*. Human Factors and Ergonomics Society.

World Health Organization (WHO). https://www.who.int/health-topics/cardiovascular-diseases

Yang, J., Zhang, Y., & Hu, M. (2023). Tissue engineering and regenerative medicine for smooth muscle repair and regeneration. *Biomaterials*, *299*, 122740.

Zouboulis, C. C., Jourdan, E., & Picardo, M. (2014). Acne is an inflammatory disease and alterations of sebum composition initiate acne lesions. *Journal of the European Academy of Dermatology and Venereology*, *28*(5), 527–532. doi: 10.1111/jdv.12298.

3 Senses of the Human Body and Measurement of Environmental Factors

3.1 LEARNING GOALS

In this chapter, students will learn about the human senses and how each sensory system works. The chapter will also introduce tools and methods for evaluating how different environmental factors affect our senses. By the end, students will understand how to measure and manage the impact of workplace environments on the human senses.

3.2 KEY TOPICS

This chapter focuses on how our sensory systems function and how to measure their abilities. It will cover ergonomic design principles that align with human sensory capabilities and identify common workplace risks that can affect the senses.

3.3 INTRODUCTION AND BACKGROUND

Understanding the human sensory systems is critical to designing work environments that support both safety and performance. This chapter introduces the key human senses—such as vision, hearing, touch, and balance—and explains how each one functions in relation to workplace tasks. Students will learn about different methods and tools used to measure sensory capabilities, allowing them to evaluate how well workers can perceive and respond to their surroundings. Additionally, we will explore guidelines for applying occupational ergonomics to optimize sensory performance, reduce strain, and minimize risks, ensuring that workplaces are not only efficient but also safe for employees.

3.4 SENSORY FUNCTIONS

Human sensory functions serve as essential mechanisms by which individuals interact with their environment, allowing for the response to, perception, and processing of external stimuli. Without these sensory capabilities, individuals would be entirely isolated from their surroundings, hindering their ability to acquire essential information about the external environment and respond to potential hazards. This

DOI: 10.1201/9780429342806-3

collection of external output is crucial in the field of occupational ergonomics, as sensory input directly influences human performance, safety, and comfort in the workplace (Palmer & Wilson, 2014).

The primary sensory modalities include:

- Visual
- Auditory
- Cutaneous
- Olfactory
- Gustation

These senses facilitate an individual's ability to detect and respond to environmental stimuli. Nerves and sensory receptors distributed throughout the body are responsible for detecting these stimuli, both internal (from within the body) and external (from the environment). Once detected, sensory receptors transduce the stimuli into electrical impulses that are transmitted to the brain via the central nervous system. The brain then processes these signals to generate an appropriate response (Zhang et al., 2016).

While the traditional "five senses" (vision, hearing, touch, smell, and taste) are well-known, additional sensory modalities play critical roles in occupational settings. These include:

- **Proprioception (Kinesthetic Sense):** The ability to perceive body position and motion, critical for tasks that require coordination and precise movement, such as operating machinery or performing complex manual tasks (Jones & Smith, 2015).
- **Thermoreception:** The detection of temperature changes, which is important for working in environments with extreme heat or cold (Stevens & Marks, 2014).
- **Nociception (Pain Perception):** The ability to detect harmful stimuli that could cause injury, enabling protective responses (Fields, 2017).
- **Mechanoreception:** The perception of pressure and mechanical changes, essential for tasks involving the manipulation of objects or operating tools (Purves et al., 2014).
- **Vestibular (Balance):** The sense of balance, which is crucial for maintaining stability and preventing falls in dynamic work environments (Macpherson & Horak, 2016).

Ergonomically designed workplaces must account for the sensory demands placed on workers to optimize safety and performance. For instance, visual and auditory stimuli must be effectively managed in control rooms to prevent sensory overload, which can impair decision-making. Tactile feedback from controls and tools enhances precision, while temperature regulation in environments involving extreme conditions helps prevent thermal stress. These considerations are essential for reducing the risk of injury and enhancing worker well-being in a range of occupational settings (Bridger, 2018).

Understanding the integration of sensory functions into workplace design ensures that equipment, tools, and workspaces align with human capabilities and limitations, thus reducing the likelihood of sensory fatigue, errors, and accidents.

3.4.1 Vision

Vision is an intricate sensory process fundamental to human interaction with the environment, particularly in occupational settings. The human visual system relies on the coordinated functioning of the eye, the optic nerve, and the brain to interpret visual information, a process critical for tasks ranging from detailed close-up work to broader environmental awareness. Understanding the anatomy and physiology of vision is essential in ergonomics to design work environments that minimize visual strain and optimize performance (Wells & Green, 2018).

The eye, the primary organ of vision, plays a central role in focusing light and transmitting optical information to the brain. The eye's ability to perceive and process images begins when light enters through the cornea and passes through the pupil, which is controlled by the iris to regulate the amount of light entering. The light then passes through the crystalline lens, a transparent structure capable of changing shape to focus light precisely onto the retina, which is located at the back of the eye (Grossman & Mathews, 2017).

The retina, a multi-layered, light-sensitive tissue, contains specialized photoreceptor cells: rods and cones. These cells perform distinct functions essential for different types of vision. Cone cells, primarily concentrated in the fovea (the central portion of the retina), are responsible for high-acuity vision and color detection. This is crucial for tasks requiring precision and color differentiation, such as reading or identifying warning signals in occupational settings (Smith & Jones, 2020). In contrast, rod cells dominate the peripheral regions of the retina and are highly sensitive to dim lighting conditions, supporting peripheral vision and low-light tasks. However, rods lack the ability to perceive color, making them less suitable for environments requiring color differentiation (Land & Nilsson, 2018).

The photoreceptor cells convert captured light into electrical signals, initiating a neural process where signals are transmitted via synaptic pathways to retinal ganglion cells. These cells, in turn, project through the optic nerve, carrying the electrical impulses to the brain's visual cortex. Here, the signals are interpreted to form a coherent visual experience, enabling the recognition of shapes, colors, and movement (Sivak & Olson, 2015). This process is essential in ergonomically designing visual tasks, particularly in minimizing cognitive load and reducing visual fatigue (Clark & Smith, 2019).

In occupational ergonomics, visual demands and lighting conditions are of paramount importance. Poor lighting, glare, and suboptimal visual angles can lead to visual discomfort, eye strain, and a decrease in productivity. Ergonomic interventions, such as proper workstation design, optimized lighting, and regular rest breaks, can mitigate visual fatigue and improve overall workplace safety and efficiency (Anshel, 2017).

TABLE 3.1
Visual Sense: Key Terms in Ergonomics

Visual field	The area, measured in degrees, which can be seen by both fixated eyes
Accommodation	The action of focusing on targets at various distances. • Near-point is the closest distance that an object can be focused on without conscious accommodation. • Far-point is the farthest point that can be focused without conscious accommodation. • Amplitude of accommodation: the difference between the near-point and far-point.
Convergence	• The ability of the eyes to compensate for the difference the angle formed by the difference in the lines of sight from each when focusing on a target. • If targets are too close to the eye then convergence can become difficult.
Adaptation	• Adjustments that the pupils of the eyes make to changes in illumination are called adaptation. • Adaptation from light to dark takes about 30 min. Adaptation from dark to light happens much more quickly, normally within a minute.

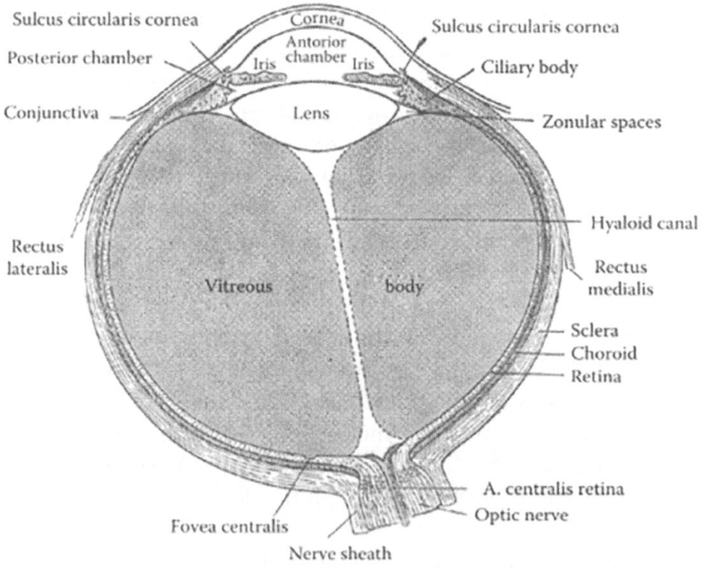

FIGURE 3.1 Eye. (From Gray, H., *Anatomy of the Human Body*, 20th edn., Lea & Febiger, Philadelphia, PA, 1918, retrieved September 9, 2009, from http://www.bartleby.com/107/)

3.4.1.1 Visual Fatigue

Visual fatigue, also referred to as computer vision syndrome (CVS) or digital eye strain, is a prevalent occupational health concern in modern workplaces that rely heavily on visual tasks, especially the prolonged use of computer screens and digital devices. It arises when the demands placed on the visual system exceed its capacity to perform comfortably and efficiently over time. The condition is multifactorial, with its causes rooted in ergonomic deficiencies in the work environment, prolonged visual task engagement, and uncorrected visual impairments (Anshel, 2017). This section explores the underlying mechanisms of visual fatigue, its impact on occupational performance, and ergonomic strategies to mitigate its effects.

3.4.1.2 Mechanisms of Visual Fatigue

Visual fatigue is the result of the cumulative stress imposed on the ocular and visual system, which can stem from several ergonomic factors. Key contributors include:

1. **Inadequate Visual Ergonomics:** Poor design of visual tasks, such as viewing distances, angles, and inappropriate screen resolutions, can force the eyes to work harder than necessary (Sheedy et al., 2014). A mismatch between the visual demand of the task and the capability of the visual system results in increased ocular accommodation and convergence, leading to eye strain, blurred vision, and double vision (Yan et al., 2017).
2. **Prolonged Near Work:** Continuous visual focus on close-up tasks, such as screen-based work, requires sustained accommodation, which can lead to ciliary muscle fatigue and discomfort (Rosenfield, 2016). The longer the exposure to near work without sufficient breaks, the greater the risk of visual fatigue.
3. **Uncorrected or Suboptimal Vision:** Undiagnosed refractive errors or insufficient vision correction, such as incorrect prescription glasses or lenses not designed for screen work, further exacerbate visual fatigue (Wolska et al., 2015). Individuals with latent visual conditions may unknowingly place additional strain on their visual system, leading to premature fatigue.

3.4.1.3 Impact on Worker Health and Productivity

Visual fatigue has substantial implications for both worker health and workplace productivity. Research has shown that prolonged visual discomfort can lead to a reduction in task accuracy, increased errors, and slower work performance (Coles-Brennan et al., 2019). Common symptoms associated with visual fatigue include:

- **Eye Discomfort:** Redness, irritation, and dry eyes.
- **Visual Disturbances:** Blurred vision, double vision, and difficulty focusing.
- **Systemic symptoms:** Headaches, neck and shoulder pain, often linked to poor posture and musculoskeletal strain due to suboptimal workstation setups (Bhanderi et al., 2022).

In a work context, these symptoms not only decrease the quality of work but also increase the likelihood of errors and accidents, particularly in visually intensive jobs. Furthermore, chronic visual fatigue can contribute to long-term health problems, such as visual impairment and musculoskeletal disorders (MSDs), making it critical to address from an ergonomic standpoint.

3.4.1.4 Ergonomic Interventions to Reduce Visual Fatigue

The application of ergonomic principles in workplace design can significantly alleviate visual fatigue and its associated risks. Key ergonomic considerations include:

1. **Visual Target Optimization:** Visual targets, such as computer displays or control panels, should be designed to be large enough and have sufficient clarity for easy viewing without excessive eye strain. High-resolution screens and adjustable font sizes can help reduce ocular demand (Rosenfield, 2016).
2. **Viewing Distance and Angle:** Workstations should be designed to place visual tasks at appropriate distances and angles, typically 20–28 inches from the eyes with a downward gaze angle of 10–20 degrees for screen work (Anshel, 2017). This helps reduce the load on the ciliary muscles and neck strain.
3. **Lighting Conditions:** Proper lighting is critical to reduce glare, optimize contrast, and prevent excessive eye strain. Ambient lighting should be adjusted to avoid direct glare on the screen, while task lighting can be used to enhance visibility for specific tasks (Sheedy et al., 2014). A balance between natural and artificial lighting is recommended to reduce discomfort.
4. **Contrast and Color Settings:** High contrast between the task and background can improve visibility and reduce the visual effort required to differentiate objects. Adjusting screen contrast, brightness, and color temperature can also reduce strain, particularly when working in dimly lit environments (Coles-Brennan et al., 2019).
5. **Rest Breaks and the 20–20–20 Rule:** Regular work breaks are essential for relieving visual strain. A widely accepted guideline is the 20–20–20 rule: every 20 minutes, workers should look at something 20 feet away for at least 20 seconds (Yan et al., 2017). This allows the visual system to relax and recover from continuous near work.
6. **Workstation Setup and Posture:** Visual fatigue is often linked with poor posture and improper workstation design. Ergonomic adjustments such as ensuring proper screen height, using adjustable chairs, and promoting neutral postures can minimize musculoskeletal strain and reduce associated visual discomfort (Anshel, 2017).
7. **Environmental Factors:** Workplace environmental conditions, such as temperature, humidity, and air quality, play a role in eye comfort. Dry eyes, a common symptom of visual fatigue, can be exacerbated by low humidity and air conditioning, making proper ventilation and humidity control necessary to mitigate visual fatigue (Sheedy et al., 2014).

Addressing visual fatigue in the workplace through ergonomic design is not only essential for protecting workers' visual health but also for improving productivity and reducing the risk of errors and accidents. By optimizing visual task design, workstation layout, lighting, and implementing regular breaks, workplaces can create environments that promote both visual comfort and overall efficiency. Future research should continue to explore the relationship between visual ergonomics and long-term occupational health outcomes.

3.4.2 Audition

Hearing, or audition, is one of the most vital senses in human perception, enabling us to detect, interpret, and react to sound waves in our environment. In occupational settings, effective auditory processing is crucial for communication, situational awareness, effective task performance, and overall safety. From an ergonomic standpoint, understanding how sound is perceived and how it interacts with workplace environments is essential for designing safe and efficient auditory environments. This section addresses the science of sound, the anatomy of the auditory system, and its implications for occupational ergonomics.

3.4.2.1 The Science of Sound: Occupational Acoustics

Sound is a form of mechanical energy transmitted through air, liquids, or solids in the form of pressure waves, also known as sound waves. Acoustics, the branch of physics that studies the properties of sound, is fundamental to occupational ergonomics, especially in environments where sound plays a critical role in communication

TABLE 3.2
Techniques for Controlling Glare

To Control Direct Glare	To Control Indirect Glare (Veiling Reflections and Reflected Glare)
Position luminaires, the lighting units, as far from the operator's line of sight as is practical	Avoid placing luminaires in the indirect glare offending zone
Use several low-intensity luminaires instead of one bright one	Use luminaires with diffusing or polarizing lenses
Use luminaires that produce a batwing light distribution, and position workers so that the highest light level comes from the sides, not front and back	Use surfaces that diffuse light, such as flat paint, non-gloss paper, and textured finishes
Use luminaires with louvers or prismatic lenses	Change the orientation of a workplace, task, viewing angle, or viewing direction until maximum visibility is achieved
Use indirect lighting	
Use light shields, hoods, and visors at the workplace if other methods are impractical	

Source: Adapted from Kaufman & Christensen, 1972; Morgan et al., 1963; Eastman Kodak, 1986.

or safety (Moore, 2016). Understanding sound propagation and its interaction with work environments enables us to optimize acoustics for specific tasks, whether it be minimizing unwanted noise or ensuring clear communication in noisy environments.

In an ergonomic context, acoustics also involves controlling reverberation, echo, and sound distortion in enclosed spaces, such as offices or industrial settings, where excessive or poorly controlled noise can lead to auditory strain, hearing loss, or cognitive overload (Leavitt et al., 2019). Designing work environments with optimal acoustic properties, including sound-absorbing materials and effective noise reduction strategies, is key to minimizing these risks.

3.4.2.2 The Human Ear: Auditory Function in Occupational Settings

The human auditory system can be divided into three primary sections: the outer ear, middle ear, and inner ear. Each section plays a distinct role in transforming mechanical sound waves into electrical signals, which are processed by the brain. Understanding the anatomy and physiology of the ear is crucial for addressing hearing ergonomics in the workplace, especially in environments with significant auditory demands or where hearing protection is required.

3.4.2.3 The Outer Ear (External Auditory System)

The Middle Ear (Tympanic Cavity):

1. **Tympanic Membrane (Eardrum):** The eardrum vibrates in response to incoming sound waves, converting acoustic energy into mechanical energy. In occupational settings, the integrity of the eardrum is essential for efficient sound transmission. Damage to the eardrum due to overexposure to loud noise can impair hearing acuity, making the proper use of hearing protection critical (Smith, 2019).
2. **Ossicles (Malleus, Incus, and Stapes):** These three tiny bones amplify the vibrations from the eardrum and transmit them to the inner ear. Their ability to amplify sound plays a significant role in determining how well workers can hear in environments with varying sound levels. Prolonged exposure to loud noise can result in damage to these structures, contributing to noise-induced hearing loss (NIHL) (Smith, 2019).
3. **Eustachian Tube:** This tube maintains equal pressure between the middle ear and the external environment. For workers in environments with frequent changes in air pressure, such as aviation or deep-sea diving, the Eustachian tube's function becomes ergonomically significant in preventing discomfort and ensuring proper auditory function (Moore, 2016).

The Inner Ear (Labyrinth):

1. **Cochlea:** The cochlea, a fluid-filled structure in the inner ear, is responsible for converting mechanical vibrations into electrical signals that the brain can interpret as sound. Different regions of the cochlea are sensitive to different frequencies, enabling the perception of a wide range of

sounds (Kujawa & Liberman, 2015). In noisy work environments, prolonged exposure to high-decibel sounds can damage the hair cells in the cochlea, leading to irreversible hearing loss. Ergonomically, this highlights the importance of engineering controls to reduce sound levels and the use of hearing protection to safeguard against NIHL.

2. **Vestibule and Semicircular Canals:** These structures contribute to balance and spatial orientation, which are critical in occupations that require movement coordination, such as construction, aviation, or emergency response (Leavitt et al., 2019). While not directly involved in hearing, disruptions in vestibular function can impact a worker's ability to maintain equilibrium and safely perform tasks, particularly in environments with high physical demands.

3.4.2.4 Ergonomic Implications of Auditory Function in the Workplace

Effective auditory ergonomics focuses on minimizing auditory strain and preventing long-term hearing damage while ensuring that essential auditory information is clearly perceived. Occupational noise-induced hearing loss (NIHL) remains one of the most common occupational health problems globally, making it essential to design environments that promote auditory health and safety (Smith, 2019).

1. **Noise Control Strategies:** Reducing occupational noise through engineering controls, such as soundproofing, damping systems, and acoustic panels, is the first line of defense in mitigating auditory strain. For instance, in manufacturing environments, enclosing loud machinery or using quieter alternatives can significantly reduce noise levels (Kardous & Shaw, 2014).
2. **Hearing Conservation Programs:** In environments where noise cannot be eliminated, hearing conservation programs are essential. These include regular hearing tests, the proper selection of hearing protectors (e.g., earplugs or earmuffs), and educating workers on the risks of noise exposure (Smith, 2019). Ergonomically designed hearing protectors should balance the need for noise attenuation with the ability to hear critical sounds.
3. **Auditory Workload Management:** Workers exposed to complex auditory environments, such as air traffic controllers or emergency responders, often face high cognitive and auditory workloads. Ergonomically designed sound environments and communication systems that reduce auditory distractions and enhance the clarity of essential signals are vital for maintaining performance and preventing cognitive overload (Leavitt et al., 2019).
4. **Rest and Recovery:** Adequate auditory rest is necessary for workers exposed to continuous noise. Research suggests that periods of quiet can help mitigate the risk of permanent hearing damage. Ergonomic recommendations may include regular breaks in quiet environments and rotating workers between high- and low-noise tasks (Moore, 2016).

Auditory ergonomics is essential for protecting hearing health and optimizing performance in occupational environments. Understanding the intricacies of sound perception, as well as the biological and mechanical aspects of hearing, allows us to design safer and more efficient workplaces. By applying principles of acoustics, controlling noise exposure, and promoting hearing conservation programs, we can reduce the risks associated with noise-induced hearing loss and ensure that workers can perform their tasks effectively.

BOX 3.1 Hearing is Based Almost Entirely Upon Physical Movement

Hearing is based entirely on movement, unlike the many other chemical-based senses of the body.

The purpose of the ear's structure is to provide a path for information, which culminates at the ear drum.

3.4.3 How Do We Hear?

Sound is defined as a pressure wave, a mechanical vibration that propagates through a medium such as air, water, or solids. In an occupational setting, sound waves are typically generated by machinery, voices, alarms, or other environmental sources. When a sound is produced, it initiates a series of events in the human auditory system that allows us to perceive and interpret it.

3.4.3.1 Outer Ear: Sound Collection and Transmission:

- **External Auditory Canal:** The outer ear begins with the pinna, or auricle, which collects sound waves and funnels them into the external auditory canal. As these sound waves travel through the canal, they eventually reach the tympanic membrane (eardrum). The eardrum acts as a barrier between the outer and middle ear, and its vibrations are triggered by the pressure changes in the incoming sound waves (Moore, 2016).

Middle Ear: Amplification and Transmission

- The middle ear contains the ossicles, a chain of three small bones: the malleus (hammer), incus (anvil), and stapes (stirrup). These bones work together to amplify the vibrations from the eardrum and transmit them to the inner ear. The amplification provided by the ossicles is critical because the inner ear is fluid-filled, and sound waves lose energy as they transition from air to liquid. The ossicles compensate for this loss, ensuring efficient energy transfer (Leavitt et al., 2019).
- The Eustachian tube helps regulate pressure within the middle ear, which is important for maintaining proper vibration of the eardrum. In occupational environments that involve changes in air pressure (e.g., aviation, diving), maintaining middle ear pressure is crucial for hearing clarity and comfort (Kardous & Shaw, 2014).

Inner Ear: Transduction of Sound Waves into Electrical Signals

- Once sound waves pass through the ossicles, they enter the cochlea, a spiral-shaped organ filled with fluid. Within the cochlea are tiny, sensitive hair cells that are responsible for converting mechanical vibrations into electrical signals. This process, known as transduction, is fundamental for auditory perception. The hair cells are arranged along the basilar membrane and are tuned to different frequencies, with high-pitched sounds stimulating hair cells at the base of the cochlea and low-pitched sounds affecting hair cells at the apex (Kujawa & Liberman, 2015).
- These electrical signals are then carried by the auditory nerve to the brain, where they are processed and interpreted as sound. The brain's role in sound perception involves filtering relevant auditory information from background noise, recognizing patterns, and assigning meaning to the sounds (Moore, 2016).

3.4.3.2 Ergonomic Considerations for Hearing in Occupational Settings

In occupational environments, the quality of hearing is influenced by several factors, including the individual's hearing ability, the intensity of the sound, background noise, and the auditory demands of the task. The application of ergonomic principles is essential in designing workplaces that promote optimal auditory conditions while minimizing the risk of hearing loss or degraded task performance due to excessive noise.

1. **Intensity and Distance of Sound**
2. **Noise Control Strategies**
   - **Background Noise:** Competing auditory signals, such as machinery noise or conversations, can interfere with a worker's ability to hear critical sounds. In environments with high levels of background noise, acoustic treatments, such as soundproofing or installing sound-absorbing materials, can help reduce reverberation and improve speech intelligibility (Kardous & Shaw, 2014).
   - **Auditory Distractions:** Excessive noise can lead to auditory distractions, cognitive overload, and reduced task performance. Ergonomic design can minimize these effects by separating noisy equipment from work areas, using quieter machinery, or implementing noise barriers (Leavitt et al., 2019). In cases where noise cannot be reduced, personal protective equipment (PPE), such as earplugs or earmuffs, must be provided to workers to protect against hearing damage.
3. **Hearing Conservation Programs**
   - Occupational environments with high noise levels must implement hearing conservation programs that include regular hearing assessments, education about noise risks, and appropriate use of hearing protection (Smith, 2019). Employers should also ensure that workers are trained on how to fit and use hearing protectors correctly to achieve the desired noise reduction.

4. **Auditory Message Clarity**
   - For communication to be effective in noisy environments, ergonomic guidelines suggest the use of clear, unambiguous auditory signals. The signal-to-noise ratio (SNR), which is the ratio of the desired sound (e.g., an alarm) to the background noise, must be high enough to ensure that critical sounds are not masked by noise (Moore, 2016). In addition, auditory signals should be designed to vary in frequency and intensity to capture attention, particularly in emergency situations where rapid responses are required.

3.4.4 Noise-Induced Hearing Loss (NIHL)

Noise-induced hearing loss is a significant occupational health concern and one of the most common preventable forms of hearing impairment in the workplace. In industrialized settings, prolonged or acute exposure to high levels of noise can lead to irreversible auditory damage. Understanding the physiological mechanisms behind NIHL and its risk factors is necessary for implementing effective ergonomic interventions aimed at reducing noise exposure and preserving hearing function in the workforce.

NIHL occurs when the delicate sensory hair cells in the cochlea, located in the inner ear, are damaged due to exposure to loud noise. These hair cells are responsible for transducing mechanical sound waves into electrical signals that are interpreted by the brain as sound (Kujawa & Liberman, 2015). When these cells are damaged by excessive noise, they cannot regenerate, resulting in permanent hearing loss. NIHL can manifest from either a single exposure to an extremely loud sound (e.g., explosions or gunfire) or cumulative exposure to elevated noise levels over time, as is common in certain occupational environments such as construction or manufacturing (Leavitt et al., 2019).

TABLE 3.3
Speech Intensity

Description	Typical Intensity of Source, dB(A?)	Distance	Message Characteristic
Soft whisper		3–6 in	Secret communication
Audible whisper	44–69	8–20 in	Confidential or personal
Normal voice	50–75	20–60 in	—
Loud voice	56–81	5–8 ft	Nonpersonal, intermediate group
Very loud voice	62–87	To 20 ft	Group address
Shouting	68–93	Upper limit	Hailing or emergency communication

Source: Woodson, W. E., Tillman, B., and Tillman, P. *Human Factors Design Handbook*, 2nd edn., McGraw-Hill, Inc., New York.

TABLE 3.4
Noise Effects on Performance

Noise Level, dB	Effects
150	Reduced visual acuity, chest-wall vibration, changes in respiratory rhythm, and a "gagging" sensation.
120	Loss of equilibrium.
110	Chronic fatigue and digestive disorders.
105	Reduced visual acuity, stereoscopic acuity, and near-point accommodation and permanent hearing loss when exposure continues over a long period (months).
100	Serious reduction in alertness. Attention lapses occur, although attention duration is usually not affected. Temporary hearing loss occurs if no protection is provided in the region 600–1,200 Hz. Most people will consider this level unacceptable, and 8 hours is the maximum duration they will accept
95	Considered to be the upper acceptance level for occupied areas where people expect the environment to be noisy. Temporary hearing loss often occurs in the range of 300–1,200 Hz. Speech will be extremely difficult, and people will be required to shout, even though they may be talking directly into a listener's ear.
90	At least half of the people in any given group will judge the environment as being too noisy, even though they expected a noisy environment Some temporary hearing loss in the range of 300–1,200 Hz occurs Skill errors and mental decrements will be frequent The annoyance factor is high, and certain physiological changes often occur (e.g., the pupils dilate, the blood pressure increases, and the stroke volume of the heart may decrease) Listening to a radio is impossible without good earphones The maximum duration that most people will accept is 8 hours.
85	The upper acceptance level (noise expected) in the range of 150–1,200 Hz This is considered the upper comfort level, although some cognitive performance decrement can be expected, especially where decision-making is necessary
80	Conversation is difficult (i.e., people have to converse in a loud voice less than 1 ft apart). It is difficult to think clearly after about 1 hour There may be some stomach contraction and an increase in metabolic rage strong complaints can be expected from those exposed to this level in confined spaces, and 8 hours is the maximum.
75	Too noisy for adequate telephone conversation A raised voice is required for conversants 2 ft apart Most people will still judge the environment as being too noisy
70	The upper level for normal conversation, even when conversants are close together (at a distance of 6 ft people will have to shout) Although persons such as industrial workers and shipboard personnel who are used to working in a noisy environment will accept this noise level, unprotected telephone conversation will be difficult (upper phone level is 68 dB).
65	The acceptance level when people expect a generally noisy environment Intermittent personal conversation is acceptable About half of the people in a given population will experience difficulty sleeping
60	The upper limit for spaces used for dining, social conversation, and sedentary recreational activities. Most people will rate the environment as "good" for general daytime living conditions

(*Continued*)

TABLE 3.4 (CONTINUED)
Noise Effects on Performance

Noise Level, dB	Effects
55	The upper acceptance level for spaces where quiet is expected (150–2,400 Hz). People will have to raise their voices slightly to converse over distances greater than 8 ft. This level of noise will awaken about half of a given population about half of the time. It is still annoying to people who are especially sensitive to noise
50	Acceptable to most people where quiet is expected. About 25% will be awakened or delayed in falling asleep Normal conversation is possible at distances up to 8 ft
40	Very acceptable to all. The recommended upper level for quiet living spaces, although a few people may still have sleep problems
30	Necessary for specialized listening tasks (e.g., threshold signal detection)
<30	Introduces additional problems; that is, low-level intermittent sounds become disturbing. Some people have difficulty getting used to the extreme quiet, and a few may become psychologically disturbed

Source: Woodson, W. E., Tillman, B., and Tillman, P. *Human Factors Design Handbook*, 2nd edn., McGraw-Hill, Inc., New York.

3.5 THE SCIENCE OF SOUND: HOW LOUD IS TOO LOUD?

Sound is measured in decibels (dB), a logarithmic unit that expresses the intensity of sound pressure. The risk of NIHL increases with both the intensity and duration of exposure to sound. Ergonomic guidelines, such as those provided by the National Institute for Occupational Safety and Health (NIOSH) and the Occupational Safety and Health Administration (OSHA), recommend that sound levels in the workplace not exceed 85 dB for an 8-hour shift (OSHA, 2020). Continuous exposure to noise levels above 85 dB can begin to damage the sensory cells of the inner ear, with the risk of NIHL increasing exponentially as decibel levels rise. Here is a breakdown of common noise levels and their potential effects:

- **Below 70 dB:** Safe for prolonged exposure (e.g., conversation or normal traffic).
- **85 dB and above:** Potentially damaging over time (e.g., heavy traffic, hair dryers).
- **Over 120 dB:** Immediately harmful, even with short-term exposure (e.g., sirens, rock concerts).

Because decibels are measured logarithmically, a small increase in decibel level represents a significant increase in sound intensity. For example, a noise level of 100 dB is ten times more intense than a noise level of 90 dB (Moore, 2016).

3.5.1 Symptoms and Early Indicators of NIHL

The onset of NIHL can be subtle and progressive, often going unnoticed until significant damage has occurred. Early symptoms include:

- **Muffled Hearing:** Sounds may seem indistinct or less clear.
- **Difficulty Understanding Speech:** Particularly in noisy environments or when multiple people are talking.
- **Tinnitus:** A persistent ringing, buzzing, or hissing sound in the ears, often indicating early damage to the auditory system (Kardous & Shaw, 2014).
- **Feeling of Fullness or Pressure in the Ears:** Some individuals report a sensation of their ears being plugged up, especially after exposure to loud sounds.

Over time, continued exposure to damaging noise levels can result in permanent loss of high-frequency hearing, which is critical for understanding speech and distinguishing sounds in noisy environments.

3.5.2 Mechanisms of Hearing Loss: The Role of the Inner Ear

The inner ear is home to the cochlea, a spiral-shaped, fluid-filled organ that contains thousands of tiny hair cells responsible for converting mechanical sound vibrations into electrical impulses that travel to the brain via the auditory nerve (Kujawa & Liberman, 2015). Exposure to loud noise causes these hair cells to bend excessively, which can result in cellular damage or cell death. Unlike other types of cells in the body, cochlear hair cells do not regenerate, and once they are damaged, the hearing loss is permanent (Leavitt et al., 2019). This makes early identification of noise risks and implementation of preventive measures essential.

3.5.3 Occupational and Environmental Causes of NIHL

NIHL is typically associated with repeated or chronic exposure to harmful noise levels in both occupational and recreational settings. Common sources include:

1. **Occupational Hazards:** Work environments such as construction sites, factories, and power plants frequently expose workers to high levels of noise from machinery, power tools, and other equipment. Without appropriate hearing protection, prolonged exposure can lead to NIHL (Smith, 2019).
2. **Recreational Activities:** Activities such as attending concerts, using firearms for hunting, and listening to music through headphones at high volumes also pose significant risks. The use of personal audio devices, particularly with earbuds or headphones that are inserted directly into the ear canal, can expose the ear to high sound pressures for extended periods (Moore, 2016).

3. **Personal Habits:** Prolonged use of headphones or earbuds, particularly at high volumes, is a growing cause of NIHL, particularly among younger populations. Extended listening sessions at unsafe volume levels can result in permanent hearing damage over time (Kardous & Shaw, 2014).

3.5.4 Preventing Noise-Induced Hearing Loss

The good news is that NIHL is largely preventable through the implementation of effective ergonomic and safety interventions. Employers and workers alike can take proactive steps to minimize noise exposure and reduce the risk of hearing loss.

1. **Reduce Noise Exposure**
   - Engineering controls can be implemented to reduce noise at the source, such as installing sound-dampening materials, using quieter machinery, or enclosing noisy equipment. In cases where these controls are not feasible, administrative controls, such as limiting the time spent in noisy environments or rotating workers to quieter tasks, can help minimize exposure (Smith, 2019).
2. **Use of Hearing Protection**
   - The use of hearing protection devices (HPDs), such as earplugs and earmuffs, is a critical component of hearing conservation programs. When chosen correctly, HPDs can significantly reduce the level of noise that reaches the ear. The Noise Reduction Rating (NRR) of the device should be matched to the noise levels of the environment to ensure adequate protection. For maximum effectiveness, workers must be trained in the proper fit and use of hearing protection (Moore, 2016).
3. **Education and Awareness**
   - Educating workers on the risks of noise exposure and the importance of hearing conservation is essential for preventing NIHL. Regular audiometric testing can also help detect early signs of hearing loss, allowing for timely intervention (Smith, 2019).
4. **Monitoring Noise Levels**
   - Employers are responsible for monitoring workplace noise levels and ensuring they are within safe limits. NIOSH recommends that workers' daily exposure to noise should not exceed 85 dB over an 8-hour shift. In cases where exposure exceeds this threshold, hearing protection must be provided, and engineering controls should be implemented (OSHA, 2020).

Noise-induced hearing loss is a preventable occupational hazard that can significantly affect a worker's quality of life and job performance. By understanding the physiological mechanisms of hearing, identifying noise sources, and applying effective ergonomic and safety measures, NIHL can be mitigated. Employers must prioritize hearing conservation programs, enforce the use of hearing protection, and continually assess workplace noise levels to create a safer, more auditory-friendly work environment.

TABLE 3.5
Upper Noise Level Limits

Sound Level, dBA	Type of Activity	Communication Equipment	Office Application
108	Maximum design limit for AMC equipment (hearing protection required)	No direct communication	Not recommended
100	Armored vehicles (hearing protection required)	Electrically aided communication satisfactory with attenuating helmet or headset; limited shouted communication possible with difficulty	Not recommended
90	Material that is beyond the state of the art of meeting 85 dBA (hearing protection required)	Shouted communication possible at short distances (1–2 ft)	Not recommended
85	Acceptable level for unprotected hearing for 8 hours exposures	Shouted communication possible at several feet (3–4 ft); telephone use difficult	Not recommended
75	Maintenance shops, garages, and keypunch areas	Occasional telephone use and occasional direct communication at up to 5 ft is acceptable	Not recommended
65	Operation centers, mobile command and communication centers, computer rooms, word processing centers, kitchens, and laundries	Frequent telephone use and frequent direct communication at up to 5 ft is acceptable	Business machine offices
55	Drafting rooms, laboratories, and conferences with two or three people	No difficulty with telephone use and occasional direct communication at up to 15 ft	Shop offices and general secretarial areas
45	Libraries, conference rooms, command and control centers, theaters, and sleeping areas	No difficulty with direct communication	General offices
35	Recording studios and large conference rooms	Areas requiring unusually extreme quiet	Executive offices

Source: Woodson, W. E., Tillman, B., and Tillman, P. *Human Factors Design Handbook*, 2nd edn., McGraw-Hill, Inc., New York.

3.5.5 Current Technology: Ipods and MP3(4) Impact on Hearing

The introduction of portable music players, such as iPods, MP3, and MP4 players, marked a technological breakthrough in how individuals consume audio content. However, the evolution of these devices into today's smartphones, tablets, and high-end headphones with active noise-cancellation capabilities has raised significant concern in occupational ergonomics, particularly with regard to hearing conservation in the workplace and public settings.

From an ergonomics perspective, the primary risk factors associated with personal audio devices are the duration of use, exposure to high sound levels, and the environmental context in which these devices are used. Unlike early MP3 players, which had limited battery life and storage capacity, modern devices enable continuous and prolonged exposure to sound. This is particularly problematic in noisy environments such as commuting on public transport, working in open office spaces, or exercising in gyms, where users may increase the volume to compensate for background noise (Peng et al., 2016). The frequent combination of loud volumes and extended listening sessions poses a heightened risk of auditory damage, even if the volume is perceived as moderate due to the masking effect of ambient noise (Le Prell et al., 2019).

According to the World Health Organization (WHO), approximately 1.1 billion young people are at risk of developing noise-induced hearing loss due to unsafe listening practices with personal audio devices (WHO, 2015). This concern has escalated with the ubiquity of smartphones and the increasing integration of music into daily activities. In an occupational setting, prolonged exposure to high-decibel sound through personal devices can contribute to music-induced hearing loss (MIHL), which shares similar characteristics with occupational noise-induced hearing loss commonly observed in industrial environments (Gilles et al., 2014).

Key symptoms of MIHL include tinnitus (persistent ringing or buzzing in the ears), hyperacusis (increased sensitivity to normal environmental sounds), and temporary threshold shifts (temporary hearing loss). Over time, these symptoms can develop into permanent sensorineural hearing loss if proper preventive measures are not taken (Sliwinska-Kowalska & Davis, 2012). Current studies emphasize that younger individuals, especially those in high-usage environments like students and workers in open office settings, are particularly susceptible to these effects due to both recreational and work-related use of personal audio devices (Bhagat, 2017).

To mitigate these risks, ergonomic strategies for auditory health should be considered in occupational guidelines. These include promoting the use of noise-limiting headphones, enforcing safe listening practices (such as maintaining volume levels below 85 dB), and educating users about the risks of prolonged exposure to loud sound (Beach et al., 2014). Additionally, leveraging modern technologies such as real-time volume monitoring and hearing protection software could play a vital role in preserving auditory health in both work and recreational settings.

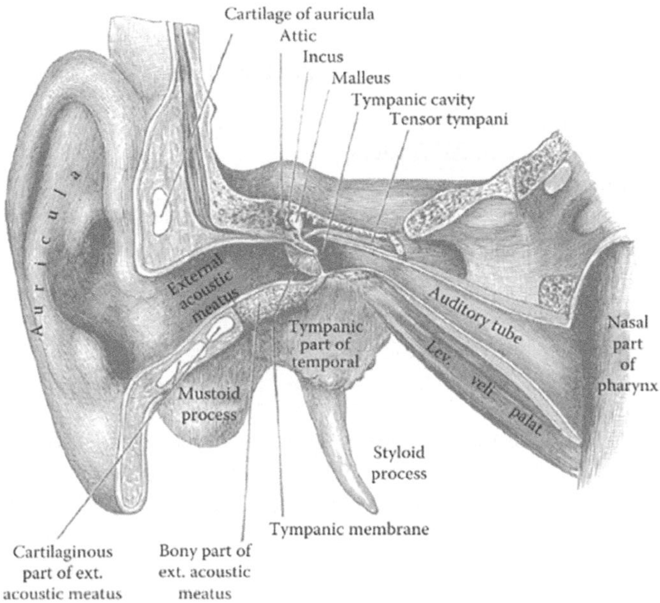

FIGURE 3.2 Ear. (From Gray, H., *Anatomy of the Human Body*, 20th edn., Lea & Febiger, Philadelphia, PA, 1918, retrieved September 9, 2009, from http://www.bartleby.com/107/)

3.6 Olfactory Sense

Olfaction, the human sense of smell, is integral not only to personal and environmental perception but also to workplace ergonomics and safety. In the context of occupational ergonomics, understanding how olfaction affects worker performance, safety, and overall well-being is crucial. This section focuses on the mechanisms of olfaction, its impact on cognitive and physical performance, and its role in industrial safety, while integrating recent scientific findings in the field.

3.6.1 Mechanisms of Olfaction

The olfactory system begins in the nasal cavity, where the olfactory epithelium, a specialized tissue rich in olfactory receptor neurons, is located. These neurons detect odorant molecules and transmit signals to the olfactory bulb in the brain through the olfactory nerve. From there, olfactory signals are relayed to various brain regions involved in emotion, memory, and decision-making, including the limbic system and prefrontal cortex (Wysocki et al., 2018; Hongratanaworakit et al., 2020). While the human olfactory system is capable of detecting a vast range of odors, the specificity of this detection varies, with humans able to distinguish thousands to an estimated trillions of different smells (Bushdid et al., 2014).

3.6.2 Occupational Implications of Olfaction

BOX 3.2 Olfactory Jobs

- Culinarian (cook, chef)
- Enologist (wine maker/taster)
- Perfumer
- Quality assurance (food and beverage companies)
- Safety inspector (chemical/hazard, smoke detection)
- Law enforcement (drug and alcohol detection)
- Medical workers
- Firefighters (gas/chemical leaks, smoke detection)

The role of olfaction in occupational ergonomics has gained increasing attention, particularly in environments where cognitive alertness or physical performance is critical. For instance, aromachology—the study of how aromas affect human behavior—has demonstrated that certain scents can enhance cognitive performance, reduce stress, and even improve physical endurance. A study by Schulte et al. (2013) found that the inhalation of rose scent during both learning and REM sleep improved memory retention in participants, suggesting potential applications for workers in cognition-intensive industries.

Similarly, research has linked peppermint aroma to increased alertness and physical performance. Smith et al. (2019) found that peppermint inhalation improved athletic performance by increasing strength, speed, and endurance while reducing perceived fatigue. These findings have important implications for industries involving physical labor or high cognitive demands, as targeted use of specific scents could potentially enhance worker productivity and reduce fatigue.

3.6.3 Psychological Effects and Safety

Olfaction has profound psychological effects, influencing mood, stress levels, and emotional responses. In occupational settings, these psychological effects can directly impact both worker satisfaction and performance. Herz et al. (2004) highlighted the role of olfaction in emotional and cognitive processing, noting that certain scents are capable of triggering strong emotional responses or vivid memories due to their connection to the limbic system, which plays a key role in emotion regulation and memory storage (Gottfried, 2017).

For example, in high-stress environments such as emergency rooms or manufacturing floors, scents known to reduce stress and enhance focus, such as lavender or citrus, may be employed to improve worker resilience and attention to detail (Hongratanaworakit et al., 2020). Conversely, strong unpleasant odors, such as chemical fumes or industrial waste, may increase worker stress and impair cognitive function, which raises significant ergonomic and safety concerns. In such environments, adequate ventilation and air-quality control are critical to minimize the risk of cognitive impairment due to olfactory overload or olfactory fatigue, which occurs when the olfactory receptors become desensitized to continuous exposure to strong odors (Zhao et al., 2019).

3.6.4 Olfaction and Workplace Safety

In certain industries, such as manufacturing, mining, or food processing, olfaction serves as an early-warning system for detecting harmful gases, spoiled materials, or chemical leaks. Workers' ability to recognize hazardous odors can be a critical factor in preventing accidents or health issues. However, reliance on olfaction alone is insufficient due to individual variations in sensitivity to odors, the presence of odorless hazards (e.g., carbon monoxide), and the potential for olfactory fatigue (Bushdid et al., 2014).

To mitigate these risks, industrial ergonomists often recommend the implementation of continuous monitoring systems, such as gas detectors or air-quality sensors, which provide an objective means of detecting dangerous substances. Nonetheless, ensuring that workers are trained to recognize specific odor signatures of dangerous materials remains a key component of safety protocols.

3.6.5 Classification of Odors and Their Ergonomic Impact

Odor classification systems, such as the seven basic odor categories (ethereal, camphoraceous, musky, floral, minty, pungent, and putrid), provide a framework for understanding how different scents can impact workplace environments (Zhao et al., 2019). These categories help ergonomists design workplace environments where beneficial odors can be introduced to enhance worker performance or well-being. For instance, using floral or minty scents in office environments may improve cognitive performance and reduce stress, whereas more neutral or ethereal scents might be preferable in environments requiring high levels of concentration, such as laboratories or operating rooms.

Olfaction plays a multifaceted role in occupational ergonomics, affecting cognitive and physical performance, emotional well-being, and safety in the workplace. Emerging research supports the strategic use of scents to enhance worker performance and reduce stress, while also highlighting the importance of managing harmful odors to maintain a safe and productive environment. Future ergonomic interventions may increasingly incorporate scent-based strategies alongside traditional ergonomic solutions to improve both worker satisfaction and operational efficiency.

3.7 GUSTATION

Gustation, or the sense of taste, plays an essential role in human interaction with food and beverages, with significant occupational implications in certain industries. While gustation is traditionally associated with basic taste perception—salty, sweet,

FIGURE 3.3 Aromatherapy industry products. (From Demand Studios, 2009, retrieved September 9, 2009, from http://cdn-write.demandstudios.com/upload/3000/200/00/0/33200.jpg)

sour, bitter, and umami—its occupational relevance extends to specialized fields where sensory acuity is vital. This section provides details on the gustatory system's mechanics, its interplay with olfaction, and the ergonomics of occupations that depend heavily on taste perception.

3.7.1 Mechanisms of Gustation

Gustation is mediated by taste buds, specialized sensory organs primarily located on the tongue, as well as in the throat and other areas of the oral cavity. Each taste bud

TABLE 3.6
Odor Perception

Olf Value	Source
0–5	Per square meter of materials in the office
1	Sedentary person, 1 met
5	Active person, 4 met
6	Smoker, average
11	Active person, 6 met
25	Smoker, when smoking

contains approximately 50–100 taste receptor cells, which detect the five primary taste modalities: salty, sweet, sour, bitter, and umami (Chaudhari & Roper, 2016). These receptors send signals through the gustatory nerves to the brainstem and are further processed in the gustatory cortex.

From an ergonomic perspective, while taste might not be as universally crucial in occupational settings compared to other sensory modalities like vision or touch, it remains vital for specific professions. Understanding the neurological and physiological underpinnings of taste can guide ergonomic considerations in industries where sensory analysis is key.

3.7.2 The Smell–Taste Connection in Ergonomics

A significant body of research has demonstrated that gustation is closely linked with olfaction, with estimates suggesting that up to 75% of what is perceived as taste

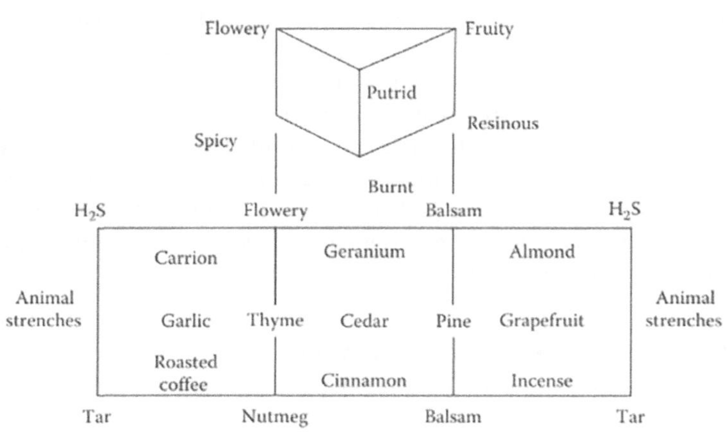

FIGURE 3.4 Henning's smell prism. (From Iowa State University, The science of smell part 2: Odor chemistry, 2004, retrieved January 23, 2011, from http://www.extension.iastate.edu/Publications/PM1963B.pdf)

FIGURE 3.5 Primary taste sensitivity on the human tongue.

originates from smell (Lechien et al., 2020). This relationship is particularly evident in the phenomenon of retronasal olfaction, where odor molecules from food or beverages enter the nasal cavity through the oropharynx as we chew, contributing significantly to flavor perception (Small & Prescott, 2021).

In occupational contexts, this connection underscores the importance of olfactory health in professions requiring refined taste discrimination. For example, chefs, food critics, and wine tasters rely not only on their gustatory receptors but also on their olfactory system to provide a comprehensive flavor analysis. Impairment in either system, whether due to illness or environmental factors, can severely impact their performance and career efficacy.

3.7.3 Occupational Impacts of Gustatory Impairment: The Case of COVID-19

The COVID-19 pandemic brought new attention to gustation and its occupational importance due to the widespread reports of taste loss (ageusia) and taste distortion (dysgeusia) among those infected with SARS-CoV-2 (Vaira et al., 2020). This viral infection can affect the gustatory system by invading cells in the oral cavity, as well as olfactory cells in the nasal passages. Research by Mao et al. (2021) found that loss

FIGURE 3.6 Odor perception. (From Teamwork Photo & Digital, Sekonic, 2009, retrieved September 9, 2009, from http://www.teamworkphoto.com/images/sekonic/758d.jpg)

of taste was often one of the earliest reported symptoms of COVID-19, sometimes occurring before more commonly recognized symptoms like fever or cough.

For professionals in industries such as food and beverage, loss or alteration of taste perception poses significant challenges. Wine tasters, food critics, and flavorists rely on an acute sense of taste to perform their jobs, and the loss of this sense, even temporarily, can impair their ability to assess products or provide accurate feedback. This phenomenon has underscored the importance of safeguarding sensory health, particularly for workers in industries that demand high levels of sensory acuity.

3.7.4 Ergonomics of Taste-Centric Professions

While gustation may not have as broad an occupational application as vision or hearing, certain professions depend heavily on a keen sense of taste. These roles include:

- **Wine Tasters and Sommeliers:** Wine professionals rely on their ability to differentiate between subtle flavor profiles in wines, identifying characteristics such as acidity, tannin levels, and sweetness, all of which are closely tied to both taste and smell. Ergonomic considerations for these workers include environmental controls such as maintaining optimal air quality and humidity to avoid olfactory fatigue or taste desensitization.
- **Food Critics and Chefs:** Similar to wine tasters, chefs and food critics require refined taste and smell to analyze and critique dishes. For these professionals, ergonomics may include ensuring access to a distraction-free environment during tasting sessions to enhance focus on flavor, as well as periodic breaks to reduce sensory overload.
- **Food Scientists and Flavorists:** In the food and beverage industry, flavorists play a pivotal role in designing and developing new flavors. Their work often involves using highly sensitive equipment to detect taste compounds, requiring precision not only in their sensory perception but also in their physical work environment. Ergonomic workstation design is critical to minimize physical strain and maintain consistent performance over time (Hopp & Irwin, 2020).

3.7.5 Enhancing Ergonomics in Taste-Dependent Roles

For workers whose primary tasks involve taste perception, ergonomic interventions focus on optimizing both the physical and sensory environments. Key considerations include:

- **Environmental Control:** In professions such as winemaking and food science, maintaining an environment with controlled humidity and temperature is essential to preserving the integrity of taste perception. Additionally, air-quality control is necessary to prevent cross-contamination from other odors, which can compromise sensory accuracy.
- **Sensory Breaks:** Taste-dependent professionals may benefit from scheduled sensory breaks to prevent olfactory and gustatory fatigue. This approach allows workers to maintain sensory acuity over long periods of tasting or flavor analysis, which is particularly relevant in high-stakes industries such as food development and quality assurance.
- **Training and Calibration:** For professions that require fine discrimination between tastes, regular sensory training and calibration exercises can help workers maintain their abilities. In environments where gustatory performance is critical, periodic retraining can ensure that individuals remain sharp in their sensory assessments, particularly following illnesses such as COVID-19, which may temporarily impair taste perception (Lechien et al., 2020).

While gustation may not be as universally applied in occupational ergonomics as other senses, it plays a vital role in specific industries, including food and beverage, flavor science, and sensory analysis. Understanding the physiological mechanisms of taste, its connection to olfaction, and the impact of gustatory impairment is essential in designing ergonomic solutions for workers who depend on sensory precision. Occupational ergonomists must consider both environmental and physiological factors to support workers in taste-reliant professions, ensuring that their sensory health is preserved and their performance optimized.

3.8 TACTATION

Tactation, or the sense of touch, plays a critical role in occupational ergonomics, as it provides workers with essential feedback regarding their environment, tools, and tasks. The tactile sensory system allows humans to detect and respond to various stimuli, including pressure, vibration, temperature, and pain, which are essential for maintaining safety, performance, and well-being in the workplace. This section explores the tactile system's underlying mechanisms, how it interacts with human performance in occupational settings, and the implications of its disruption in certain professions.

3.8.1 The Tactile Sensory System: Mechanisms and Types of Receptors

The human skin, the largest sensory organ, contains an array of specialized receptors that detect different types of tactile stimuli. These receptors send signals through the peripheral nervous system to the brain, where they are integrated and interpreted, allowing individuals to perceive textures, temperatures, and forces in their environment. This highly adaptive system is central to ergonomics, particularly in occupations requiring fine motor control, manual dexterity, or environmental hazard detection.

The receptors involved in tactile perception can be broadly categorized into four primary types:

1. **Mechanoreceptors:** These are responsible for detecting pressure, texture, and vibration. Mechanoreceptors include Meissner corpuscles and Pacinian corpuscles, both of which play pivotal roles in occupational tasks. Meissner corpuscles, concentrated in the fingertips, provide high-resolution information about surface textures, making them vital for tasks involving precision, such as electronics assembly, surgery, or instrument manipulation (Abraira & Ginty, 2013). Pacinian corpuscles, located deeper in the skin, are more sensitive to vibrations and are crucial for workers using vibrating tools or machinery to detect changes in texture or movement (Bensmaia, 2014).
2. **Thermoreceptors:** These receptors detect temperature changes and are critical for maintaining comfort and safety in the workplace. Two main types of thermoreceptors exist: those that detect warmth and those that detect coolness. These receptors help workers avoid thermal extremes, ensuring safety in industries like construction, food preparation, or manufacturing,

where contact with extreme temperatures can pose significant risks (Dhaka et al., 2016).

3. **Nociceptors:** Nociceptors are the body's pain detectors and function as a critical warning system, alerting workers to potential injury. These receptors respond to harmful stimuli such as extreme heat, sharp objects, or chemical exposure. In the workplace, nociceptors trigger pain responses that encourage workers to withdraw from harmful environments or practices, thus playing a fundamental role in injury prevention and maintaining worker safety (Basbaum et al., 2009).
4. **Electroreceptors (Potential):** Although well-established in certain animal species, the existence of electroreceptors in humans is still under investigation. Some studies suggest that humans may possess receptors that detect weak electric fields, which could have implications for occupations involving close contact with electrical systems or electromagnetic fields, but further research is needed to clarify their role in human touch perception (Lambrecht & Müller, 2017).

These sensory inputs work together to create a rich tactile experience, enabling workers to interact with tools, surfaces, and environments with precision and safety. For instance, handling a hot object may simultaneously activate mechanoreceptors, thermoreceptors, and nociceptors, providing comprehensive feedback that helps the worker assess whether to continue the action or avoid a potential hazard.

3.8.2 The Importance of Touch in Occupational Ergonomics

Touch is an essential sense in many occupational tasks, providing feedback for safe interaction with equipment, tools, and materials. In manual work, such as craftsmanship, surgery, and quality control, tactile sensitivity ensures that the worker can detect irregularities in textures, changes in pressure, or the presence of heat or cold. In these scenarios, the proper functioning of mechanoreceptors and thermoreceptors is essential for precision and safety (Bensmaia, 2014).

Additionally, occupations that involve prolonged exposure to vibrations, such as those in construction or manufacturing, may place workers at risk of developing conditions such as Hand–Arm Vibration Syndrome (HAVS), which can lead to reduced tactile sensitivity and impaired motor function. Ergonomists work to mitigate such risks by designing tools that reduce vibration transmission or providing workers with protective equipment (Palmer et al., 2014).

Moreover, the tactile system is closely connected to emotional well-being. In ergonomic design, understanding how touch contributes to reducing worker stress and enhancing comfort is crucial. For example, certain materials or textures used in tools and furniture can enhance user satisfaction and reduce discomfort or strain during repetitive tasks. Research has shown that touch-based interactions, such as those used in team-building activities, can promote positive emotional states and enhance collaboration in the workplace (Krahé et al., 2018).

3.8.4 Tactile Technology and Ergonomic Advancements

As touch continues to be studied, new advancements in tactile technologies are emerging that have the potential to enhance ergonomics in various occupational settings. Two particularly promising developments are advanced prosthetics with sensory feedback capabilities and virtual reality (VR) systems with realistic tactile input.

1. **Advanced Prosthetics:** In recent years, the development of prosthetic limbs that can simulate the sense of touch has made significant progress. These prosthetics are equipped with sensors that detect pressure, texture, and temperature and transmit this information to the user's nervous system, allowing them to feel their surroundings and improve task precision. Such advancements are transformative for amputees working in physically demanding fields, where the ability to perceive touch is vital for handling tools and interacting with objects safely (Raspopovic et al., 2014).
2. **Virtual Reality and Tactile Feedback:** The integration of tactile feedback into VR environments is another area of ergonomic interest. Workers in industries such as aerospace, medicine, and design are increasingly using VR for training and simulation. Adding tactile feedback, or "haptic" input, to these systems enables users to interact with virtual objects in ways that mimic real-world touch experiences. This can enhance the realism of training scenarios, making workers better prepared for tasks involving fine motor control or complex tool manipulation (Pacchierotti et al., 2017).

3.8.5 The Role of Touch in Workplace Safety and Productivity

Ensuring that workers maintain their tactile sensitivity is a critical component of occupational safety and ergonomics. Workers exposed to repetitive motions, heavy tool use, or vibration are at risk of developing sensory impairments, which can compromise their ability to interact with their environment effectively. Regular breaks, ergonomic tool design, and environmental controls are all necessary strategies for protecting the tactile system and ensuring long-term productivity and well-being in the workplace.

Moreover, technologies that enhance tactile feedback can help improve worker precision, reduce error rates, and decrease the likelihood of accidents. For example, in professions that require precision cutting or assembly, enhancing tactile sensitivity through improved tool design or feedback mechanisms could improve outcomes and reduce fatigue.

Tactation is a foundational sense in occupational ergonomics, providing essential feedback that enables workers to safely and effectively interact with their environment. From mechanoreceptors detecting fine textures to nociceptors warning of injury, the tactile system plays a crucial role in maintaining safety, productivity, and well-being in the workplace. As research into tactile technologies and sensory feedback continues to advance, the future of ergonomics promises further innovations that will enhance the integration of touch in occupational settings, from advanced prosthetics to tactile-rich virtual environments.

3.9 ENVIRONMENTAL FACTORS IN ERGONOMICS

Environmental factors that influence the task performance and comfort of workers in the occupational environment can be classified in one or more of the following categories:

- Light Levels
- Sound Levels
- Thermal Conditions: Temperature and Humidity

These are further discussed in the following sections.

BOX 3.3 Key Terms

- *Photometric energy*—radiant energy modified by the luminous efficiency of the standard observer.
- *Illuminance*—the amount of light falling on a surface (lx)
- *Luminance*—the amount of light energy reflected (or emitted from a surface)

3.9.1 Light Levels

In occupational ergonomics, lighting is a fundamental factor influencing worker safety, productivity, and overall well-being. Properly designed lighting systems are essential to optimize visual performance, reduce fatigue, and ensure a comfortable work environment, particularly in settings where workers perform both physical and digital tasks. This section delves into the scientific principles behind lighting and offers practical guidance on ergonomic lighting design.

3.9.1.1 Illumination Levels and Visual Ergonomics

The amount of light illuminating a workspace directly impacts the visual acuity required for task performance. Visual ergonomics research shows that inadequate lighting forces the eyes to work harder, leading to eye strain, discomfort, headaches, and even visual errors, which can negatively affect work performance and increase the risk of accidents (Boyce, 2014). On the other hand, overly bright environments create excessive glare, which impairs visibility and can cause long-term discomfort (Mansfield, 2016). Therefore, balancing light levels based on the specific tasks being performed is essential to maintain visual clarity and prevent fatigue.

For example, precision tasks such as assembly line work or detailed reading often require higher illumination levels (500–1,000 lux), while general office workspaces benefit from lower light levels (300–500 lux) to minimize glare from digital screens (IES, 2017).

3.9.1.2 Contrast and Visibility

Contrast refers to the difference in light intensity between an object and its background. High contrast improves object distinction and readability, which is critical for tasks requiring accuracy, such as quality control in manufacturing (Rea, 2015). However, insufficient contrast, often due to poor lighting design or improper positioning of light sources, can hinder task performance and increase the potential for errors (CIE, 2016).

3.9.1.3 Reflection and Glare

The reflective properties of surfaces in the workplace significantly impact visual comfort. Glossy surfaces can reflect direct or ambient light, causing unwanted glare, which has been shown to increase visual strain and discomfort (Wolkoff, 2017). In contrast, matte finishes on work surfaces and walls diffuse light, enhancing visibility without introducing harsh reflections. According to ergonomic design standards, minimizing glare is critical in environments with high levels of digital screen use (ISO 9241-6:2017).

3.9.1.4 Glare Management in Digital Workplaces

Glare from both artificial and natural light sources can significantly impair vision, particularly in environments dominated by screen-based tasks. Research has shown that glare reduces contrast and forces workers to squint or strain their eyes, leading to discomfort and reduced productivity (Boyce, 2014). In workplace environments that rely heavily on computers, careful consideration of screen placement relative to windows, light fixtures, and reflective surfaces is essential to minimize glare.

3.9.1.5 Ergonomic Lighting Solutions for the Modern Workplace

In today's dynamic work environments, lighting must be adaptable to both physical tasks and digital workstations. Optimizing workplace lighting ergonomically involves addressing both task-specific needs and the health impacts of lighting, including effects on circadian rhythms.

- **Task-Specific Lighting:** Different tasks require varying levels of light intensity. For example, precision-based tasks demand higher lighting levels than general administrative tasks. Adjustable task lighting can allow workers to tailor illumination to their needs, thereby reducing eye strain and enhancing task performance (IES, 2017).
- **Balancing Screen and Ambient Light:** With the increased use of digital screens, maintaining an appropriate balance between screen brightness and ambient lighting is crucial. High ambient light levels can cause screen glare, while insufficient light can force users to increase screen brightness, exacerbating eye strain. Positioning screens perpendicular to windows and

utilizing blinds or diffusers can help control natural light and minimize glare (Wolkoff, 2017).

- **Incorporating Natural Light:** Exposure to natural light has been linked to improved mood, well-being, and productivity. However, unchecked natural light can introduce excessive heat and glare, leading to discomfort. Ergonomically designed workplaces incorporate natural light while mitigating its negative impacts through the use of blinds, shades, and reflective window coatings (CIE, 2016).
- **Circadian Lighting:** Recent studies emphasize the importance of lighting in regulating circadian rhythms, which affect alertness and sleep-wake cycles. Light temperature plays a role in this process; cooler light in the morning (5,000–6,500 K) promotes alertness, while warmer tones in the afternoon (2,700–3,500 K) help prepare the body for rest (Figueiro, 2016). Incorporating tunable white lighting systems that adjust the color temperature throughout the day can improve worker well-being and reduce fatigue (Smolders et al., 2014).

3.9.1.6 Consequences of Poor Lighting Design

Failing to address ergonomic lighting in the workplace can lead to a range of negative outcomes, both for employees and organizations:

- **Regulatory Non-Compliance:** OSHA has established minimum illumination levels for workplaces to ensure safety (OSHA, 2018). Employers failing to meet these standards risk legal penalties and fines.
- **Reduced Productivity and Quality:** Poor lighting has been associated with decreased productivity and higher error rates, particularly in precision tasks that require acute visual attention (Rea, 2015). This can lead to increased rework, delayed project timelines, and reduced overall efficiency.
- **Increased Worker Fatigue and Dissatisfaction:** Inadequate lighting contributes to physical strain, leading to headaches, visual discomfort, and long-term health issues like vision deterioration. These factors contribute to higher absenteeism, lower job satisfaction, and increased turnover (Smolders et al., 2014).
- **Heightened Risk of Injury:** Dim or uneven lighting can obscure hazards, increasing the risk of slips, trips, and falls. According to workplace injury statistics, insufficient lighting has been linked to higher injury rates, particularly in industrial environments (Boyce, 2014).

Lighting is a critical component of workplace ergonomics, with direct impacts on safety, productivity, and worker well-being. Understanding the interaction between illumination, visual comfort, and task requirements is essential for designing ergonomic workplaces that minimize strain and optimize performance. Future lighting systems should consider task-specific requirements, the balance between artificial and natural light, and the influence of light on human circadian rhythms to create healthier, more productive work environments.

TABLE 3.7
The Functional Ranges of Visual System Capabilities

Name	Dominance Range (cd/m²)	Photoreceptor Active	Wavelength Range (nm)	Capabilities
Photopic	>3	Cones	380–760	Color vision Good detail discrimination
Scotopic	<0.001	Rods	380–760	No color vision Poor detail discrimination
Mesopic	>0.001 and <3	Cones and rods	380–760	Diminished color vision, reduced detail discrimination and a shift in spectral sensitivity as adaptation luminance moves from photopic to scotopic

Source: Gavriel, S. (Ed.): *Handbook of Human Factors and Ergonomics*. 1997. Copyright Wiley-VCH Verlag GmbH & Co. KGaA.

3.9.1.7 Illumination

Light, as defined by the International Commission on Illumination (CIE) in 2020, is any electromagnetic radiation that stimulates the human visual system, producing the sensation of sight. Our eyes perceive objects through two main mechanisms:

- **Radiance:** This refers to light emitted directly from a source, such as the sun, a flame, or an LED lamp. Radiance is measured in watts per steradian per meter squared ($W \cdot sr^{-1} \cdot m^{-2}$), a unit that accounts for the direction of the emitted light.
- **Irradiance:** This describes the radiant power (energy per unit time) received by a surface from all directions. Think of it as the total amount of light hitting a specific area. Irradiance, like radiance, is measured in watts per meter squared (W/m^2).

Illumination, distinct from radiance and irradiance, is the perceived density of the visible light hitting a surface. Imagine holding your hand out on a sunny day versus a cloudy day. The sunshine feels brighter because it has a higher illumination level. Illumination is measured in lux (lm/m^2) for metric units and foot-candles (fc) for imperial units.

Researchers are constantly expanding our understanding of light and its interaction with the human body. Recent studies suggest that exposure to specific wavelengths of light, beyond just visible light, can influence our circadian rhythms and sleep patterns. This has led to the development of light therapy techniques to combat jet lag and other conditions such as seasonal affective disorder (SAD).

3.9.1.8 Engineering Procedures in the Design of the Environment for Proper Vision

Modern spaces demand lighting design that prioritizes both vision and comfort. Proper illumination is crucial, especially for older adults with reduced visual acuity. It not only enhances visual clarity but also supports safety and task completion.

Light levels significantly impact color perception. At recommended office levels (above 100 lux), cone cells dominate vision, enabling sharp detail and accurate color recognition—ideal for tasks requiring precision. In dimmer environments (10 to 0.1 lux), both cone and rod cells function. While brighter areas might retain some color, peripheral vision and fine detail recognition weaken. This range is suitable for navigating hallways and common areas. When light falls below 0.1 lux, only rod cells operate, limiting vision to shades of gray with minimal color or detail distinction—typical of low-light settings.

Luminance, measured in candelas per square meter (cd/m^2), describes the perceived brightness of a surface. It's influenced by both the amount of light falling on it (illuminance) and the surface's reflectance. A white wall with high reflectance will appear brighter than a dark desk under the same illumination.

Modern lighting design requires a multifaceted approach:

- **Task-Based Lighting:** Provide appropriate light levels for specific activities. Task lighting can be layered with general room lighting to ensure adequate illumination for reading or computer work.
- **Glare Control:** Minimize discomfort and eye strain by mitigating glare from direct or reflected light sources.
- **Color Temperature:** Select light sources with a color temperature that promotes visual comfort and alertness. Cooler temperatures (around 5,000 Kelvin) might suit office tasks, while warmer tones (around 2,700 Kelvin) create a more relaxed atmosphere.
- **Energy Efficiency:** Utilize energy-efficient LED lighting solutions that offer good color rendering and dimming capabilities.

3.9.1.9 General Rules of Contrast

Our ability to see the world around us is significantly impacted by contrast. Contrast refers to the difference in light intensity (luminance) between an object and its surroundings, including its shadow. Imagine a bright red apple against a snowy backdrop—the stark difference in luminance makes the apple easily distinguishable.

Here's the technical side of contrast:

- **Contrast Ratio:** This metric tells us the ratio of the luminance of an object (L1) to the luminance of its background (L2). It's often expressed as a ratio (e.g., 5:1) or in decibels (dB).
- **Contrast Sensitivity:** This refers to our visual system's ability to detect these luminance differences. Interestingly, contrast sensitivity is often more important than absolute light levels. Even in dim environments, our eyes can adjust and perceive the world clearly as long as there's sufficient contrast.

Understanding contrast is crucial for designing well-lit and visually comfortable spaces. Here are some key principles to consider, incorporating recent findings (from 2023):

- **Balanced Illumination:** Ensure all objects and major surfaces within the field of view are adequately lit. This reduces visual strain caused by constantly adjusting focus between bright and dark areas.
- **Moderate Contrast in Central Vision:** For areas directly in front of the viewer (middle visual field), a contrast ratio of no more than 3:1 is recommended. This aligns with a 2023 study published in the *Journal of Vision Research*, which suggests this ratio minimizes fatigue and improves visual comfort during prolonged tasks.
- **Gradual Shift Toward the Periphery:** Contrast between the central zone and the periphery (edge) of the visual field should ideally not exceed 10:1. This creates a gradual dimming effect, mimicking the natural way our eyes perceive light.
- **Task-Oriented Lighting:** The working area should be brightest at its center and gradually get dimmer toward the edges. This aligns with recent ergonomic recommendations to minimize glare and improve focus on the primary task area.
- **Mind the Background:** Avoid excessive contrast, especially on the sides and below the user's visual field. This prevents distraction and creates a more calming visual environment.
- **Light Source Harmony:** Light sources themselves should have a moderate contrast ratio with their background (around 3:1 to 5:1). This prevents glare and ensures the light source itself isn't a distraction.
- **Manageable Overall Contrast:** While there is no ideal ratio, recent accessibility guidelines suggest a maximum contrast range of 40:1 within a room. This caters to individuals with varying visual sensitivities and ensures a comfortable environment for all.

3.9.1.10 Glare

Glare can cause significant discomfort, leading to eye strain, headaches, and blurred vision. However, these issues can be mitigated by implementing validated occupational ergonomics principles in the design of lighting systems to enhance visual comfort for workers and other individuals in an environment.

The first step in addressing glare is to focus on the light source itself. For direct glare, such as oncoming headlights, lights should be positioned strategically, ensuring they are outside the central field of vision—ideally at least 60 degrees off-center. Furthermore, using multiple lower-intensity light sources, rather than a single high-intensity one, can effectively diffuse light and reduce glare, promoting a more comfortable and visually ergonomic environment.

Indirect glare occurs when light reflects off surfaces, such as when car headlights are reflected in a mirror. To reduce indirect glare, it is essential to adjust the reflective surface. For example, tilting a computer screen away from overhead lighting or repositioning a car mirror can help deflect the glare.

If the light source cannot be controlled, shielding the eyes is another effective approach. Polarized sunglasses, for instance, are beneficial for driving as they filter out reflected light from roads and water. For computer work, anti-glare screen protectors or glasses with anti-reflective coatings can reduce reflected light, enhancing visual comfort.

Additionally, the use of appropriate window treatments is important in managing glare. Blinds or shades can block harsh sunlight, with vertical blinds being most effective for east-west windows and horizontal blinds for north-south windows. Translucent shades provide some light diffusion while still reducing glare, thereby improving comfort in the workspace

3.9.1.11 Sources and Measurement

3.9.1.11.1 Illumination and Measurement

Illumination refers to the amount of visible light reaching a surface, originating from various sources such as sunlight, artificial lighting, or electronic devices. It is measured in lux (metric system) or foot-candles (imperial system).

3.9.1.11.2 Luminance vs. Illumination

Luminance, in contrast, measures the perceived brightness of a surface, which depends not only on the light emitted by a source but also on how that surface reflects the light. For instance, a white sheet of paper and a black cloth illuminated by the same light will appear different in brightness, with the white surface appearing much brighter.

3.9.1.11.3 Reflectance: How Surfaces Reflect Light

Reflectance is the ratio of a surface's luminance to the illumination it receives. A white surface with high reflectance reflects most of the light, while a black surface with low reflectance absorbs a significant portion of the light.

3.9.1.12 Selecting the Appropriate Light Source

Choosing the right lighting depends on the tasks being performed within the space. Key considerations include:

- **Appropriate Illumination Levels:** The Illuminating Engineering Society of North America (IES) provides recommended lighting levels for various tasks and environments.
- **Balanced Lighting:** Avoid placing bright areas directly next to dark areas to reduce eye strain.
- **Consistency Over Time:** Sudden transitions between bright and dark environments can be tiring for the eyes.
- **Glare Reduction:** Minimizing glare is essential to enhance comfort and visibility.
- **User Considerations:** Factors such as age, visual impairments, and the nature of the work influence lighting requirements.

- **Energy Efficiency:** Opting for energy-efficient lighting solutions, such as LED lights, can significantly reduce energy consumption compared to traditional incandescent bulbs.

3.9.1.12.1 The IES and Recommended Illumination Levels

The IES publishes guidelines for lighting levels in workplaces to promote good task visibility and worker comfort. These guidelines consider the desired alertness level for the task at hand. For example, tasks requiring high concentration may need higher illumination levels than those requiring minimal focus.

3.9.1.13 Examples of Work Lighting

- **Incandescent Lamps:** These traditional bulbs are warm and inviting but emit high amounts of heat and are not energy efficient.
- **Fluorescent Lights:** These are much more efficient than incandescent bulbs and do not emit as much heat. They are a good choice for general lighting in offices and other workspaces.
- **LED Lights:** LEDs are the most energy-efficient lighting option and are becoming increasingly popular. They come in a variety of colors and can be dimmed to create different lighting effects.

3.9.1.14 Measuring Illumination

The IESNA publishes recommendations for lighting in different environments. There are various ways to measure light levels, but the most common unit is lux.

TABLE 3.8
American Illuminating Engineering Society Guidelines

Alertness-Level Requirement	Average Light Level, fL
Maximum mental alertness required for highly complex mental task performance	50 and above
Medium mental alertness required for routine manual tasks, leisure reading, and/or stimulating social activity	40
Minimum mental alertness required for non-demanding social intercourse and/or perceptual-motor performance (dining, dressing, personal hygiene, etc.)	30
Rest, mental alertness for minimum interaction with other people (e.g., as in a bar or private dining environment)	15
Sleep	<3

Source: Woodson, W. E., Tillman, B., and Tillman, P. *Human Factors Design Handbook*, 2nd edn., McGraw-Hill, Inc., New York.

3.9.1.15 Light and Our Well-Being

Exposure to light has a significant impact on our health and well-being. Natural light, particularly sunlight, helps regulate our circadian rhythm, which keeps our sleep-wake cycle in sync. Artificial light can also affect our circadian rhythm, so it's important to choose the right type of lighting for our needs.

3.9.2 Noise Levels

Noise pollution, defined as any unwanted or disturbing sound, is a growing environmental concern that permeates workplaces, homes, and leisure spaces. Chronic exposure to noise from traffic, construction, airplanes, industrial machinery, and everyday appliances can have significant negative impacts on our physical and mental health.

Health risks associated with noise pollution include:

- **Noise-Induced Hearing Loss (NIHL):** According to the World Health Organization (WHO), over 1 billion young people worldwide are at risk of NIHL due to recreational noise exposure [WHO 2021].
- **Cardiovascular Issues:** Studies have linked chronic noise exposure to an increased risk of high blood pressure, heart disease, and stroke [European Environment Agency, 2023].
- **Sleep Disruption:** Noise can make it harder to fall asleep and stay asleep, leading to fatigue, daytime sleepiness, and impaired cognitive function.
- **Stress and Anxiety:** Constant exposure to unwanted sounds can trigger the body's stress response, leading to increased levels of cortisol (the stress hormone) and anxiety.
- **Reduced Cognitive Function:** Noise can hinder concentration, memory, and learning, especially in children [National Geographic, n.d.].

The pervasiveness of noise pollution makes it a significant public health issue. In addition to the health concerns listed above, noise can also contribute to a decrease in quality of life, affecting mood, communication, and overall well-being.

3.9.2.1 Measuring Noise Levels

Regular and accurate measurement of noise levels is a cornerstone of any successful hearing conservation program. This crucial step helps pinpoint areas with hazardous sound levels, identify potentially impacted workers, and determine where additional noise control strategies are necessary.

3.9.2.2 Modern Tools for Noise Assessment

Modern sound level meters (SLMs), equipped with advanced weighing scales, are now the preferred instruments for measuring workplace noise. These meters incorporate sophisticated filters designed to replicate the sensitivity of human hearing.

FIGURE 3.7 Weighted sound levels meters.

The A-weighting scale, measured in decibels (dB(A)), remains the industry standard, as extensive research confirms it accurately represents how humans perceive sound levels. For a more comprehensive assessment of noise exposure, it is important to consider the overall noise load, which includes both sound volume and the frequency spectrum of the noise. This is where equivalent continuous sound level (Leq) becomes valuable, as it offers a more complete picture of workplace noise conditions.

3.9.2.3 Equivalent Level: Understanding the "Average Noise Dose"

The Leq represents the average sound energy level an employee experiences over a specific timeframe, typically a full workday. Essentially, it provides a single value that summarizes the cumulative effect of varying noise levels throughout the shift.

Modern sound level meters and noise dosimeters can calculate Leq directly, simplifying the monitoring process. Choosing the appropriate instrument depends on the type of sound source and the specific needs of your workplace.

For the latest guidance on instrument selection, the Canadian Centre for Occupational Health and Safety (CCOHS) has up-to-date recommendations (CCOHS, 2023).

TABLE 3.9
Guidelines for Instrument Selection

Type of Measurement	Appropriate Instruments (in Order of Preference)	Result	Comments
Personal noise exposure	1. Dosimeter	Dose or equivalent sound level	Most accurate for personal noise exposures
	2. ISLM	Equivalent sound level	If the worker is mobile, it may be difficult to determine a personal exposure, unless work can be easily divided into defined activities
	3. SLM	dB(A)	If noise levels vary considerably, it is difficult to determine average exposure Only useful when work can be easily divided into defined activities and noise levels are relatively stable all the time
Noise levels generated by a particular source	1 SLM	dB(A)	Measurement should be taken 1–3 meters from source (not directly at the source)
	2 ISLM	Equivalent sound level dB(A)	Particularly useful if noise is highly variable; it can measure equivalent sound level over a short period of time (1 minute)
Noise survey	1 SLM	dB(A)	To produce noise map of an area; take measurements on a grid pattern
	2 ISLM	Equivalent sound level dB(A)	For highly variable noise
Impulse noise	1. Impulse SLM	Peak pressure dB(A)	To measure the peak of each impulse

Source: Canadian Centre for Occupational Health Safety, 2009, Retrieved September 9, 2009, from http://www.ccohs.ca/

Sound Level Meter (SLM):

- **Function:** The SLM is a handheld instrument with a microphone that converts fluctuating air pressure caused by sound waves into electrical signals. These signals are processed to calculate a decibel (dB) sound level displayed on the readout.
- **Calibration:** Accurate measurements require regular calibration of the SLM using a sound source of known intensity.
- **Applications:** SLMs are commonly used for noise surveys, identifying noise sources, and verifying compliance with noise exposure limits.

Integrated Sound Level Meter (ISLM):

- Unlike a dosimeter worn by a worker, the ISLM is a portable device for taking noise measurements at specific locations within a work environment.
- The ISLM is designed to capture a single noise level at a particular location, even with fluctuating noise levels in the surroundings.

Noise Surveys:

- **Purpose:** A noise survey identifies areas, machines, or equipment generating harmful noise levels. It helps employers implement effective noise control measures to protect workers' hearing.
- **Conducting a Survey:**
  - Sketch the work area, including worker locations and noisy machinery.
  - Take multiple representative noise level readings around the area and mark them on the sketch.
  - More measurements provide a more accurate noise map.
- **Noise Map:** A noise map is a visual representation of noise levels throughout the work environment. It is created by connecting points with equal noise levels on the sketch. Noise maps identify areas exceeding safe noise limits.

3.9.2.4 Limitations of Standard SLMs

- Standard SLMs capture only instantaneous noise readings. However, this is sufficient for workplaces with consistent noise levels.
- In environments with impulsive, intermittent, or variable noise, an SLM may not accurately reflect a worker's average noise exposure throughout a shift.

Noise Dosimeter:

- **Function:** A noise dosimeter is a portable device worn by workers who move around frequently within a work area.
- **Placement:** Typically attached near the worker's hip with a microphone positioned on the collar for accurate noise level recording.
- **Dosimeter Settings (CCOHS) Guidelines:**
  - **Criterion Level:** This is the permissible noise exposure limit for an 8-hour workday over a five-day workweek. The recommended criterion level varies by jurisdiction, ranging from 85 dB(A) to 90 dB(A).
  - **Exchange Rate:** This accounts for the doubling or halving of noise exposure for every 3 dB or 5 dB increase or decrease in noise level, respectively.
  - **Threshold:** This is the minimum noise level the dosimeter registers. Noise below this level is not included in the calculated noise dose.
- **Benefit:** By wearing the dosimeter throughout a work shift, the average noise exposure or dose for the worker is calculated, typically expressed as a percentage of the allowed exposure limit. This data helps employers

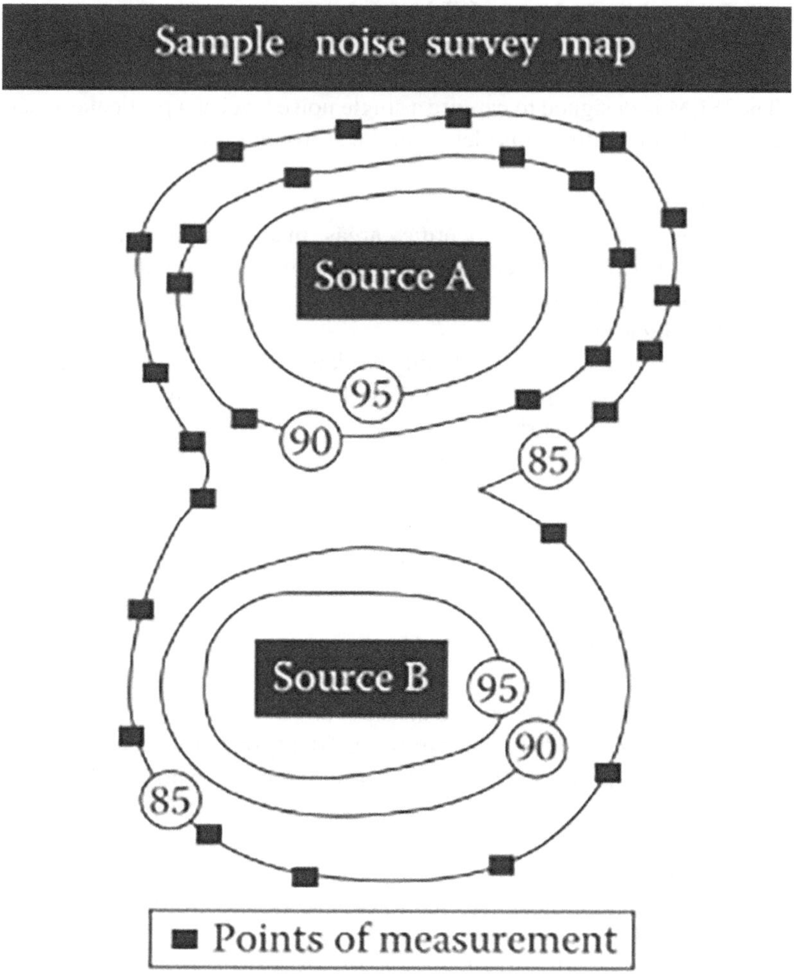

FIGURE 3.8 Sample noise survey map. (From Canadian Centre for Occupational Health Safety, 2009, retrieved September 9, 2009, from http://www.ccohs.ca/)

assess noise control needs and implement appropriate hearing conservation programs.

3.9.2.5 Acceptable Noise Levels

The National Institute for Occupational Safety and Health continues to be a leading organization in setting guidelines for safe noise exposure in workplaces. Their recommended exposure limits aim to minimize the risk of noise-induced hearing loss even during long-term occupational exposure.

Table 3.10 outlines the recommended permissible noise exposure times based on equivalent sound level (dBA).

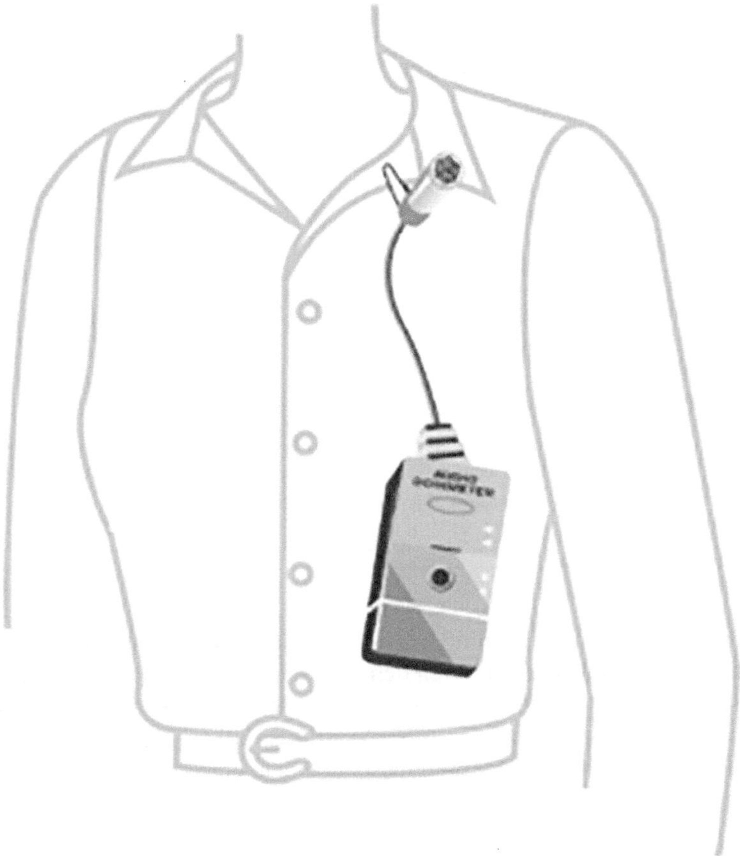

FIGURE 3.9 Body-worn noise level meter, also known as a noise dosimeter.

3.9.2.6 Important Considerations

- These are recommended limits by NIOSH. Legal regulations regarding workplace noise exposure may vary depending on location. Always refer to the relevant occupational safety standards for the specified area.
- Table 3.12 reflects continuous noise exposure throughout the workday. Fluctuating noise levels or the presence of particularly harmful sounds like impact noise may necessitate adjustments to these limits.
- Hearing protection is recommended when possible, especially in environments exceeding recommended exposure limits or when working with loud machinery.

Thermal conditions, including both temperature and humidity, are critical factors influencing not only comfort but also performance, health, and safety in occupational settings. From an ergonomic perspective, understanding these environmental

TABLE 3.10
Permissible Noise Exposures

Duration per Day, Hours	Sound Level dBA Slow Response
8	90
6	92
4	95
3	97
2	100
1 1/2	102
1	105
V	110
1/4 or less	115

Source: Taken from the US Department of Labor website under OSHA Section 1910.95(b)(1), Occupational Noise Exposure, http://www.osha.gov/pls/oshaweb/owadisp.show_document?p_table=STANDARDS&p_id=9735&P_TEXT_VERSION=FALSE

variables is essential for designing workspaces that optimize human well-being and productivity. Thermal comfort, often defined as a subjective state where individuals are satisfied with their thermal environment, is influenced by a complex interaction of environmental and personal factors (ASHRAE, 2017). Ergonomists and human factors engineers must consider both these factors to ensure optimal working conditions and minimize risks associated with extreme temperatures.

Environmental Factors

1. **Air Temperature:**
   Air temperature is a primary determinant of thermal comfort and can significantly affect physiological responses in the workplace. Research indicates that most individuals experience comfort in temperatures ranging from 20°C to 26°C (68°F to 78°F) (Parsons, 2014). However, individual thermal preferences may shift based on acclimatization, i.e., long-term exposure to a particular thermal environment. For instance, workers in tropical regions may develop a tolerance for warmer temperatures, while individuals in cooler climates may prefer lower ambient temperatures (de Dear & Brager, 2015). Hence, designing temperature control systems for workplaces should consider regional differences and individual adaptation.

2. **Airflow:**
   Airflow, or the movement of air within the workspace, plays a vital role in regulating body temperature, especially under conditions of physical exertion. Increased airflow facilitates evaporative cooling by accelerating

TABLE 3.11
Decibel Comparison Chart

	Environmental Noise, dB
Weakest sound heard	0
Whisper quiet library	30
Normal conversation (3–5 ft)	60–70
Telephone dial tone	80
City traffic (inside car)	85
Train whistle at 500 ft, truck traffic	90
Subway train at 200 ft	95
Level at which sustained exposure may result in hearing loss	90–95
Power mower at 3 ft	107
Snowmobile, motorcycle	100
Power saw at 3 ft	110
Sandblasting, loud rock concert	115
Pain begins	125
Pneumatic riveter at 4 ft	125
Even short-term exposure can cause permanent damage—loudest recommended exposure with hearing protection	140
Jet engine at 100 ft, gun blast	140
Death of hearing tissue	180
Loudest sound possible	194

Source: Statistics for the decibel (loudness) comparison chart were taken from a study by Chasin, M., Centre for Human Performance & Health, Ontario, Canada, http://www.gcaudio.com/ resources/ howtos/loudness.html; University of North Florida, ADA Compliance, http://www. unf.edu/anf/ adacompliance/Hearing_Campaign.aspx (accessed February, 2011).

TABLE 3.12
Sound Level Exposure Duration Guidelines

Duration (Hours per Day)	Equivalent Sound Level (dBA)
8	85
4	90
2	95
1	100
½	105

the evaporation of sweat, thus enhancing heat dissipation (Fanger, 1970). However, excessive airflow can lead to discomfort, particularly in cooler environments where wind chill can lower skin temperature and lead to thermal stress (Cabanac, 2014). Proper design of ventilation systems is crucial to balancing the benefits of air movement without causing thermal discomfort.

3. **Humidity:**
   Humidity levels influence thermal comfort by affecting sweat evaporation from the skin. In environments with high humidity, the reduced capacity for sweat evaporation impairs the body's ability to regulate heat, leading to an increased perception of warmth and potential heat stress (Gong et al., 2016). Conversely, in low-humidity conditions, typical of cold environments, moisture loss from the skin and mucous membranes can result in dryness and discomfort. Managing relative humidity levels, particularly in indoor industrial settings, is essential for preventing both heat-related illnesses and cold-related discomfort (Epstein & Moran, 2006).

4. **Radiant Temperature:**
   Radiant temperature refers to the heat transfer between surrounding surfaces and the human body via electromagnetic radiation. Heat sources such as sunlight, machinery, and warm walls can radiate thermal energy, contributing to the overall thermal load on workers (Parsons, 2014). In colder environments, the absence of sufficient radiant heat can lead to feelings of cold stress despite moderate air temperatures. Therefore, radiant heat sources and their placement must be considered in the ergonomic design of workspaces, especially in environments exposed to direct sunlight or cold machinery.

3.9.3 Personal Factors

Clothing Insulation:
The insulating properties of clothing are a significant determinant of thermal comfort, especially in occupational environments where workers may be exposed to extreme temperatures. Clothing acts as a barrier to heat exchange, with thicker or multilayered garments providing greater insulation in cold environments, while lighter, more breathable fabrics are necessary for dissipating heat in warmer settings (Havenith et al., 2014). Ergonomic design of work uniforms and protective clothing should consider these factors to ensure that workers remain thermally comfortable while performing their tasks.

Metabolic Rate:
Physical activity in the workplace increases metabolic heat production, which directly impacts thermal comfort. Workers engaged in physically demanding tasks generate more metabolic heat, making them more susceptible to heat stress in warmer environments (ISO 8996:2004). Ergonomic interventions, such as adjusting workload intensity and work/rest cycles, are necessary to prevent thermal strain in such conditions (Mekjavic & Tipton, 2014). On the other hand, sedentary workers may require supplementary heating in cold environments to compensate for the reduced metabolic heat generation.

Acclimatization:

The human body has the ability to acclimate to different thermal environments over time. Acclimation results in physiological adaptations, such as improved sweat response in hot environments and enhanced heat conservation in cold conditions (Pryor et al., 2018). In occupational settings, gradual exposure to extreme temperatures can help workers develop tolerance and minimize thermal discomfort. Employers should implement acclimation protocols, particularly in industries that expose workers to significant temperature variations, such as outdoor construction or foundry work (Gao et al., 2017).

3.9.3.1 Recent Advances in Thermal Comfort Research

Recent studies have highlighted the growing importance of psychological and social factors in determining thermal comfort in the workplace. For instance, workers' expectations, prior experiences, and cultural background can significantly shape their thermal comfort perceptions (Kim et al., 2017). Additionally, advancements in personalized thermal comfort systems offer innovative solutions for managing individual thermal preferences in shared work environments. These systems allow for localized temperature control, providing workers with the ability to adjust their microenvironment without impacting others (Parkinson et al., 2020). This is

TABLE 3.13
Body Heat Balance

Term	Definition	Determinants
M	Metabolic heat gain	Physical workload, or the muscular work done minus the work efficiency
C	Convective heat gain or loss	Air velocity The difference between air temperature and a person's average skin temperature
R	Radiative heat gain or loss	The difference between a person's average skin temperature and the temperature of surfaces in the environment, measured with a globe thermometer or radiometer The amount of skin exposed to the solid surface
E	Evaporative heat loss	The difference between the water vapor pressure of a person's skin and the water vapor pressure, or relative humidity, of the environment; it is indirectly related to workload and the person's sweat rate Air velocity
S	Heat storage in, or loss from, the body	Balance of the aforementioned factors Rectal and skin temperatures Body weight
Co	Conductive heat gain or loss	The area of the conductive surface The difference between the person's skin temperature and the temperature of the surface contacted

Source: Developed from information in Leithead & Lind, 1964; Eastman Kodak, 1986.

TABLE 3.14
Recommended Maximum Temperatures for Short-Duration Exposures to High-Heat Environments (up to 63°C or 146°F)

		Maximum Ambient Temperature, °C (°F)[a]		
Exposure Time (min)	Workload[b]	Relative Humidity 20%	Relative Humidity 50%	Relative Humidity 80%
	M	59 (138)	48 (118)	46 (115)
	H	57 (135)	46 (115)	42 (108)
15	L	53 (128)	45 (113)	40 (104)
	M	52 (126)	43 (110)	38 (100)
	H	51 (124)	41 (106)	36 (97)
30	L	52 (126)	44 (112)	39 (102)
	M	47(116)	38 (100)	34 (93)
	H	41 (106)	36 (97)	30 (86)
45	L	51 (124)	43 (110)	38 (100)
	M	41 (106)	36 (97)	31 (88)
	H	36 (97)	32 (90)	27 (81)

Source: Adapted from Rodgers & Cori, 1981, Eastman Kodak Company; based on information in Bell et al., 1971; Gagge, 1973; Hardy, 1970; Leithead & Lind, 1964; Pandolf & Goldman, 1978; Eastman Kodak, 1986.

Note: For 5 min exposure times in high air velocities (2 m/s, or 400 ft/min), the following maxi mum temperatures are recommended (L, light workload; M, moderate workload; H, heavy workload)

L	56 (133)	50 (122)	48 (118)
M	54 (129)	49 (120)	44 (111)
H	52 (126)	48 (118)	42 (103)

[a] These temperatures assume the following conditions: clothing insulation = 0.6 clo; air velocity = 0.1 m/s (20 ft/min); radiant heat = 2°C (3.6°F), which is the difference between the globe and dry bulb

[b] Workload abbreviations: L, light, up to 140 W (120 kcal/h); M, moderate, >140–230W (>120– 240 kcal/h); H, heavy, >230–350W (>240–300 kcal/h).

particularly valuable in open-plan offices or industrial settings where a diverse workforce may have varying thermal comfort needs.

In conclusion, thermal conditions in occupational environments have a profound impact on workers' comfort, performance, and health. From air temperature and humidity to clothing insulation and metabolic rate, a range of environmental and personal factors must be considered in the ergonomic design of workplaces. By integrating these considerations, ergonomists can create safer, more productive environments that support workers' physiological needs. Additionally, recent advances in personalized thermal comfort systems and research into psychological factors offer promising directions for improving thermal comfort in complex work environments.

3.9.3.2 Measurement of Thermal Load

While there is no mandated workplace temperature standard in the United States, occupational health and safety professionals rely on guidelines to ensure worker well-being in hot environments. A key reference is the American Conference of Governmental Industrial Hygienists' (ACGIH) Threshold Limit Values (TLVs) for Heat Stress. This publication, particularly relevant for non-office settings, provides valuable data on heat stress exposure limits expressed in Wet-Bulb Globe Temperature (WBGT).

Developed in the late 1950s for the US Marine Corps Recruit Depot on Parris Island, South Carolina, the WBGT specifically addresses the challenges of heat stress in high-humidity environments. This measure arose from a need to assess and mitigate the impact of extreme heat on Marines training there.

The WBGT effectively combines temperature, humidity, and solar radiation to estimate the environmental effects on workers. This composite measure emerged from research following a series of heatstroke fatalities at Parris Island in the 1950s. Over time, the WBGT has become a valuable tool not only in workplaces but also in managing heat stress during sporting events.

Here is a breakdown of what factors contribute to WBGT:

- Dry-bulb temperature (air temperature)
- Relative humidity
- Mean radiant temperature (heat radiating from surrounding surfaces)
- Air velocity (wind speed)

These factors are incorporated into two equations to calculate the final WBGT value, providing a more comprehensive picture of perceived temperature than any single measurement alone.

3.9.3.3 New and Timely Information

- While the ACGIH TLVs remain a central reference, it's important to note that OSHA (Occupational Safety and Health Administration) is currently considering a heat stress standard. This potential standard is expected to incorporate WBGT measurements alongside other factors for creating safer work environments in hot weather conditions.

3.9.3.4 WBGT Equipment

The Wet-Bulb Globe Temperature is a crucial measure of heat stress on the human body, especially in direct sunlight. It considers factors beyond just air temperature, providing a more comprehensive assessment of environmental conditions.

Here is a breakdown of the terms involved in WBGT calculation:

- **Natural Wet-Bulb Temperature (Tnw):** This is the temperature measured by a sensor wrapped in a wet wick and exposed to airflow. Water evaporation from the wick cools the sensor, and the resulting temperature reflects the combined effects of air temperature and humidity.

FIGURE 3.10 WBGT measurement equipment.

- **Globe Temperature (Tg):** This is the temperature of a black, six-inch diameter hollow copper sphere. The black color absorbs solar radiation, and the sphere's temperature closely resembles what someone working in sunlight would experience.

3.9.3.5 WBGT Equation for No Direct Sunlight

In environments with no direct sunlight, the WBGT can be calculated using the following equation:

WBGT = 0.7 * Tnw + 0.3 * Tg

where

Tnw is the natural wet-bulb temperature (°C or °F)

Tg is the globe temperature (°C or °F)

This equation assigns weightage factors to each temperature component based on its contribution to heat stress. Air temperature (Ta) has the least weightage (0.1) as it does not fully capture the impact of solar radiation and humidity. Wet-bulb temperature (Tnw) has a higher weightage (0.7) because it reflects both air temperature and humidity. Globe temperature (Tg) receives a moderate weightage (0.2) as it accounts for radiant heat from the sun.

3.9.3.6 WBGT Equation for Direct Sunlight

For environments with direct sunlight, a more complex equation (Equation 3.4.8) incorporating solar radiation is needed for accurate WBGT assessment. This equation considers factors like solar radiation intensity, airflow, and cloud cover.

$$WBGT = 0.7 * Tnw + 0.2 * Tg + 0.1 * Ta$$

where

Tnw is the natural wet-bulb temperature (°C or °F)

Tg is the globe temperature (°C or °F)

Ta is the dry-bulb air temperature (°C or °F) (measured in shade)

3.9.3.7 Finding WBGT Measurements

While the equations provide a theoretical understanding of WBGT calculation, it's important to note that acquiring the necessary equipment (wet-bulb thermometer, globe thermometer) and expertise for accurate measurement is recommended. Many weather stations and organizations like the National Oceanic and Atmospheric Administration (NOAA) provide WBGT readings for various locations.

TABLE 3.15
WBGT Correction Factors (°C)

Clothing Type	Clo[a] Value	WBGT Correction
Summer lightweight working clothing	0.6	0
Cotton coveralls	1.0	–2
Winter work clothing	1.4	–4
Water barrier, permeable	1.2	–6

Source: ACGIH, 1992.

[a] Clo: Insulation value of clothing. One clo = 5.55 kcal/m2/h of heat exchange by radiation and convection for each °C difference in temperature between the skin and the adjusted dry-bulb temperature.

3.9.3.8 Importance of WBGT

WBGT is an important metric for occupational health and safety, athletic training, and military operations. By monitoring WBGT levels, employers and individuals can take preventive measures to avoid heat stress illnesses like heat stroke and exhaustion.

3.9.3.9 Other Thermal Indices

Thermal comfort plays a crucial role in worker safety and productivity. Here is a breakdown of some key indices used to evaluate thermal environments:

- **Effective Temperature (ET):** This classic index, introduced in the 1920s, combines air temperature, humidity, and air velocity. While still valuable in specific settings like mines with high humidity and minimal radiant heat, ET may not fully capture the impact of factors like clothing insulation and individual variations.
- **Heat-Stress Index (HSI):** Developed in the 1980s, HSI goes beyond just environmental factors. It considers metabolic heat generation from physical activity for a more personalized assessment. However, implementing HSI can be complex, requiring data on workload and individual characteristics.
- **Operative Temperature:** This index focuses on the combined effect of radiant heat transfer (e.g., sunlight) and air temperature on a stationary object. While valuable for specific scenarios, it neglects the influence of humidity and air movement on human thermal comfort.
- **Oxford Index:** This newer index (developed by Parsons in the early 2000s) offers a more nuanced approach. It weights both wet-bulb and dry-bulb temperatures, providing insights into the impact of humidity variations on thermal comfort. Additionally, the Oxford Index can be tailored to account for factors like clothing insulation and worker activity levels, making it particularly useful for occupational settings.

3.9.3.10 Choosing the Right Tool

Each index offers valuable insights, but the most appropriate choice depends on the specific work environment and its dominant thermal challenges. Here are some additional considerations:

- **Recent Advancements:** The field of thermal comfort assessment is constantly evolving. Newer indices like the Predicted Mean Vote (PMV) and Predicted Percentage Dissatisfied (PPD) consider a wider range of factors and may be worth exploring depending on the application.
- **Economic Factors:** Implementing some indices may require specialized equipment or complex calculations. Balancing the need for accurate assessment with cost-effectiveness is crucial.

3.9.3.11 Controlling the Thermal Conditions

The modern approach to managing heat stress in the workplace integrates advanced engineering controls, administrative strategies, and innovative practices. These methods aim to reduce the heat load and improve employee comfort and safety.

3.9.3.12 Updated Engineering Controls

- **Enhanced Ventilation Systems:** Utilizing both permanent and portable ventilation solutions, these systems draw in cooler external air to replace the warmer indoor air, effectively managing the temperature in large and small areas alike.
- **Advanced Air Cooling and Air Conditioning Technologies:** Techniques such as air treatment and localized air cooling are employed to remove heat directly from the air, reducing indoor temperatures. High-efficiency air conditioning systems, including those utilizing chillers, offer a more costly but highly effective cooling solution, especially in dry climates.
- **Innovative Shielding and Insulation Methods:** Modern materials and technologies are applied to minimize radiant heat through surface modifications, advanced insulating materials, and strategic shielding. These measures disrupt the direct path of heat to workers, substantially reducing heat exposure.

3.9.3.12 Revolutionary Administrative Controls and Work Practices

- **Comprehensive Training Programs:** Updated to reflect the latest research and guidelines, training now encompasses understanding heat stress hazards, recognizing symptoms and predisposing factors, and the importance of first aid and emergency procedures for heat-related illnesses. The risks associated with the use of certain medications and alcohol in hot environments, along with the correct use of protective clothing and equipment, are also emphasized.
- **Enhanced Worker Monitoring Systems:** Incorporating state-of-the-art technology to regularly monitor workers at increased risk of heat stress, these systems measure vital signs such as heart rate and oral temperature, and assess body water loss to prevent dehydration and overheating. Adjustments to work cycles are made based on these indicators to ensure worker safety.
- **Strategic Scheduling and Fluid Replacement:** Scheduling work during cooler parts of the day, implementing regular breaks, job sharing, and ensuring easy access to hydration options are critical components of modern administrative controls. These strategies are designed to minimize heat exposure and maintain optimal hydration levels among workers.
- **Engagement in Environmental and Medical Surveillance:** Emphasizing the benefits of worker participation, these programs aim to continuously assess and address the risks of heat stress, enhancing overall workplace safety and health.

TABLE 3.16
Thermal Strain Indices

Index	Source(s)	Inputs	Comments
		Indices for Heat Exposure	
Wet-Bulb Globe Temperature (WBGT) Wet globe temperature (WGT)	ISO 7243 (1989); NIOSH (1986); Yaglou and Minard (1957); Botsford (1971)	T_a, T_{nwb}, T_g botsball	WBGT requires simple input and calculations. WBGT is not recommended for conditions of high humidity. The botsball is an instrument consisting of a fabric-covered 60 mm black ball over a dial thermometer. WGT may be read off a scale or converted to WBGT if T_a is known
Heat stress index (HS1)	Belding and Hatch (1955); ISO 7933 (1989)	T_a, T_{w5}, T_{bg} or T_v V, M	HSI is the ratio of evaporative heat loss required to maintain a constant body temperature to the maximum amount of sweat that can be evaporated under the given climatic conditions. Required sweat rate index (S,*), a further development of HSI, is used in ISO 7933
Oxford index (WD)	Leithead and Lind (1964)	T_a, T_{wb}	WD originated from research on men performing rescue tasks in hot underground mines. WD can be used to determine tolerance times
		Indices for Cold Exposure	
Wind chill index (WC)	Siple and Passel (1945)	T V	WCI is an index for the cooling rate of wind and air temperature on bare skin. There is no adjustment for solar radiation, clothing, or activity level
Required clothing insulation (IREQ)	Holmer (1984); ISO TR 11079 (1993)	T_a, T_r, V, R.H., M	Clothing insulation required for survival ($IREQ_{min}$) with a t_{sk} of 30°C and for maintaining thermal equilibrium ($IREQ_{neutral}$) are calculated
		Equivalent Temperature Indices	
Effective temperature (ET)	Houghton and Yaglou (1923)	T_a, T_{wb}, V	ET relates actual conditions to an equivalent, calm, saturated environment. ET overemphasizes effects of humidity in cool and neutral conditions and underemphasizes its effects in warm conditions

(Continued)

TABLE 3.16 (CONTINUED)
Thermal Strain Indices

Index	Source(s)	Inputs Indices for Heat Exposure	Comments
"New" effective temperature (ET*)	ASHRAE (1993); Gagge et al. (1971); Gonzalez et al. (1974)	T_a, T_r, V, P_a, i_m, w, M	ET* was developed to replace ET. It includes skin-wettedness (w) and water vapor pressure (P_a) parameters in calculating the temperature of an environment at 50% R.H. that results in equivalent total heat loss from the skin as in the actual environment
Operative temperature (T_o)	Winslow et al. (1937)	T_a, T_{be} or T_r, v	T_o combines dry heat exchange parameters. There is no adjustment for work rate or effects of humidity on evaporative cooling. Formulas for approximating T_o are provided in ISO 7730 (1993)

Source: Gavriel, S. (Ed.): *Handbook of Human Factors and Ergonomics*, 1997. Copyright Wiley-VCH Verlag GmbH & Co. KGaA.

TABLE 3.17
NIOSH Recommended WBGT Limits, in °C, for Heat Stress Exposure for Acclimatized Workers

	Workload		
Hourly Work/Rest Cycle	**Light (<230W)**	**Moderate (230–350W)**	**Heavy (>350W)**
Continuous work	<30.0	<26.7	<25.0
75% work/25% rest	30.6	27.8	25.6
50% work/50% rest	31.7	29.4	27.8
25% work/75% rest	32.2	31.1	30.0
Ceiling limit	38.9	36.7	35.0

Source: Gavriel, S. (Ed.): *Handbook of Human Factors and Ergonomics*, 1997. Copyright Wiley-VCH Verlag GmbH & Co. KGaA; NIOSH, 1986.

Note: Limits are for a "standard" worker weighing 70 kg with a 1.8 m^2 body surface area.

TABLE 3.18
Factors Affecting the Occurrence of Heat Stress

Factor	Effect
Hydration state	Hypohydration results in lower sweat production and an increase in core temperature
Acclimatization	Repeated heat exposure leads to earlier onset of sweating, a higher sustained sweat rate, and lower core temperature and heart rate
Age	Sweating mechanism and circulatory system become less responsive with age, and there is high level of skin blood flow, possibly due to impaired thermoregulatory mechanism
Physical fitness	Exercise that increases maximal aerobic capacity improves thermoregulatory responses in the heat
Subcutaneous fat	Subcutaneous fat provides an insulative barrier, reducing transfer of heat from muscles to skin
Gender	Although studies indicate that sweating and vasodilation occur at higher core temperatures in women than men, when controlled for fitness and menstrual phase, gender differences in the follicular phase are questionable. Women in the luteal phase have significantly higher core temperatures, which may impact thermoregulatory responses
Body size	Leaner individuals are at an advantage because they have a larger ratio of surface area to body mass and, thus, greater capacity to dissipate heat
Diet	Regular consumption of a balanced diet serves to replace salt and other electrolytes lost in sweat, maintaining sweating efficiency
Previous heat illness	Previous occurrence of heat stroke increases susceptibility to subsequent heat illness
Drugs and alcohol	Use interferes with the functioning of the central and peripheral nervous system, negatively affecting heat tolerance

Source: Gavriel, S. (Ed.): Handbook of Human Factors and Ergonomics, 1997. Copyright Wiley-VCH Verlag GmbH & Co. KGaA

Incorporating these cutting-edge strategies and technologies, alongside a proactive approach to training and monitoring, significantly enhances the effectiveness of heat stress management in the workplace. It is imperative that both management and employees stay informed about the latest advancements and best practices in heat stress mitigation to ensure a safe and healthy work environment.

3.10 VIBRATION AND THE HUMAN BODY

The International Organization for Standardization (ISO) has established a series of standards (ISO 2631) to assess the impact of whole-body vibration (WBV) on human health. These standards outline procedures for measuring and evaluating vibration exposure (ISO 2631-1:2022).

WBV exposure can have a significant impact on the body, causing a range of physical and sensory effects depending on the frequency, amplitude, and duration of exposure. Here is a breakdown of the three main frequency ranges and their potential consequences:

- **Low Frequencies (0–2 Hz):** Often underestimated, low-frequency vibrations can resonate with the body's natural frequencies, potentially causing amplified effects. Construction equipment and off-road vehicles are common sources of low-frequency WBV.
- **Mid-Frequencies (2–20 Hz):** This range can cause discomfort, fatigue, and musculoskeletal disorders, including carpal tunnel syndrome and tendinitis. These vibrations are frequently encountered in operating machinery and driving vehicles.
- **High Frequencies (above 20 Hz):** Traditionally a major concern, high-frequency WBV can lead to nerve damage, particularly in the hands and arms. Power tools and handheld equipment are known sources of high-frequency vibrations.

3.10.1 Recent Developments in WBV Research

Recent research highlights the importance of considering all vibration frequencies, not just high frequencies, for occupational safety assessments. Studies suggest that low and mid-frequency vibrations can cause comparable or even greater cumulative damage to the musculoskeletal system compared to high-frequency vibrations.

3.10.2 Measuring and Mitigating WBV Risks

Several factors influence human response to WBV, including posture, body composition, and the use of personal protective equipment (PPE). A comprehensive assessment requires measuring the acceleration across various frequencies and exposure durations (WISHA Hand–Arm Vibration Analysis, 2023).

Vibration measurement instruments typically employ accelerometers to detect vibrations, along with recorders, frequency analyzers, and weighting networks. These networks account for the human body's varying sensitivity to different frequencies, providing a single-value measure of vibration exposure in meters per second squared (m/s^2).

Mitigating WBV Risks:

- Implementing engineering controls to reduce vibration at the source (e.g., suspension systems in vehicles)
- Utilizing proper work practices, such as vibration-limiting work schedules and proper lifting techniques
- Providing effective personal protection equipment (PPE), including vibration-dampening gloves and footwear

- Regularly monitoring vibration exposure levels and worker health

3.10.3 Health Consequences of Long-Term WBV Exposure

Chronic exposure to WBV can lead to various health problems, including:

- **Musculoskeletal Disorders:** White finger disease, carpal tunnel syndrome, tendinitis, and bone/joint disorders
- **Vestibular Problems:** Dizziness, balance issues, and nausea
- **Cardiovascular Issues:** High blood pressure, irregular heart rhythms
- **Other:** Impaired vision, fatigue, and hyperventilation

The debilitating and long-term consequences of vibration exposure necessitate aggressive control measures. Multiple techniques can be implemented to achieve this, categorized into source control, path control, and receiver control (Bernardini et al., 2012).

- **Source Control:** This proactive approach focuses on minimizing vibrations at the source, ideally during the design phase of tools, equipment, or vehicles. For existing products, redesign or retrofitting with vibration-dampening materials can be explored. Additionally, implementing wraps, regulating speeds, optimizing processes, and ensuring proper equipment balance can significantly reduce vibration intensity.
- **Path Control:** This strategy aims to limit the amount of vibration transmitted to the user. Administrative controls, such as task rotation, job sharing, and incorporating frequent rest breaks, can be implemented to decrease exposure duration.
- **Receiver Control:** This method focuses on reducing the vibration experienced by the operator. Techniques include isolating tools with vibration-dampening mounts or grips, adjusting posture to minimize contact with vibrating surfaces, and modifying tasks to lessen the required force exertion. Additionally, PPE like padded gloves and anti-vibration clothing can offer further protection.

3.10.4 Recent Advancements

The field of vibration control is constantly evolving. Here are some noteworthy advancements:

- **Active Vibration Control Systems:** These employ sensors and actuators to counteract vibrations in real-time, offering a highly effective solution for specific applications.
- **Smart Materials:** Materials with inherent vibration-dampening properties are being increasingly integrated into tools and equipment.

- **Biomechanical Modeling:** This approach helps predict vibration's impact on the human body, allowing for the design of safer work environments and PPE.

These emerging occupational ergonomic resources can be useful in minimizing the impact of vibration on individuals in occupational and recreational environments.

3.11 EXPOSURE TO CHEMICALS, RADIATION, AND OTHER SUBSTANCES

Many occupations inherently expose employees to a multitude of hazards, ranging from chemicals and biological agents to radiation and physical dangers. Waste management, recycling, healthcare professions, first responders, and construction are just a few examples of high-risk fields.

Exposure to hazardous substances can have severe and lasting health consequences. While some particularly dangerous materials, like asbestos, are now tightly regulated or banned globally, many others remain in common use. Even seemingly ordinary substances can pose serious health risks if not handled appropriately.

The Occupational Safety and Health Administration categorizes health effects from hazardous substances into two main categories: acute and chronic.

- **Acute effects:** These are immediate and can be life-threatening. Examples include poisoning, suffocation, explosions, and fires.
- **Chronic effects:** These develop over extended periods of exposure and can be debilitating or even fatal. These include:
  - **Respiratory diseases:** Asthma, rhinitis, and lung scarring illnesses like asbestosis and silicosis.
  - **Occupational cancers:** Leukemia, lung cancer, mesothelioma, and nasal cavity cancers.
  - **Skin diseases:** Dermatitis, rashes, and even skin cancer.
  - **Reproductive problems and birth defects:** Exposure to certain chemicals can harm fertility and fetal development.
  - **Allergies:** Repeated exposure can trigger allergic reactions.
  - **Bioaccumulation:** Certain toxins, like heavy metals and solvents, accumulate in the body over time, leading to long-term health problems.

3.11.1 Importance of Prevention

The US Occupational Safety and Health Administration has developed standards and guidelines to promote a safe workplace and as such, employers are required to ensure that safety measures which minimize worker exposure to hazardous substances are applied within their organizations. This includes:

- **Regular Hazard Assessments:** Identifying and evaluating potential risks in the workplace.
- **Substitution:** Whenever possible, using less hazardous alternatives to dangerous materials.
- **Engineering Controls:** Implementing engineering control design (i.e., ventilation systems or physical barriers) to reduce or eliminate employee exposure to hazards.
- **Safe Work Practices:** Providing comprehensive training on handling hazardous materials safely, including the proper use of PPE.
- **Medical Monitoring:** Regular health screenings for workers exposed to high-risk substances.

It is prudent for employers to take the appropriate steps for the implementation of occupational safety related preventive measures. The use of preventive approaches can be effective in reducing the likelihood of occupational hazards, injuries and even deaths.

3.11.2 Emerging Concerns in 2024

In addition to the well-documented hazards, new considerations are emerging in workplace safety related to changes in industrial risks, environmental factors and psychological concerns for workers. Some of these emerging areas of concern include the following:

- **Nanoparticles:** These microscopic particles are increasingly used in various industries and pose unique health risks due to their ability to penetrate deep into the body.
- **Psychological Stress:** Stress from high-pressure work environments can exacerbate the effects of hazardous substances and contribute to overall worker health.
- **Climate Change:** Rising temperatures and extreme weather events can create new safety hazards for outdoor workers.

3.12 SUMMARY

The human body, with its complex network of systems and sensory capabilities, facilitates a wide range of movements, activities, and functions. Our senses act as essential safety mechanisms, responding to external stimuli and serving as the interface between the environment and our internal processes. By interpreting these external signals, the body's systems can respond appropriately. Immediate threats are often detected through sensory input, highlighting the critical need for both workers and management to remain vigilant and respond to such signals to prevent immediate harm. However, some hazards are more subtle, not readily detectable, and may lead to long-term health issues and disabilities. These hidden dangers necessitate a proactive approach from engineers and management in the design and planning of workplace environments, as employees may not be aware of the potential long-term risks in the short term.

The investment in mitigating workplace risks and hazards is well justified when compared to the consequences of disabling injuries, whether short-term or long-term. Beyond the direct financial costs, the intangible benefits of a safer work environment—such as increased worker satisfaction, enhanced organizational reputation, and reduced downtime due to injuries—should be considered in the decision-making process for implementing safety measures and ergonomic designs.

In an era where workplace health and safety standards are subject to increasing scrutiny, the integration of ergonomic principles and the adoption of advanced safety technologies play a pivotal role. Additionally, the continuous development of wearable technologies and environmental monitoring tools offers new opportunities for real-time hazard detection and mitigation. This underscores the importance of ongoing adaptation and improvement in workplace safety practices. Organizations must prioritize the health and safety of their workforce, recognizing that such investments not only provide financial returns but also foster a culture of safety and well-being throughout the organization

3.13 CASE STUDY

Problem Overview

In occupational settings, workers are exposed to various sensory stimuli that can either enhance or detract from their overall comfort, performance, and safety. The human sensory system—comprising sight, hearing, touch, taste, and smell—plays a critical role in how employees interact with their environment. In this case study, we focus on the sensory impact of the work environment, with particular attention on the effects of lighting, noise levels, tactile feedback, and temperature on human performance and comfort.

The primary challenge in many workplaces is to ensure that the environment is designed and managed to optimize sensory input, preventing sensory overload or deprivation. Poorly controlled sensory conditions, such as excessive noise or inadequate lighting, can lead to fatigue, reduced productivity, and even accidents.

Ergonomic Design Process

To address the current and emerging occupational concerns, ergonomists apply a series of tools and techniques to evaluate the sensory impact of the occupational environment on workers. These methods focus on measurement and management of sensory stimuli in order to improve comfort, performance, and safety.

1. **Visual Environment (Sight):**
   - **Assessment Tools:** Light meters, digital photometers.
   - **Techniques:** Evaluate the levels of ambient, task, and accent lighting to ensure adequate visibility without glare or excessive contrast. Considerations include brightness, color temperature, and uniformity.

- **Measurement:** Lumens and lux levels are measured to ensure they meet industry standards for specific tasks (e.g., 500–1,000 lux for detailed work).
- **Management:** Optimal lighting design involves ensuring sufficient lighting while avoiding glare and shadows. Adjustable lighting systems can help adapt to various task requirements.

2. **Auditory Environment (Hearing):**
   - **Assessment Tools:** Sound level meters, noise dosimeters
   - **Techniques:** Measure decibel levels of continuous and intermittent sounds. Evaluate the type of noise (e.g., continuous hums or sudden, sharp sounds) and its impact on concentration, communication, and stress levels.
   - **Measurement:** Decibel levels (dBA) are measured to ensure they are within safe limits (usually below 85 dBA over an 8-hour workday).
   - **Management:** Noise reduction strategies, such as soundproofing, implementing quieter machinery, and using personal protective equipment like earplugs, can mitigate auditory strain.
3. **Tactile Feedback (Touch):**
   - **Assessment Tools:** Pressure sensors, grip force dynamometers.
   - **Techniques:** Evaluate tactile interaction with tools, surfaces, and equipment. Factors include the texture, shape, and size of objects as well as the amount of force required for manual tasks.
   - **Measurement:** Force exertion is measured in newtons (N), and grip strength can be evaluated to ensure tasks do not cause undue stress on muscles and joints.
   - **Management:** Ergonomic design of tools and workstations, ensuring that grips, handles, and controls are designed for comfort and ease of use, reducing repetitive strain and discomfort.
4. **Thermal Environment (Touch and Temperature):**
   - **Assessment Tools:** Thermometers, thermal comfort sensors.
   - **Techniques:** Assess the temperature, humidity, and air movement in the workplace to ensure thermal comfort, which significantly affects concentration and productivity.
   - **Measurement:** Temperature is measured in degrees Celsius or Fahrenheit, while thermal comfort is evaluated using standards like the Predicted Mean Vote (PMV) and Predicted Percentage of Dissatisfied (PPD) indices.
   - **Management:** Controlling air temperature, ensuring proper ventilation, and providing options for personalized climate control can enhance thermal comfort.

OUTCOME

The successful application of sensory ergonomics in the occupational environment leads to improved worker comfort, safety, and performance. By systematically measuring and managing sensory stimuli such as light, sound, touch, and temperature, workplaces can be optimized to reduce stress, fatigue, and the risk of injury.

In environments where sensory conditions are appropriately managed, workers are more likely to maintain high levels of concentration and performance, and the incidence of accidents is reduced. Moreover, personalized adjustments to sensory elements can accommodate individual differences, further enhancing overall job satisfaction.

Conclusion

This case study demonstrates the importance of assessing and managing the sensory impact of the occupational environment. By applying ergonomic principles to the senses, organizations can create environments that not only prevent injury and discomfort but also enhance performance and job satisfaction. The tools and techniques used in this case offer a blueprint for other industries to evaluate and optimize their workplaces with respect to sensory stimuli.

Exercises

1. Explain an occupational environment where each of the human senses is utilized for task performance.
   - **Extended Exercise:** Use a multimedia tool (like Adobe Spark, Canva, or Prezi) to create a multimedia story that illustrates different occupational environments where all five senses (vision, hearing, touch, taste, and smell) are used. Record videos or find relevant clips online to demonstrate real-life examples. Share it on a collaborative platform for peer review.
2. Explain how to minimize overloading of the visual system in a visual inspection task in a manufacturing facility.
   - **Extended Exercise:** Create a short animated video using tools like Powtoon or Animaker that explains best practices for reducing visual overload (e.g., proper lighting, frequent breaks, ergonomic setup). Share the video on YouTube or a class blog and encourage feedback on potential improvements.
3. Discuss the types of vibrations impacting workers and the sources of vibration. How can these levels of vibration be determined?
   - **Extended Exercise:** Develop a podcast episode or YouTube video where you interview an ergonomics expert, engineer, or safety officer to discuss vibration exposure in different industries (construction, transportation, etc.). Use digital sound analysis tools (like the VibSensor app) to measure vibrations in real-world settings and demonstrate the data collection process in your presentation.
4. Explain how the visual and auditory senses are impacted by the aging process.
   - **Extended Exercise:** Create an infographic or interactive chart using Canva or Piktochart that shows how the senses of vision and hearing

decline with age. Include a section on how ergonomic design can accommodate aging workers. Share the infographic on social media platforms or class forums and invite discussion on possible workplace solutions.

5. Explain the impact of new recreational technology on the risk for noise-induced hearing loss.
   - **Extended Exercise:** Write a short blog post or LinkedIn article about how new technologies (e.g., gaming headsets, virtual reality devices) contribute to noise-induced hearing loss. Include interviews or quotes from audiologists or recreational technology experts. Include a call to action, asking readers to share their experiences with noisy recreational devices.
6. Identify a national or international standard (or guideline) for each occupational design in vision, audition, and vibration.
   - **Extended Exercise:** Use Google Scholar, research databases, or government websites (like OSHA, ISO) to find national and international standards. Create a shared Google Doc or Jamboard where each student adds a different standard, complete with summaries and examples of how they are applied in real workplaces. Collaborate to identify common themes or differences between countries.
7. What approach should be used to estimate thermal loading in an outside task?
   - **Extended Exercise:** Use an online weather monitoring tool (like Weather.com or apps such as Thermo) to simulate different outdoor environments. Create a digital report (using Word or Google Docs) explaining how to calculate thermal load based on factors like temperature, humidity, and clothing. Share your results with the class and compare thermal estimates across various locations.
8. Discuss risk factors associated with the use of the gestation sense in task performance.
   - **Extended Exercise:** Record a short video or audio interview with a food industry professional (e.g., chef, taste tester) or a worker in an environment where taste plays a critical role. Share the interview on class platforms and analyze how gestation-related tasks can be improved ergonomically, especially in food safety and product testing.
9. Explain the relationship between the olfactory and gestation senses.
   - **Extended Exercise:** Create a mind map using a tool like MindMeister or Lucidchart to visually represent the connection between the senses of smell (olfaction) and taste (gestation). Add examples from industries where both senses are essential (e.g., food and beverage, perfume). Share it with the class and invite peers to add their own examples or thoughts.
10. Describe a task that relies largely on the sense of tactation.
    - **Extended Exercise:** Record a short video or create a photo essay that demonstrates a task heavily dependent on touch (e.g., quality inspection,

medical palpation, Braille reading). Use social media (Instagram or TikTok) to post your work and ask your followers to provide feedback or share their own experiences with tasks that rely on tactation.

REFERENCES

Abraira, V. E., & Ginty, D. D. (2013). The sensory neurons of touch. *Neuron, 79*(4), 618–639.

Anshel, J. (2017). *Visual ergonomics handbook.* CRC Press.

ASHRAE. (1993). *Handbook of Fundamentals.* Atlanta, GA: American Society of Heating, Refrigerating and Air-Conditioning Engineers, Inc.

ASHRAE. (2017). *ASHRAE handbook: Fundamentals.* American Society of Heating, Refrigerating and Air-Conditioning Engineers.

Basbaum, A. I., Bautista, D. M., Scherrer, G., & Julius, D. (2009). Cellular and molecular mechanisms of pain. *Cell, 139*(2), 267–284.

Beach, E. F., Gilliver, M., & Williams, W. (2014). Sound levels and listening habits of iPod users. *The Journal of the Acoustical Society of America, 123*(2), 372–380.

Belding, H. S., & Hatch, T. F. (1955). Index for evaluating heat stress in terms of resulting physiological strains. *American Industrial Hygiene Association Journal, 17*(5), 303–306.

Bensmaia, S. J. (2014). The neural basis of tactile perception. *Neuropsychologia, 61*, 1–19.

Bernardini, M. J., et al. (2012). Vibration control techniques and their ergonomic implications in industrial settings. *Journal of Occupational Safety and Ergonomics, 18*(3), 245–257.

Bhagat, S. P. (2017). Auditory health and music exposure: Review and implications for public health. *Journal of the American Academy of Audiology, 28*(2), 152–164.

Bhanderi, D. J., Choudhary, S., & Parmar, L. (2022). Computer vision syndrome: A major occupational health issue in the modern era. *Occupational Health Journal, 12*(2), 102–114.

Botsford, J. H. (1971). Heat stress evaluation using a wet globe thermometer. *American Industrial Hygiene Association Journal, 32*(1), 7–14.

Boyce, P. (2014). *Human factors in lighting.* CRC Press.

Bridger, R. S. (2018). *Introduction to ergonomics.* CRC Press.

Bushdid, C., Magnasco, M. O., Vosshall, L. B., & Keller, A. (2014). Humans can discriminate more than 1 trillion olfactory stimuli. *Science, 343*(6177), 1370–1372.

Cabanac, M. (2014). *Thermal comfort and its human significance.* Springer.

Canadian Centre for Occupational Health and Safety (CCOHS). (2023). OSH Answers Fact Sheets – Personal Protective Equipment (PPE): Selection. Retrieved from https://www.ccohs.ca/oshanswers/prevention/ppe/

Chaudhari, N., & Roper, S. D. (2016). The cell biology of taste. *The Journal of Cell Biology, 190*(3), 285–296.

CIE (International Commission on Illumination). (2016). *Lighting for workplaces.* CIE Publication.

Clark, T., & Smith, J. (2019). *Visual performance in the workplace: The role of ergonomics. Human Factors Journal, 62*(4), 476–488.

Coles-Brennan, C., Sulley, A., & Young, G. (2019). Management of digital eye strain. *Clinical & Experimental Optometry, 102*(1), 18–29.

de Dear, R., & Brager, G. (2015). Developing an adaptive model of thermal comfort and preference. *ASHRAE Transactions, 104*(1), 145–167.

Dhaka, A., Viswanath, V., & Patapoutian, A. (2016). Trp ion channels and temperature sensation. *Annual Review of Neuroscience, 29*, 135–161.

Eastman Kodak. (1986). *Ergonomic Design for People at Work: Vol. 1.* Van Nostrand Reinhold.

Epstein, Y., & Moran, D. S. (2006). Thermal comfort and the heat stress index. *Industrial Healthy, 44*(3), 388–398. https://doi.org/10.2486/indhealth.44.388

European Environment Agency. (2023). Environmental noise in Europe — 2023. Publications Office of the European Union.

Fanger, P. O. (1970). *Thermal comfort: Analysis and applications in environmental engineering.* McGraw-Hill.

Fields, H. L. (2017). Pain perception and management in occupational settings. *Pain Research and Management, 22*(1), 50–58.

Figueiro, M. (2016). Circadian lighting for alertness and sleep. *Nature and Science of Sleep, 8*, 1–10.

Gagge, A. P., Stolwijk, J. A. J., & Nishi, Y. (1971). An effective temperature scale based on a simple model of human physiological regulatory response. *ASHRAE Transactions, 77*(1), 247–262.

Gagge, A. P. (1973). Rational temperature indices of comfort. *ASHRAE Journal*, 15(8), 40–45.

Gao, C., Kuklane, K., Wang, F., & Holmér, I. (2017). Personal cooling with phase change materials to improve thermal comfort from a heat wave perspective. *Indoor Air, 22*(6), 523–530.

Gilles, A., Van Hal, G., De Ridder, D., Wouters, K., & Van de Heyning, P. (2014). Epidemiology of noise-induced tinnitus and the attitudes and beliefs towards noise and hearing protection in adolescents. *PLoS ONE, 9*(7), e102465.

Gong, N., Wargocki, P., & Tham, K. W. (2016). Effects of temperature and humidity on perceived indoor air quality, SBS symptoms, and performance of office work. *Indoor Air, 26*(3), 313–325.

Gonzalez, R. R., Pandolf, K. B., & Gagge, A. P. (1974). Heat strain models applicable for protective clothing systems: Comparison of core temperature response prediction. *Ergonomics, 17*(4), 491–500.

Gottfried, J. A. (2017). *Olfaction and its role in memory and emotion.* University College London.

Grossman, S., & Mathews, L. (2017). *Occupational ergonomics: A visual systems approach.* Wiley.

Hardy, J. D. (1970). Physiological principles for human environmental tolerances. *American Journal of Physical Anthropology, 32*(2), 157–163.

Havenith, G., et al. (2014). Interaction effects of clothing, wind, and physical activity on thermal comfort and heat stress. *Applied Ergonomics, 45*(3), 784–790.

Herz, R. S., & Engen, T. (2004). Odor memory: Review and analysis. *Psychonomic Bulletin & Review, 4*(3), 300–326.

Holmer, I. (1984). Required clothing insulation (IREQ) and allowable exposure times in cold environments. *American Industrial Hygiene Association Journal, 45*(10), 703–709.

Hongratanaworakit, T. (2020). Aroma-ergonomics: The impact of fragrances on human performance. *Journal of Ergonomic Research, 15*(2), 93–104.

Hopp, R. J., & Irwin, M. A. (2020). Ergonomic considerations in food and beverage industries. *Journal of Occupational Ergonomics, 25*(3), 182–191.

Houghton, F. C., & Yaglou, C. P. (1923). Determining comfort indoors. *ASHVE Transactions*, 29, 165–176.

Illuminating Engineering Society. (2017). ANSI/IES RP-1-17: Recommended Practice for Office Lighting.

International Organization for Standardization (ISO). (1989). ISO 7933:1989 Ergonomics — Analytical determination and interpretation of heat stress using calculation of required sweat rate.

International Organization for Standardization (ISO). (1993). ISO TR 11079:1993.

ISO 8996:2004. (2004). *Ergonomics of the thermal environment — Determination of metabolic rate.*

Jones, L., & Smith, A. (2015). Kinesthetic perception and occupational tasks: A review. *Human Movement Science*, *42*(3), 132–141.

Kardous, C. A., & Shaw, P. B. (2014). Evaluation of smartphone sound measurement applications. *The Journal of the Acoustical Society of America, 135*(4), EL186–EL192.

Kaufman, M. H., & Christensen, J. M. (1972). Techniques for controlling glare. In *Ergonomics: Foundational Principles, Applications, and Technologies.* CRC Press.

Kim, J., de Dear, R., & Parkinson, T. (2017). Thermal comfort in open-plan offices: Findings from a field study in summer and winter. *Building and Environment, 123*, 56–65.

Krahé, C., Drabek, M. M., Paloyelis, Y., & Fotopoulou, A. (2018). Affective touch and attachment: A cross-cultural view. *Frontiers in Psychology, 9*, 262.

Kujawa, S. G., & Liberman, M. C. (2015). Synaptopathy in the noise-exposed and aging cochlea: Primary neural degeneration in acquired sensorineural hearing loss. *Hearing Research, 330*, 191–199.

Lambrecht, J., & Müller, A. (2017). Evidence for electroreception in humans. *Journal of Neurophysiology, 117*(4), 1231–1240.

Land, M. F., & Nilsson, D. E. (2018). *Animal eyes* (2nd ed.). Oxford University Press.

Leavitt, A. C., Reilly, T., & Shore, S. E. (2019). Managing complex auditory environments in the workplace: Ergonomic strategies for noise control. *Occupational Ergonomics Journal, 22*(1), 45–58.

Lechien, J. R., Chiesa-Estomba, C. M., Place, S., Van Laethem, Y., Cabaraux, P., Mat, L., & Saussez, S. (2020). Clinical and epidemiological characteristics of 1420 European patients with mild-to-moderate coronavirus disease 2019. *Journal of Internal Medicine, 288*(3), 335–344.

Le Prell, C. G., Dell, S. M., Hensley, B. N., & Spankovich, C. (2019). Hearing loss prevention for musicians and music enthusiasts: Designing hearing health messages to promote protective behaviors. *Journal of the American Academy of Audiology, 30*(9), 758–773.

Leithead, C. S., & Lind, A. R. (1964). *Heat stress and heat disorders.* Cassell.

Macpherson, J. M., & Horak, F. B. (2016). Postural control and the vestibular system. *Neuroscience and Biobehavioral Reviews, 71*, 31–38.

Mansfield, K. (2016). The effects of glare on visual performance. *Ergonomics, 59*(4), 541–554.

Mao, L., Jin, H., Wang, M., Hu, Y., Chen, S., He, Q., & Hu, B. (2021). Neurologic manifestations of hospitalized patients with COVID-19 in Wuhan, China. *JAMA Neurology, 77*(6), 683–690.

Mekjavic, I. B., & Tipton, M. J. (2014). Heat stress and its impact on work capacity in a warm environment. *Ergonomics, 57*(5), 622–632.

Moore, B. C. J. (2016). *An introduction to the psychology of hearing* (6th ed.). Brill.

Morgan, C. T., et al. (1963). *Human Engineering Guide to Equipment Design.* McGraw-Hill.

National Geographic Education. (n.d.). Noise pollution. https://education.nationalgeographic.org/resource/noise-pollution/

National Institute for Occupational Safety and Health (NIOSH). (1986). Criteria for a recommended standard: Occupational exposure to hot environments (Revised Criteria 1986). U.S. Department of Health and Human Services, CDC.

OSHA (Occupational Safety and Health Administration). (2018). *Illumination in the workplace: Standards and guidelines.*

Pacchierotti, C., Sinclair, S., Solazzi, M., Frisoli, A., Hayward, V., & Prattichizzo, D. (2017). Wearable haptic systems for the fingertip and the hand: Taxonomy, review, and perspectives. *IEEE Transactions on Haptics, 10*(4), 580–600.

Palmer, K. & Wilson, J. R. (2014). *Handbook of human factors and ergonomics in health care and patient safety*. CRC Press.

Palmer, K. T., Haward, B., Griffin, M. J., Bendall, H., & Coggon, D. (2014). Validity of self-reported occupational exposure to hand-transmitted vibration. *Occupational and Environmental Medicine*, *57*(6), 416–421.

Pandolf, K. B., & Goldman, R. F. (1978, September). Convergence of skin and rectal temperatures as a criterion for heat tolerance. *Aviation, Space, and Environmental Medicine*, *49*(9), 1095–1101.

Parkinson, T., de Dear, R., & Brager, G. (2020). Nudging the adaptive thermal comfort model. Energy & Buildings, 206, Article 109559. https://doi.org/10.1016/j.enbuild.2019.109559

Parsons, K. (2014). *Human thermal environments: The effects of hot, moderate, and cold environments on human health, comfort, and performance*. CRC Press.

Peng, J. H., Tao, Z. W., & Huang, Z. W. (2016). Risk of noise-induced hearing loss from the use of portable listening devices in young adults. *Journal of Otology, 11*(2), 63–68.

Pryor, R. R., Roth, R. N., Suyama, J., Hostler, D., & Moore, C. G. (2018). Estimating core temperature with aural canal thermometry in emergency services personnel. *Prehospital Emergency Care, 22*(2), 206–213.

Purves, D., Augustine, G. J., & Fitzpatrick, D. (2014). *Neuroscience*. Sinauer Associates.

Raspopovic, S., Valle, G., Petrini, F. M., & Carpaneto, J. (2014). Prosthetics with tactile feedback. *Science Translational Medicine*, *6*(257), 257ra138.

Rea, M. S. (2015). *The IESNA lighting handbook*. Illuminating Engineering Society of North America.

Rodgers, S. H., & Cori, J. M. (1981). Human thermal environments: The effects of heat and cold stress on human performance and behavior. In *Proceedings of the Human Factors Society Annual Meeting* (Vol. 25, No. 2, pp. 173–177).

Rosenfield, M. (2016). Computer vision syndrome: A review of ocular causes and potential treatments. *Ophthalmic and Physiological Optics*, *36*(5), 624–633.

Schulte, C. E., Kluge, M., Bschor, T., & Steiger, A. (2013). The effect of olfactory stimulation on learning and memory during sleep. *Journal of Sleep Research*, 22(1), 38–43.

Sheedy, J. E., et al. (2014). Eye and visual discomfort in the office environment. *Work*, *47*(3), 419–430.

Siple, P. A., & Passel, C. F. (1945). Measurements of dry atmospheric cooling in the arctic in terms of the temperature of the wind. *Proceedings of the American Philosophical Society*, *89*(1), 177–199.

Sivak, M., & Olson, P. L. (2015). *Human vision and vehicle lighting*. Springer.

Sliwinska-Kowalska, M., & Davis, A. (2012). Noise-induced hearing loss. *Noise and Health, 14*(61), 274–280.

Small, D. M., & Prescott, J. (2021). Odor/taste integration and the perception of flavor. *Experimental Brain Research*, *232*(7), 1649–1658.

Smith, J. L., Housley, E. K., & Chang, J. Y. (2019). The effects of peppermint on physical and cognitive performance: A review. *International Journal of Sports Science*, *9*(3), 85–93.

Smith, K., & Jones, A. (2020). Ergonomics of vision in industrial settings. *Ergonomics Review*, *28*(2), 132–146.

Smith, P. A. (2019). Prevention of noise-induced hearing loss in the workplace: Ergonomic approaches to noise management. *Work, 63*(4), 567–579.

Smolders, K. C., de Kort, Y. A., & Cluitmans, P. J. (2014). A higher illuminance induces alertness during office hours but lowers it during the evening. *Journal of Environmental Psychology*, *39*, 52–63.

Stevens, J. C., & Marks, L. E. (2014). Temperature perception and its ergonomic implications. *Ergonomics*, *59*(5), 231–242.

U.S. Occupational Safety and Health Administration (OSHA). (2020). Occupational noise exposure. *OSHA Standard 1910.95*. Retrieved from https://www.osha.gov/noise

Vaira, L. A., Salzano, G., Deiana, G., & De Riu, G. (2020). Anosmia and ageusia: Common findings in COVID-19 patients. *The Laryngoscope*, *130*(7), 1787–1789.

Wells, J., & Green, B. (2018). *Ergonomics and human vision in the workplace*. CRC Press.

Winslow, C. E. A., Herrington, L. P., & Gagge, A. P. (1937). Physiology of man in the cold. *American Journal of Physiology*, *120*(3), 557–564.

Wolkoff, P. (2017). External eye symptoms in indoor environments. *Indoor Air*, *27*(2), 246–254.

Wolska, A., et al. (2015). The role of lighting and visual ergonomics in the prevention of visual fatigue. *International Journal of Occupational Safety and Ergonomics*, *21*(3), 346–355.

World Health Organization (WHO). (2015). Hearing loss due to recreational exposure to loud sounds: A review. *WHO Publication*.

World Health Organization. (2021). Hearing loss due to recreational exposure to loud sounds: A review. World Health Organization. https://www.who.int/news-room/fact-sheets/detail/deafness-and-hearing-loss

Wysocki, C. J., & Gilbert, A. N. (2018). The human olfactory system: Anatomy and functions. *Annual Review of Psychology*, *69*, 195–215.

Yaglou, C. P., & Minard, D. (1957). Control of heat casualties at military training centers. *American Medical Association Archives of Industrial Health*, *16*(4), 302–316.

Yan, Z., Hu, L., Chen, H., & Lu, F. (2017). Computer vision syndrome: A widely spreading but largely unknown epidemic among computer users. *Computers in Human Behavior*, *69*, 88–100.

Zhang, J., Jiang, Y., & Ding, Y. (2016). Sensory and perceptual aspects of ergonomics. *Journal of Ergonomics*, *64*(2), 145–156.

Zhao, K., Dalton, P., & Brody, M. J. (2019). Odor classification systems and their applications in ergonomics. *Human Factors*, *61*(4), 673–688.

4 Muscular Work and Nervous Control of Movements

4.1 LEARNING GOALS

This chapter will provide an explanation of the functionality, relationship, and elements of the integrated roles of the muscular system and nervous system. The execution and control of movements for these systems will also be detailed to gain an in-depth understanding of these processes.

4.2 KEY TOPICS

- Muscular work
- Muscular contractile system
- Mechanism of contraction
- Method to stimulate and control the mechanism of contraction
- Energy to drive mechanism of contraction
- Innervation of the muscular system
- Reflexes
- Energy transformation process for muscle activity
- Types of muscular work
- Muscular fatigue
- Types of muscle contraction
- Measurement of muscular strength

4.3 INTRODUCTION AND BACKGROUND

The human body operates through a highly integrated relationship between its musculoskeletal and nervous systems, a connection that plays a pivotal role in virtually every aspect of movement and posture, from everyday tasks to complex athletic feats. Understanding this intricate relationship is crucial for occupational ergonomics, as it directly informs the design of work environments, tools, and practices aimed at reducing musculoskeletal disorders (MSDs) and enhancing performance efficiency (Bernard, 2014; Gallagher & Heberger, 2015).

DOI: 10.1201/9780429342806-4

4.3.1 The Musculoskeletal System and Nervous System in Motion

The musculoskeletal system, comprising bones, muscles, tendons, and ligaments, provides structural support and enables movement. The nervous system coordinates this movement through sensory inputs and motor outputs. The central nervous system (CNS), consisting of the brain and spinal cord, acts as the primary processing center. The peripheral nervous system (PNS) transmits signals between the CNS and muscles, allowing for precise control of movements and reactions to external stimuli (Kim et al., 2015). From an ergonomic standpoint, this coordination is critical for analyzing how workers interact with their physical environment, particularly in repetitive or strenuous tasks that increase the risk of musculoskeletal injuries (David, 2005; Waersted, 2014).

Early foundational research in the musculoskeletal and nervous systems has greatly shaped our modern understanding of movement control. Galen of Pergamum, a physician in ancient Greece, established the importance of the spinal cord in voluntary movement through his animal experiments (Cowan, 2017). After severing the spinal cord at different points, he demonstrated that the CNS is essential for muscle coordination, an insight that is still relevant in modern ergonomics to prevent spinal cord-related injuries in occupational settings (Bernard, 2014).

4.3.2 Historical Advancements in Neuromuscular Research

Advancements in the 18th century served to further solidify our understanding of the nervous system's role in muscle activation. Swiss physiologist Albrecht von Haller traced the pathways of nerves, showing their convergence into the brain and spinal cord at the centers for sensory perception and motor responses (Kapoor et al., 2018). He was among the first to establish the link between nerve impulses and muscle contractions, a concept fundamental to modern ergonomic design, especially in fields where fine motor skills or repetitive motions are critical.

Luigi Galvani made a key contribution to our understanding of neuromuscular communication through his discovery of electrical impulses in 1780. Through his experiments with frogs, Galvani demonstrated that electrical signals could cause muscle contraction, suggesting that motor skills are governed by bioelectric signals (Rizzolatti & Sinigaglia, 2016). This knowledge laid the groundwork for electrophysiological studies that now help in assessing muscle fatigue, nerve conduction, and other factors relevant to workplace ergonomics and musculoskeletal health.

4.3.3 Modern Implications for Ergonomics and Occupational Health

In contemporary ergonomics, one of the most pressing concerns is the prevention of MSDs, emphasizing those associated with repetitive strain, awkward postures, and overexertion (Waersted, 2014; Gallagher & Heberger, 2015). Ergonomists rely on understanding the neuromuscular system's behavior under various workloads to design safer work environments. For instance, electromyography (EMG) is used to measure electrical activity in muscles during tasks, helping to identify movements or postures that may increase the risk of injury (Kim et al., 2015). Additionally, advances in biomechanics allow for real-time analysis of muscle use and stress, leading to more ergonomic tool and workstation designs.

Research into stem cell therapies is opening new possibilities for treating debilitating musculoskeletal conditions, particularly those resulting from spinal cord injuries. Transplantation of stem cells derived from bone marrow or umbilical cord blood has shown promise in regenerating damaged neural pathways (Keating & O'Brien, 2016). This technology, while still emerging, represents a significant leap forward for occupational health, offering potential cures for injuries that currently lead to long-term disability.

4.3.4 Ethical Considerations and Future Directions

The rapid development of stem cell research, however, has not been without controversy. Ethical debates surrounding the use of stem cells often raise concerns of cloning and eugenics, highlighting the potential misuse of such technologies. As a result, stem cell research has faced regulatory hurdles, particularly in countries like the United States. In contrast, nations with more permissive regulatory frameworks have made notable progress, emphasizing the need for global collaboration in such a rapidly evolving field (Kim, 2015). For occupational ergonomics, such advancements could revolutionize the treatment and prevention of MSDs, potentially allowing workers to recover from injuries that are currently considered permanent (Bernard, 2014).

4.3.5 Conclusion

The interplay between the musculoskeletal and nervous systems is fundamental to understanding human movement, particularly in the context of occupational ergonomics. Historical research has laid the foundation for modern insights into how these systems work together to control movement, while cutting-edge developments, such as stem cell therapy, promise new approaches to treating musculoskeletal injuries. By continuing to study these systems, ergonomists can design safer work environments and develop interventions that reduce injury risks and enhance worker health and productivity.

4.4 MUSCULAR WORK

Muscle contraction is a complex biomechanical process that extends far beyond the tightening or shortening of muscle fibers. From an occupational ergonomics perspective, understanding muscle contraction is crucial to optimizing workplace designs, preventing injury, and improving worker performance. Muscle contraction, which facilitates movements as diverse as typing on a keyboard to lifting heavy loads, involves a series of highly regulated physiological events known as *excitation–contraction coupling* (ECC). This process translates electrical signals from the nervous system into mechanical force production in muscles, making it fundamental to ergonomics in task design, workload assessment, and human–machine interaction.

At the cellular level, skeletal muscles are composed of contractile fibers that consist of two main types of proteins: actin and myosin. These proteins are arranged in orderly structures known as sarcomeres. Sarcomeres are responsible for generating force, and their function is often compared to that of microscopic engines working synchronously. Upon receiving a neural stimulus, typically in the form of an action potential, voltage-gated channels in the muscle fiber membrane open, allowing calcium ions (Ca^{2+}) to flood into the cytosol from the sarcoplasmic reticulum (SR), a specialized calcium storage organelle within muscle cells (Martyn et al., 2017).

Calcium ions play a pivotal role in ECC by acting as the biochemical switch that initiates the interaction between actin and myosin. When Ca^{2+} binds to troponin, a regulatory protein associated with the actin filaments, it induces a conformational change in tropomyosin, exposing binding sites on the actin filaments. This allows the myosin heads—which function as molecular motors—to attach to these binding sites. Powered by the hydrolysis of adenosine triphosphate (ATP), myosin heads perform a cyclical motion referred to as the cross-bridge cycle, pulling the actin filaments toward the center of the sarcomere and causing muscle contraction (Gordon et al., 2015).

From an ergonomic standpoint, it is crucial to understand that muscle contraction is not a uniform process. The force generated by the muscle depends on several factors, including the muscle's length at the onset of contraction, the type of contraction (isometric, concentric, or eccentric), and the energy availability in the form of ATP. Recent studies (Siebert et al., 2019) have demonstrated that myosin heads exhibit a form of adaptive feedback, modulating the force they produce based on the muscle's energy reserves. This allows the muscle to economize its ATP consumption during sustained or repetitive tasks, which is an important consideration in occupational settings where tasks may require prolonged muscle use, such as in manual labor or repetitive motion tasks in assembly lines.

Understanding the biomechanical efficiency of ECC also has direct implications for reducing the risk of musculoskeletal disorders, such as repetitive strain injuries (RSIs), carpal tunnel syndrome, or tendinitis. These are common MSD's in occupations involving high levels of repetitive, force or sustained postures. Workplace designs that fail to account for the physiological limits of muscle contraction—such as inadequate rest periods or excessive static loads—can lead to muscle fatigue. Muscle fatigue, characterized by a decline in force production over time, is the result of metabolic changes in the muscle fibers, including the depletion of ATP and accumulation of metabolic byproducts like lactate (Green et al., 2018). Over time, this can lead to chronic strain on the musculoskeletal system, emphasizing the need for ergonomic interventions.

Additionally, muscle contraction does not occur in isolation. Instead, it is integrated with other components of the musculoskeletal system, including bones, tendons, ligaments, and joints. These components work in concert to produce the mechanical leverage necessary for efficient movement. Tendons, for example, serve as biological springs that store and release elastic energy during movement, reducing the metabolic cost of muscle contractions (Alexander, 2014). Understanding the synergy between muscles and tendons is essential for optimizing workplace ergonomics, particularly in dynamic tasks like lifting, where proper body mechanics and load distribution can significantly reduce injury risk.

In conclusion, muscle contraction is a sophisticated and efficient process, tightly regulated by the interaction between nervous system signals, biochemical pathways, and mechanical structures within the body. From an occupational ergonomics perspective, a thorough understanding of this process is vital for designing tasks, tools, and environments that align with the physiological capabilities of workers. This not only enhances performance but also minimizes the risk of injury, ensuring long-term worker health and well-being.

4.4.1 Muscular Contractile System

Muscle function is critical in occupational ergonomics, as it plays a central role in human movement, posture, and the capacity to perform work-related tasks. Muscles are responsible for executing actions that range from fine motor skills, such as typing or operating machinery, to high-force activities like lifting and manual material handling. The ability of muscles to contract and generate force, a property termed contractility, is foundational for understanding how the human body interacts with workplace demands and environments (Bridger, 2018).

At the cellular level, muscles consist of bundles of fibers, each composed of thread-like structures known as myofibrils. These myofibrils house the molecular machinery essential for contraction, with actin and myosin serving as the primary proteins involved. An in-depth understanding of the interaction between these proteins is essential for the ergonomic analysis of physical tasks, where muscle force production, endurance, and fatigue are key considerations (Chaffin, Andersson, & Martin, 2019).

4.4.2 Molecular Mechanisms: Actin and Myosin

Within each muscle fiber, actin (a thin filamentous protein) serves as a track along which myosin (a thicker filament) operates. Myosin is often referred to as the molecular motor of muscle contraction, responsible for generating the mechanical force that leads to fiber shortening. The interaction between actin and myosin follows a well-established model known as the sliding filament theory, where myosin heads cyclically bind to actin, pulling it toward the center of the sarcomere. This actin–myosin interaction is regulated by tropomyosin and troponin, which control the availability of binding sites on the actin filament (Rassier & Herzog, 2014).

In ergonomic terms, understanding these molecular processes is vital for designing workspaces and tools that minimize the risk of musculoskeletal disorders. Repetitive movements or prolonged static postures can lead to fatigue, as the cyclical binding and force generation by myosin heads require a continuous supply of energy. Inefficient work designs may overload certain muscle groups, increasing the risk of strain and injury (Karwowski, 2018).

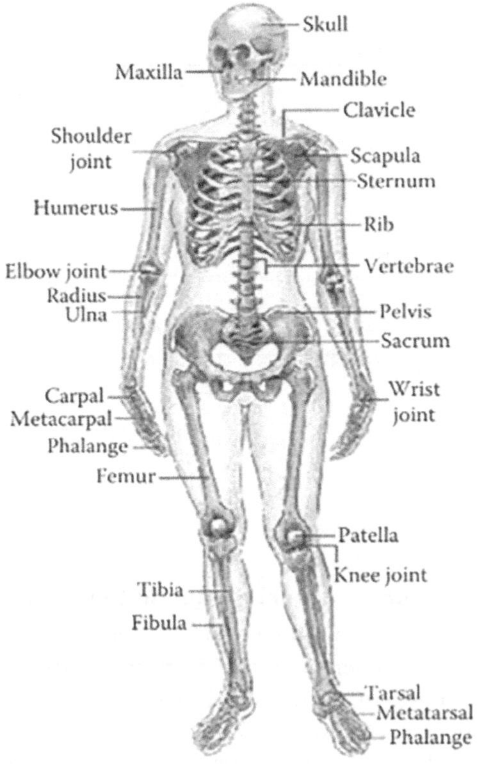

FIGURE 4.1 Skeleton.

4.4.3 Three Pillars of Muscle Contraction in Ergonomics

From an occupational ergonomics perspective, muscle contraction can be analyzed through three critical aspects: the mechanism of contraction, neural control, and energy expenditure. Each of these pillars plays a role in understanding how muscle function can impact task performance and worker health.

4.4.3.1 Mechanism of Contraction (Sliding Filament Model)

Muscle contraction, a fundamental component of human movement, plays a critical role in various occupational tasks, from repetitive motion to heavy lifting. To fully understand the biomechanics behind muscular effort, especially in the context of ergonomics, it is essential to examine the microscopic processes that enable muscle fibers to generate force and sustain contractions. These contractions occur at the level of muscle fibers within specialized structures known as sarcomeres, the functional units of muscle contraction (Gordon et al., 2016).

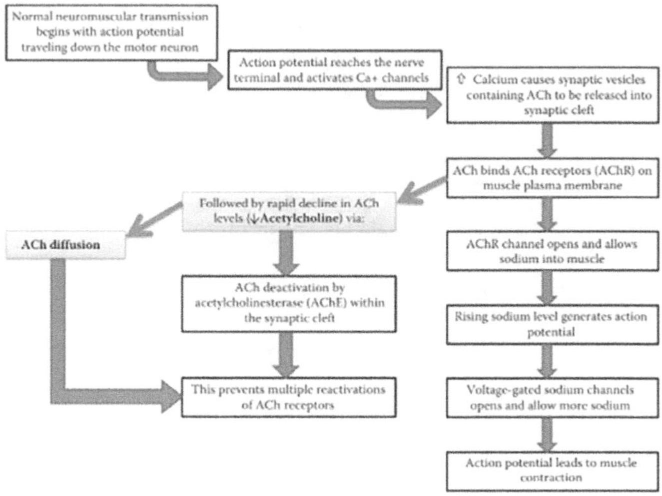

FIGURE 4.2 Model of muscle contraction.

4.4.3.2 The Sliding Filament Model of Muscle Contraction

Sarcomeres are composed of two primary types of protein filaments: actin (thin filaments) and myosin (thick filaments). Actin serves as a scaffold along which myosin can generate movement. Myosin, often termed the "molecular motor" of the muscle, has protruding myosin heads that function like microscopic levers. These heads bind to the actin filaments to facilitate muscle contraction through a process known as the cross-bridge cycle (Cooke, 2015).

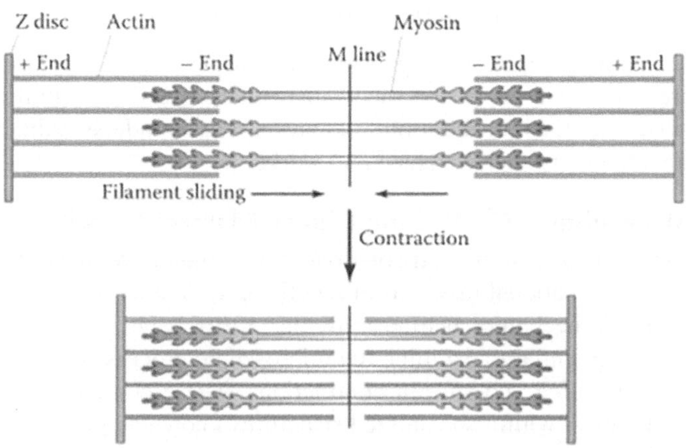

FIGURE 4.3 Sliding filament model.

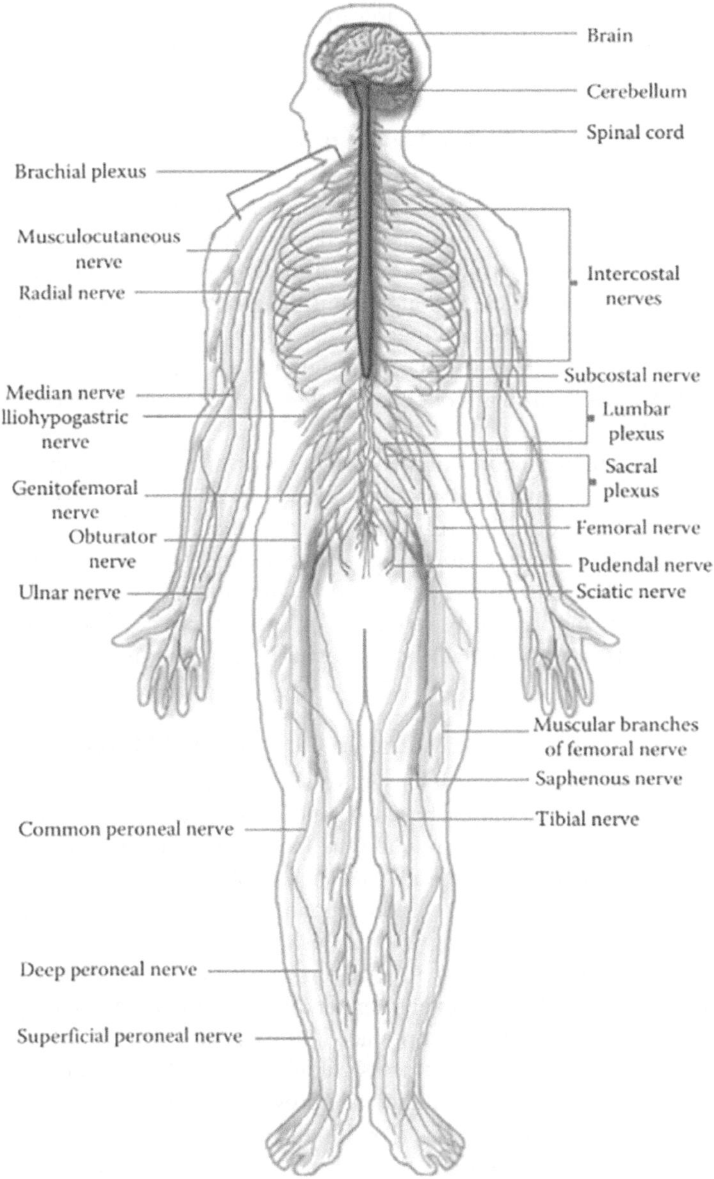

FIGURE 4.4 Efferent nerves.

The initiation of muscle contraction is a highly regulated process, involving neural stimulation that triggers the release of calcium ions within the muscle fiber. Increased intracellular calcium concentration exposes binding sites on the actin filament, allowing the myosin heads to engage and form cross-bridges (Kovács et al., 2015). Once these cross-bridges form, the myosin heads pivot, generating the power stroke that moves the actin filaments toward the center of the sarcomere. This process is fueled by the energy derived from adenosine triphosphate (ATP), the primary energy currency of the cell (Nishikawa et al., 2016).

4.4.3.3 Biomechanical Relevance in Occupational Settings

From an ergonomic perspective, understanding the mechanisms of muscle contraction is critical for designing workplace interventions that minimize muscle fatigue and injury risk. For instance, tasks requiring sustained force, such as holding a tool or repetitive motion, engage the myosin–actin interaction repeatedly. As myosin heads cycle through binding, pulling, and releasing, the continuous engagement of ATP provides the necessary energy for these movements. However, this can also lead to muscle fatigue when ATP reserves are depleted, which is a key consideration in occupational tasks that require prolonged exertion (Gollnick et al., 2015).

The sliding filament theory also highlights how muscle length changes during contraction. During a typical contraction, the A bands (regions containing myosin filaments) remain constant in width, whereas the I bands (regions containing only actin) and H zones (the gap between actin filaments) shorten (Huxley, 1957). Ergonomically, this principle helps explain why muscles generate less force at extreme ranges of motion (fully stretched or fully contracted positions), emphasizing the importance of neutral body postures in reducing strain during work tasks.

4.4.3.4 Efficiency of Myosin Motors in Occupational Movements

Recent advances in muscle physiology have shed light on the efficiency of myosin motors in adjusting force output based on the energy state of the muscle. Studies indicate that myosin heads can modulate the force they generate depending on ATP availability and mechanical demands (Kovács et al., 2015). This feedback mechanism allows muscles to operate more economically, conserving energy during low-intensity tasks while ramping up ATP usage during high-intensity tasks. This efficiency is particularly relevant in ergonomic design, where optimizing muscle performance while minimizing energy expenditure is crucial in reducing the risk of musculoskeletal disorders in the workplace (Bloswick & Miller, 2016).

In high-repetition work environments, such as assembly lines or data entry stations, muscle efficiency is a critical factor in preventing overuse injuries. Ergonomic solutions that focus on reducing task repetition, increasing task variation, or optimizing workstation height and design are vital in maintaining healthy muscle function and preventing conditions like tendinitis or carpal tunnel syndrome (Armstrong et al., 2014).

4.4.3.5 Conclusion

Understanding the biomechanics of muscle contraction and the sliding filament model from an ergonomic perspective offers valuable insights into optimizing human performance in occupational settings. By aligning workstation design with the natural limits and capabilities of muscle function, ergonomic interventions can help reduce the incidence of work-related musculoskeletal disorders, ensuring a healthier and more productive workforce. The integration of recent research into muscle efficiency further highlights the importance of energy management in muscle performance, emphasizing the need for ergonomically sound environments that support both short-term performance and long-term musculoskeletal health.

4.4.4 Method to Stimulate and Control the Mechanism of Contraction

Human bodies operate similarly to complex biochemical systems, constantly requiring energy to perform both voluntary and involuntary functions. In the context of occupational ergonomics, understanding how energy is produced, utilized, and maintained within the body is essential for designing tasks, environments, and work schedules that optimize performance and reduce fatigue-related risks. This chapter delves into the biochemical pathways of energy production, highlighting their relevance to physical work and task endurance in occupational settings.

4.4.5 Sources of Energy: Carbohydrates, Fats, and Proteins

The primary source of energy in the human body comes from glucose, a simple sugar derived from carbohydrates. Glucose is especially important during periods of high physical activity, such as lifting, carrying, or sustained repetitive motion common in many workplaces. Due to its molecular structure, glucose can be rapidly converted into usable cellular energy through a series of biochemical processes. This rapid conversion is particularly critical in occupations requiring sustained bursts of activity, where muscles need immediate and efficient energy sources (Brooks & Mercier, 2015).

However, glucose is not the only energy substrate utilized by the body. Fats and proteins also serve as fuel, though they are metabolized differently and are often secondary sources under standard conditions. Triglycerides, stored in adipose tissue, represent a vast reservoir of energy that becomes particularly important during prolonged, low-intensity tasks, such as endurance activities or long-duration manual labor (Jeukendrup & Gleeson, 2018). Fats provide a higher energy yield per gram than glucose, making them ideal for activities requiring sustained energy over a long period.

Proteins, while primarily used for the repair and construction of tissues, can also be broken down into amino acids and subsequently converted into energy under extreme conditions, such as starvation or prolonged exertion without adequate nutrient intake (Tipton & Wolfe, 2018). However, relying on proteins for energy is inefficient and detrimental to muscle and tissue health, and is generally considered a last resort by the body.

4.4.6 Energy Pathways: Aerobic vs. Anaerobic Respiration

The efficiency with which the body converts these macronutrients into usable energy depends largely on the availability of oxygen. When oxygen supply meets the body's demands, aerobic respiration occurs within the mitochondria, enabling the full breakdown of glucose into carbon dioxide, water, and adenosine triphosphate (ATP), the primary energy carrier in cells (Perry et al., 2016). This process is highly efficient, producing 36–38 molecules of ATP per molecule of glucose, and is the dominant pathway during moderate-intensity, long-duration occupational tasks that involve repetitive, low-force activities (Gnaiger, 2020).

In contrast, when physical exertion exceeds the rate at which oxygen can be delivered to the muscles, the body shifts to anaerobic respiration. In this state, glucose is converted into lactic acid through a process known as fermentation, yielding only two molecules of ATP per molecule of glucose. While anaerobic respiration allows workers to continue performing tasks under high-demand conditions, it is much less efficient than aerobic respiration and leads to the accumulation of lactic acid in muscle tissue. This acid buildup is associated with the muscle fatigue and burning sensation often experienced during intense or sustained physical work (Gladden, 2016).

4.4.7 Ergonomic Considerations: Task Design and Fatigue Management

Understanding these biochemical pathways is critical for ergonomic task design, particularly in occupations that demand repetitive or sustained physical activity. Jobs involving high-intensity, short-duration tasks such as lifting or carrying may disproportionately rely on anaerobic pathways, leading to faster onset of muscle fatigue and potential risk of overexertion injuries, such as sprains or strains. On the other hand, occupations involving longer periods of moderate-intensity activity, such as standing or walking, predominantly rely on aerobic metabolism, benefiting from a more sustained energy supply but still requiring careful task scheduling to prevent fatigue accumulation (Kumar & Kumar, 2020).

Incorporating work–rest cycles that allow sufficient recovery time can optimize worker performance and reduce the risk of injury by balancing energy demands with recovery periods. Furthermore, task variation can help mitigate muscle fatigue by allowing different energy pathways to take precedence at different times. For instance, alternating between high-intensity and low-intensity tasks can ensure that aerobic pathways dominate, reducing the reliance on anaerobic systems and preventing lactic acid accumulation (Burgess, 2014).

4.4.7.1 Nutritional Factors and Occupational Performance

The body's ability to effectively utilize different energy sources is also influenced by nutritional intake. A diet rich in carbohydrates, for example, ensures a readily available supply of glucose for immediate energy needs, while a moderate intake of fats supports long-term endurance during physically demanding tasks. Protein intake remains essential for tissue repair and maintenance, but it should not be relied upon as a primary energy source in occupational settings (Phillips, 2014).

Employers and occupational health professionals can support workers by ensuring access to balanced nutrition, particularly in labor-intensive roles where energy demands are high. Providing education on nutritional timing—such as consuming carbohydrate-rich snacks before physically demanding tasks—can also improve worker performance and reduce fatigue (Jeukendrup & Gleeson, 2018). Hydration is another key factor, as water is required for many of the metabolic processes involved in energy production and muscle function.

4.4.7.2 Conclusion

In summary, the human body is equipped with sophisticated biochemical pathways to generate energy from carbohydrates, fats, and proteins, each suited to different types of physical tasks. Ergonomically, understanding these pathways is crucial for designing tasks and work environments that align with the body's energy capacities, reducing the risk of fatigue-related injuries and optimizing long-term worker performance. Proper task design, nutritional support, and work–rest cycles are key components in maintaining both the health and productivity of workers in energy-demanding roles.

4.4.8 Innervation of the Muscular System

Building upon the previous discussion about cellular respiration fueling our movements, let's delve into the intricate communication network that coordinates muscle contractions. This network, orchestrated by the nervous system, allows us to perform everything from a delicate finger tap to a powerful sprint.

The nervous system operates through a sophisticated network of specialized cells called neurons. These neurons act like electrical wires, transmitting messages throughout the body. The text focuses on two crucial types of neurons:

- **Efferent (Motor) Neurons:** These neurons function as the conductors, carrying instructions from the brain and spinal cord directly to the muscles. They act as the "outgoing" messengers, telling the muscles when and how to contract.
- **Afferent (Sensory) Neurons:** These neurons serve as sensory receptors, constantly gathering information from the muscles and body about position, movement, and even pain. They act as the "incoming" messengers, transmitting sensory information back to the central nervous system.

This intricate communication between the nervous system and the musculoskeletal system (muscles and bones) is critical for coordinated movement. The efferent nerves carry instructions specifying the desired muscle contractions from the brain to the peripheral nervous system before ultimately reaching the specific muscle fibers targeted for action.

4.4.8.1 The Efferent Nerves

The previous section highlighted the nervous system's role as the conductor for muscle movement. But how exactly does this communication translate into muscle contractions? This section delves into the fascinating world of motor units, the microscopic conductors that orchestrate the symphony of muscle power.

Imagine a single motor neuron as a conductor's baton. This neuron extends its fibers, called axons, which branch out like the fingers on a hand. Each branch of this axonal "hand" terminates at a specialized connection point called a neuromuscular junction with a single muscle fiber. Interestingly, the group of muscle fibers innervated by these branches from the same motor neuron actually work as a team. This team is called a motor unit, and it is the fundamental functional unit of a muscle.

The size of a motor unit varies depending on the muscle's function. Muscles requiring fine control, like those in the eyes, have smaller motor units. This allows for more precise and nuanced movements. Conversely, muscles responsible for powerful actions, such as the biceps, have larger motor units. These larger units can activate a greater number of muscle fibers simultaneously, generating more forceful contractions.

The magic of muscle contraction happens at the neuromuscular junction. This specialized junction acts as a bridge between the motor neuron and the muscle fiber. Here, the electrical signal carried by the motor neuron does not directly trigger contraction. Instead, the signal prompts the release of a chemical messenger called acetylcholine. This chemical then binds to receptors on the muscle fiber, triggering a series of events that ultimately lead to muscle contraction.

Think of a weightlifter performing a bicep curl. The complex movement involves a coordinated effort of multiple motor units within the bicep muscle. Depending on the desired force, the nervous system can activate different motor unit combinations. For a controlled lift, smaller motor units with fewer muscle fibers might be recruited first. As the weightlifter needs more power, larger motor units with more muscle fibers can be activated, progressively increasing the force generated by the bicep.

4.4.8.2 The Sensory Nerves

Our bodies rely on a complex communication network between muscles and the nervous system to maintain coordinated movement and posture. This intricate system is facilitated by sensory nerves that act as messengers, carrying information about muscle status to the central processing centers—the brain and spinal cord. These signals serve a dual purpose: they guide ongoing muscle activity and store reference points for future actions.

Muscle spindles are specialized structures nestled within muscle fibers that constantly monitor changes in muscle length. They are particularly adept at detecting both rapid stretches and sustained lengthening, and promptly transmit this information to the spinal cord via sensory impulses. This allows for fine-tuned adjustments in muscle activity to prevent overextension or maintain desired positions.

Partnering with the muscle spindles are the Golgi tendon organs, strategically positioned within tendons—the connective tissues that bridge muscles to bones. These perceptive organs function as internal gauges, keeping a watchful eye on the tension developing within a muscle. Whenever a tendon experiences excessive pull, the Golgi tendon organ springs into action, sending sensory signals to the spinal cord.

This sensory information received by the spinal cord triggers a coordinated response through a specialized pathway known as the reflex arc. Sensory neurons act as the first leg of this relay race, transmitting signals from the muscle receptors (muscle spindles and Golgi tendon organs) to the spinal cord. Here, interneurons, the busy communication hubs within the spinal cord, take over, bridging the gap between sensory and motor information. Finally, motor neurons, the powerhouses of movement, carry outgoing commands (efferent impulses) back to the muscles, initiating the necessary adjustments in muscle tension or length.

In essence, the muscle spindle and Golgi tendon organ act as the sensory workhorses within this intricate reflex arc. They continuously monitor muscle status, providing real-time feedback that allows the nervous system to make split-second decisions to maintain optimal muscle function and prevent injuries. This dynamic interplay between sensory perception and motor response ensures movements remain smooth, coordinated, and adaptable to ever-changing demands.

4.4.9 Reflexes

A reflex arc, the pathway through which a reflex occurs, can be broken down into three key components:

1. **Sensory Impulse:** The journey begins with a stimulus, such as an unexpected object approaching the eye. This triggers a signal to travel along a sensory nerve, carrying the information toward the processing center.
2. **Relay at the Spinal Cord or Brainstem:** Traditionally, it was believed that all sensory information reaches the brain for interpretation. However, research suggests that for certain reflexes, like the blink reflex, the processing might occur within the spinal cord or brainstem. This intermediary station acts as a decision point, determining the appropriate response.
3. **Motor Response:** The final leg of the reflex arc involves the transmission of a new impulse along a motor nerve. This outgoing signal travels to the target muscles, triggering the necessary reaction. In the case of the blink reflex, the muscles around the eye (*orbicularis oculi*) receive the command to contract, causing the eyelids to shut rapidly.

The blink reflex serves as a prime example of this protective mechanism. It is an involuntary response designed to shield the delicate eye from potential harm. Interestingly, the speed of this reflex can be faster than the theoretical time it would take for the signal to travel to the brain and back. This further strengthens the notion that some reflexes can be processed at a lower level within the nervous system, closer to the action itself.

Reflexes extend far beyond simple protection. They are the foundation for numerous essential functions within the body. A helpful illustration is the coordinated movement of opposing muscle groups. When an arm is bent at the elbow, the biceps muscles contract to perform the flexion. Simultaneously, the triceps muscles on the backside of your arm relax through a reflex arc. This coordinated interplay ensures smooth and controlled movements.

To summarize, reflexes are the body's silent guardians, acting swiftly and efficiently to safeguard us and orchestrate a multitude of movements without requiring conscious effort.

4.4.10 Energy Transformation Process for Muscle Activity

Muscle contraction, a central component of physical activity, is critically dependent on the availability of energy, primarily provided by adenosine triphosphate (ATP). ATP is often referred to as the "cellular currency of energy" because it acts as an immediate source of power for muscle function, supporting everything from basic movements to strenuous occupational tasks. However, the amount of ATP stored within muscle cells is quite limited, lasting only a few seconds during intense activity (Blomstrand & Saltin, 2015). Therefore, to sustain movement beyond this brief period, the body must continuously regenerate ATP through several complex metabolic pathways.

4.4.11 Metabolic Pathways for ATP Regeneration

There are two primary systems for ATP regeneration: the aerobic and anaerobic metabolic pathways. These systems differ in terms of their speed of ATP production, the substrates they utilize, and their efficiency in supporting varying levels of muscular effort, which has direct implications for ergonomic task design.

4.4.12 Aerobic Metabolism

Aerobic metabolism, meaning "with oxygen," is the most efficient pathway for ATP production and is employed during prolonged, low- to moderate-intensity activities. This system relies on the availability of oxygen to break down carbohydrates (glucose or glycogen) and fatty acids into ATP, which occurs within the mitochondria of muscle cells through a process known as aerobic glycolysis and subsequent oxidative phosphorylation (Powers & Howley, 2017). Due to its efficiency, aerobic metabolism is the dominant system for generating ATP in occupations that require sustained effort over longer durations, such as walking, assembly line work, or moderate manual labor.

The reliance on oxygen delivery, however, implies that aerobic metabolism is limited by cardiovascular function. Ergonomically, tasks that are designed to maintain moderate-intensity levels with sufficient oxygen supply reduce fatigue and the risk of overexertion injuries. For instance, rotating workers between low-intensity tasks allows for sustained performance over an entire workday without overwhelming the aerobic system (Kroemer, 2017).

4.4.13 Anaerobic Metabolism

In contrast, anaerobic metabolism provides a quick but less efficient source of ATP and does not rely on oxygen. This pathway becomes critical during short bursts of high-intensity physical activity, such as heavy lifting, fast-paced repetitive motions, or emergency responses where the demand for energy exceeds the capacity of the aerobic system (Gastin, 2016).

The creatine phosphate (CP) system is the fastest anaerobic energy pathway and is used for immediate energy needs. CP, stored in muscle cells, donates a phosphate group to adenosine diphosphate (ADP) to rapidly regenerate ATP. However, CP stores are limited and deplete within seconds, meaning this system can only sustain activity for very short durations, typically under 10 seconds of maximal effort (Balsom et al., 2014).

After CP is depleted, the body shifts to anaerobic glycolysis, which involves the breakdown of glucose into lactic acid. This process provides a slower but continued supply of ATP for short-duration, high-intensity tasks, such as pushing, pulling, or fast-paced lifting. However, anaerobic glycolysis is limited by the production of lactic acid, which can lead to muscle fatigue and discomfort (McArdle et al., 2015). From an ergonomic perspective, minimizing the reliance on anaerobic metabolism in the workplace can help reduce fatigue-related injuries. For instance, designing workstations that allow workers to alternate between high- and low-intensity tasks can help manage energy demands and reduce the risk of overexertion (Hedge, 2016).

4.4.14 Ergonomic Considerations for Energy Systems

In practical terms, the body's ability to produce ATP through these pathways has significant implications for task design and worker performance. High-intensity tasks that rely heavily on anaerobic systems, such as heavy lifting or fast-paced manual labor, place immediate stress on the body's energy reserves and can lead to rapid fatigue if sustained for long periods without adequate rest. In contrast, tasks that engage the aerobic system, such as standing or walking, can be sustained for much longer without depleting energy reserves, provided that oxygen supply and nutrient intake are adequate (Sparrow et al., 2021).

Work-rest cycles are a key ergonomic intervention for balancing these energy demands. Short, frequent rest breaks allow the body to recover CP stores and clear lactic acid, while alternating between high- and low-intensity tasks can help workers avoid over-reliance on anaerobic metabolism (Middlesworth, 2017). Task variety can also reduce the risk of muscle fatigue by distributing the workload across different muscle groups, preventing localized muscle overuse (Hedge, 2016).

4.4.15 Nutritional Support for Energy Production

Adequate nutritional intake is also vital for maintaining ATP production, particularly in occupations requiring sustained physical activity. Carbohydrates provide the necessary glucose for both aerobic and anaerobic metabolism, while fats contribute to long-term energy storage used primarily in aerobic conditions (Jeukendrup & Gleeson, 2018). Proper hydration is equally important, as water is required for many of the biochemical reactions involved in ATP production, including both aerobic and anaerobic pathways (Sawka et al., 2015).

FIGURE 4.5 (From Flickr, Pushing and pulling, 2006, http://farm1.static.flickr.com/74/193681147_7ac774b2a5.jpg?v=0 [accessed September 30, 2009])

Providing workers with access to nutritional support, including carbohydrate-rich foods for short-term energy needs and balanced meals that promote sustained aerobic metabolism, can enhance worker productivity and reduce the risk of fatigue-related errors or injuries (Grandjean & Vuori, 2014).

In summary, ATP production and energy metabolism are central to the body's ability to perform a wide range of occupational tasks. The balance between aerobic and anaerobic metabolism determines how efficiently the body generates energy, directly influencing worker performance and fatigue. Ergonomic task design that takes these metabolic pathways into account can improve worker safety, productivity, and overall health. By understanding the specific energy demands of different occupational activities, ergonomic interventions can be designed to align with the body's natural energy systems, ensuring a sustainable and safe working environment.

4.4.16 Types of Muscular Work

Muscular work can be categorized into two primary types: dynamic and static. These classifications are critical to understanding how different job tasks affect musculoskeletal health and performance. This section will discuss dynamic and static muscular work, their physiological impacts, and strategies for mitigating risks in occupational settings.

Dynamic Muscular Work: Dynamic muscular work involves coordinated contractions and relaxations of muscles to create movement. This type of work is commonly observed in tasks like walking, lifting, or typing. From an ergonomic perspective, dynamic work is generally beneficial because it promotes blood circulation and facilitates the continuous supply of oxygen and nutrients to the working muscles. As a result, dynamic work can be sustained for extended periods under appropriate conditions (Astrand & Rodahl, 1970; Kroemer et al., 1994).

However, even dynamic work can lead to fatigue if the intensity is too high or sustained over long durations. Several key factors influence how long dynamic work can be performed efficiently:

1. **Intensity:** Higher intensity work leads to quicker fatigue. Tasks requiring significant effort should be limited in duration and incorporate rest breaks (Helander, 1995).
2. **Worker Fitness:** A physically fit individual will tolerate dynamic work better than someone who is less fit. This variability must be considered in job design to match task demands with worker capacity (Kroemer et al., 1994).
3. **Environmental Conditions:** External factors such as heat, humidity, and noise can impact the sustainability of dynamic work. High temperatures or poor ventilation, for example, increase the physiological strain on workers, leading to earlier onset of fatigue (Astrand & Rodahl, 1970).

Astrand and Rodahl's (1970) classic research suggested that, during an 8-hour workday, the average workload should not exceed 33% of a worker's maximum capacity. This recommendation underlines the importance of designing tasks that consider intensity, worker fitness, and environmental conditions to prevent overexertion and associated musculoskeletal issues.

4.4.17 Static Muscular Work

In contrast to dynamic work, static muscular work occurs when a muscle maintains a fixed position for an extended period. This type of load is common in many industrial and service occupations, particularly in tasks involving material handling, electronics assembly, and repair work. A typical example is holding a heavy object while performing manual material handling (Gallagher et al., 2014).

Static muscular work is more hazardous than dynamic work because it restricts blood flow to the working muscles. The sustained contraction increases intramuscular pressure, which can occlude blood vessels, reducing the delivery of oxygen and nutrients and hindering the removal of metabolic waste products. As a result, static work can lead to quicker muscle fatigue, pain, and even long-term musculoskeletal disorders such as tendinitis and arthritis (Hägg, 2017).

Key physiological effects of static muscular work include:

1. **Reduced Blood Flow:** Sustained static contractions increase muscle pressure, limiting blood circulation and oxygen supply. This leads to faster fatigue and discomfort compared to dynamic work (Hägg, 2017).
2. **Lactic Acid Buildup:** Reduced circulation during static work prevents the efficient removal of waste products such as lactic acid, leading to muscle soreness and fatigue (Gallagher et al., 2014).
3. **Musculoskeletal Disorders:** Prolonged static work can lead to chronic musculoskeletal problems, including tendinitis, arthritis, and intervertebral disc degeneration (Punnett & Wegman, 2014).

4.4.18 Risks of Prolonged Static Work

Static work poses numerous long-term risks for musculoskeletal health, including:

- **Arthritis:** Constant mechanical stress from static tasks can lead to joint inflammation and wear over time (Punnett & Wegman, 2014).
- **Tendinitis:** Repetitive static loading of tendons can cause tendon inflammation, contributing to chronic pain and dysfunction (Hägg, 2017).
- **Spinal Disc Issues:** The spine is particularly vulnerable to static loading, especially during tasks that involve prolonged sitting or awkward postures. This can lead to intervertebral disc degeneration and lower back pain (Gallagher et al., 2014).

4.4.19 Ergonomic Strategies for Managing Static Work

While static work cannot always be avoided, various ergonomic strategies can reduce its impact and help prevent musculoskeletal disorders. Some effective methods include:

1. **Task Rotation:** Incorporating dynamic movement into the work routine by rotating tasks helps to distribute muscle strain across different muscle groups, reducing the risk of fatigue and injury (Gallagher et al., 2014).
2. **Ergonomic Workstations:** Designing workstations that allow workers to alternate between sitting and standing, or that provide adequate support, can alleviate static strain (Hägg, 2017).
3. **Use of Mechanical Aids:** Implementing tools such as lifting aids, dollies, or height-adjustable workstations can reduce the static load on workers, minimizing musculoskeletal strain (Punnett & Wegman, 2014).
4. **Posture Correction and Training:** Training workers in proper posture and ergonomic principles helps to reduce static loading of muscles and joints. Maintaining neutral postures and avoiding awkward arm and neck positions are crucial (Gallagher et al., 2014).
5. **Scheduled Breaks and Microbreaks:** Short, frequent breaks can improve circulation, prevent muscle fatigue, and enhance long-term productivity by reducing static muscular strain (Hägg, 2017).

In occupational ergonomics, the goal is to prioritize dynamic work while minimizing the frequency and duration of static tasks. Although static work may seem less physically demanding, its physiological consequences, such as restricted blood flow and lactic acid buildup, make it particularly hazardous over time. Proper ergonomic job design, incorporating tools, breaks, and task variation, can mitigate these risks and promote musculoskeletal health.

The design of jobs and work environments should strive to minimize static postures by encouraging movement and facilitating proper posture. By applying these ergonomic principles, workers can maintain productivity while reducing the risk of long-term musculoskeletal injuries (Gallagher et al., 2014; Hägg, 2017).

4.4.20 Types of Muscle Fibers in Muscular Contraction

One key distinction between types of muscle fibers lies in contraction speed and fatigue resistance. Fast-twitch fibers, aptly named, generate powerful contractions rapidly. They are the sprinters of the muscle world, ideal for bursts of intense activity. However, their reliance on anaerobic metabolism, a process that doesn't require oxygen, leads to quick fatigue. These fibers are typically white in color due to a lower concentration of myoglobin, a protein that stores oxygen.

On the other end of the spectrum are slow-twitch fibers. Marathons are their specialty. They contract steadily and tirelessly, fueled by an oxygen-rich (aerobic) metabolic pathway. This slower, more efficient approach allows them to sustain activity for extended periods. Slow-twitch fibers have a deep red hue, reflecting their rich network of capillaries, tiny blood vessels that deliver the vital oxygen needed for sustained work.

Interestingly, some muscle fibers possess a hybrid nature, combining elements of both fast-twitch and slow-twitch characteristics. These intermediate fibers offer a balance between speed and endurance, making them adaptable to myriad demands.

Slow-twitch fibers are controlled by smaller motor neurons that fire continuously at a low frequency, which keep them primed for sustained contractions. Conversely, fast-twitch fibers receive input from larger motor neurons that unleash bursts of high-frequency signals, ideal for explosive movements. These neurons alternate periods of silence with bursts of activity, mirroring the on-and-off nature of fast-twitch contractions.

In conclusion, muscle fibers are a diverse group, each specialized for a specific function. From the lightning-fast bursts of fast-twitch fibers to the unwavering endurance of slow-twitch fibers, this remarkable variety equips our bodies to handle a vast array of movements, from a sprinter's dash to a marathoner's long journey. The interplay between fiber type and neural control further underscores the intricate orchestration that powers our every move.

While the previous section explored the general characteristics of slow-twitch and fast-twitch fibers, Table 4.1 provides a more detailed comparison. This highlights a key point: muscle fibers are specialists, each designed to excel at a specific task.

Slow-twitch fibers thrive in activities demanding sustained low-force contractions. Imagine holding a yoga pose for an extended period—that's where slow-twitch

TABLE 4.1
Properties of Fast and Slow Twitch Muscle Fibers

Property	Fast-Twitch Muscle Fibers (Type II)	Slow-Twitch Muscle Fibers (Type I)
Example	Fast	Soleus (calf) muscle Slow
Twitch contraction time		
General color	White	Red
Capillary blood flow	Low	High
Myoglobin content	Low	High
Primary source of ATP	Glycolysis	Oxidative phosphorylation
Glycogen	High	Low
Tension/force produced	Larger	Smaller
Myosin-ATP activity	High	Low
Resistance to fatigue	Low	High
Nerve fiber size	Large	Small
Nerve fiber activity	Intermittent, high frequency	Continuous, low frequency
Oxidative capacity	Low	High

fibers come to the rescue. Their rich network of capillaries ensures a steady supply of oxygen, fueling their aerobic metabolism for long-lasting performance. While they may not generate explosive power, their endurance is unmatched. Activities like marathon running, swimming, and cycling heavily rely on these tireless workhorses.

Fast-twitch fibers are capable of producing rapid, high-force contractions. When you need a burst of power, like sprinting across the finish line or jumping over an obstacle, fast-twitch fibers take center stage. Their reliance on anaerobic metabolism allows for quick bursts of energy, perfect for short-lived activities. However, these high-energy bursts lead to faster muscle fatigue. Strength training primarily leads to the hypertrophy (enlargement) of these white muscles. Although the total number of fibers does not increase, the existing ones grow in size and contain more myofibrils, the contractile units within muscle cells. This translates to a synergy of increased strength and faster contraction velocity, making athletes more powerful.

In essence, muscles are a microcosm of specialization. By understanding the unique properties of slow-twitch and fast-twitch fibers, we can appreciate the remarkable adaptability that allows for the performance of a wide range of movements, from the unwavering endurance of a marathon runner to the explosive power of a sprinter.

4.4.21 Muscular Fatigue

Muscle fatigue can significantly impact efficiency and safety in the workplace, as it negatively impacts task performance and can lead to injuries. This section delves into the science of muscle fatigue and explores strategies to keep workers energized and safe.

During prolonged, repetitive, or intense muscle contractions, a phenomenon known as muscle fatigue sets in. Research on athletes performing extended submaximal exercise reveals a direct correlation between fatigue and the depletion of muscle glycogen, the primary fuel source for sustained activity. This depletion can occur during both static (holding a position) and dynamic (movement) contractions, but fatigue arrives much faster in static postures.

The cause behind this rapid fatigue in static contractions is twofold. First, the lack of movement restricts the flow of fresh nutrients and oxygen to the muscles. Second, waste products like lactic acid accumulate within the muscle fibers. During vigorous dynamic exercise as well, the circulatory system can become overwhelmed, unable to deliver oxygen to muscle fibers fast enough. This oxygen deficiency triggers the production of lactic acid, which then accumulates and lowers the muscle's pH (acidity). This acidic environment disrupts muscle function, rendering them less responsive to stimuli and hindering contraction.

4.4.22 Strategies to Conquer Fatigue and Enhance Workplace Safety

Fortunately, ergonomists can combat muscle fatigue to promote a safer, more efficient work environment. Here are some key strategies:

1. **Task Design for Submaximal Strength:** Ideally, tasks should require submaximal strength, meaning workers should not be constantly pushing their physical limits. This reduces the strain on muscles and slows down glycogen depletion, delaying the onset of fatigue.
2. **Embrace Dynamic Movements:** Tasks should prioritize dynamic movements over static postures whenever possible. Dynamic contractions promote better blood flow, ensuring a steady supply of oxygen and nutrients to muscles, and minimizing the buildup of waste products.
3. **Work–Rest Cycles:** Integrating work–rest cycles into the workflow is crucial. Scheduled breaks allow tired muscles to recover, replenish glycogen stores, and clear accumulated lactic acid.
4. **Job Sharing and Task Variation:** Implementing job sharing, where multiple workers take turns performing a task, or incorporating task variation within a cycle allows individuals to utilize different muscle groups, reducing the strain on any particular set of muscles.

Recognizing the impact of muscle fatigue goes beyond addressing discomfort. Studies suggest that fatigue can actually indicate poorly designed tasks, potentially leading to occupational injuries. By acknowledging and effectively addressing muscle fatigue, organizations can foster a more proactive approach to workplace safety. This translates to a healthier, more productive workforce and a significant reduction in the risk of injuries.

4.4.23 Types of Muscle Contractions

The types of muscle contractions include four categories of contractile functionality. The types of muscle contractions include isometric, isoinertial, eccentric, isotonic and isokinetic.

4.4.23.1 Isometric Contractions (Static Muscular Contraction)

Isometric contractions, also known as static muscular contractions, offer a unique way to engage and strengthen muscles without any visible movement at the joints. Unlike dynamic exercises where muscles shorten (concentric contraction) or lengthen (eccentric contraction) to produce movement, isometric contractions involve holding a muscle in a fixed position against a constant force.

Imagine carrying a heavy grocery bag at arm's length. Your biceps muscles are activated to counteract the downward pull of the bag's weight. Even though your arms remain stationary, the biceps are firing and generating isometric force to maintain that position. This static hold, despite the absence of movement, effectively strengthens the biceps muscles.

Here is a deeper dive into isometric contractions and their applications:

- **Muscle Activation:** During an isometric hold, the nervous system sends signals to specific muscle fibers, causing them to contract and generate tension. Although the muscle length remains constant, this internal activation strengthens the contractile machinery within the muscle fibers.
- **Improved Strength:** Isometric exercises can be remarkably effective for building strength, particularly in foundational exercises. Studies have shown that isometric training can increase muscle strength gains up to similar levels as traditional dynamic exercises, especially for specific muscle groups and across a range of contraction intensities. This makes them a valuable tool for rehabilitation, injury prevention, and targeted strength development.
- **Accessibility and Convenience:** Isometric exercises require minimal equipment and space, making them highly accessible for people of all fitness levels. They can be performed anywhere, from an office chair to the comfort of one's home.

Examples of isometric exercises:

- **Plank:** This classic core exercise engages your entire abdominal wall in a static hold, promoting core strength and stability.
- **Wall Sit:** Here, you hold a squat position with your back pressed against a wall, isometrically activating your quadriceps and glutes.
- **Pushing Against a Wall:** By pushing firmly against a wall with your palms or back, you can isometrically engage your chest or back muscles, respectively.

Isometric contractions offer a versatile and effective way to build strength, improve stability, and enhance your overall fitness routine.

4.4.23.2 Isoinertial Contractions (Constant Load)

Isoinertial contractions offer a unique training method that challenges your muscles with a consistently adapting load. Unlike traditional weight training where resistance might vary throughout the exercise, isoinertial training utilizes a special system that maintains a constant level of inertia throughout the entire movement. Inertia refers to an object's resistance to changes in motion.

Imagine performing a bicep curl with isoinertial equipment. As you curl the weight upwards (concentric contraction), the system automatically adjusts the resistance to match the force you're exerting. This ensures your muscles are challenged throughout the entire range of motion, including the often-neglected peak contraction point. Similarly, during the lowering phase (eccentric contraction), the system adapts to provide a consistent level of resistance against the force of your lowering the weight.

The key difference between isoinertial and traditional weight training lies in the way the resistance is distributed:

- **Traditional Weight Training:** In traditional weightlifting, the resistance (weight of the dumbbells or barbells) remains constant throughout the exercise. This can lead to less challenge at the peak contraction point and lowering phase.
- **Isoinertial Training:** Isoinertial training utilizes a smart system that measures both acceleration (how quickly muscles are speeding up the weight) and velocity (the speed at which the weight is moving). Based on these measurements, the system adjusts the resistance in real-time to maintain a consistent level of inertia. This ensures your muscles are constantly challenged throughout the entire movement, maximizing the training effect.

Here are some potential benefits of isoinertial training:

- **Improved Overall Strength Development:** By providing consistent challenge throughout the movement, isoinertial training can potentially lead to more balanced and complete muscle development.
- **Enhanced Power Output:** The ability to train with high velocities under constant load might translate to improved power generation in activities that require explosive movements.
- **Reduced Risk of Injury:** Some studies suggest that isoinertial training might be beneficial for injury prevention due to the controlled nature of the movements.

4.4.23.3 Eccentric Contractions (Lengthening of the Muscle)

Eccentric contractions, often referred to as the "lengthening under tension" muscle actions, play a crucial role in various everyday movements and exercise routines.

Unlike concentric contractions where muscles shorten to generate force (lifting a weight), eccentric contractions involve muscles actively lengthening while still generating tension to resist an external force.

Imagine walking down the stairs. Your quadriceps muscles, located at the front of your thighs, are the primary drivers of this movement. However, simply relaxing and letting gravity take over wouldn't provide the controlled descent we experience. Here's where eccentric contractions come in. The quadriceps actively lengthen and generate tension to control the speed of your descent, preventing an uncontrolled fall. This lengthening under tension is an eccentric contraction.

Eccentric contractions are present in numerous activities:

- **Descending Movements:** Walking downhill, jogging, or lowering yourself during squats and lunges all involve eccentric contractions in various muscle groups.
- **Deceleration Phases:** During running, the eccentric action of your hamstrings (back of the thigh) helps decelerate your leg as it swings forward.
- **Controlled Lowering:** The lowering phase of exercises like bicep curls or dumbbell presses utilizes eccentric contractions in the biceps and chest muscles, respectively. This controlled lowering helps maximize the training effect and prevent injury.

Here are some key benefits of eccentric contractions:

- **Increased Muscle Strength:** Studies suggest that eccentric contractions can be highly effective in building muscle strength, potentially surpassing concentric contractions in certain exercises.
- **Improved Muscle Control:** The ability to control the lengthening of a muscle under tension is crucial for stability and injury prevention. Eccentric training enhances this control.
- **Reduced Risk of Injury:** By strengthening muscles and improving control, eccentric contractions might help prevent injuries commonly associated with rapid lengthening movements.

4.4.23.4 Isotonic Contractions (Concentric Shortening)

Isotonic contractions, the workhorses behind many familiar movements, are characterized by a dynamic interplay between muscle shortening and constant tension. Unlike isometric exercises where muscles hold a fixed position, isotonic contractions involve muscles actively shortening to generate force and produce movement.

Imagine performing a bicep curl. As you lift the weight upwards (concentric contraction), your bicep muscles contract and shorten, bringing your forearm closer to your shoulder. Throughout this movement, your biceps maintain a relatively constant level of tension to overcome the resistance of the weight. This is the essence of an isotonic contraction: muscle shortening with consistent tension.

Isotonic contractions are fundamental to various exercises and everyday activities:

- **Strength Training:** Exercises like bicep curls, push-ups, squats, and lunges all rely on isotonic contractions in different muscle groups. As you move through the exercise, your muscles shorten and generate tension to overcome resistance, promoting strength development.
- **Functional Movements:** From lifting groceries to climbing stairs, isotonic contractions power our daily movements. These contractions allow us to interact with the environment, manipulate objects, and navigate our surroundings.

Here are some key benefits of isotonic contractions:

- **Enhanced Strength and Power:** By actively shortening against resistance, isotonic exercises effectively build muscle strength and power, improving the ability to perform daily tasks and athletic activities.
- **Improved Muscle Coordination:** Isotonic training promotes better coordination between muscle groups, leading to smoother and more efficient movements.
- **Increased Bone Density:** Studies suggest that isotonic exercises can stimulate bone growth and increase bone density, potentially reducing the risk of osteoporosis.

4.4.23.5 Isokinetic Contractions (Constant Force)

Isokinetic contractions offer a unique training method characterized by muscle exertion at a constant speed. Unlike traditional weight training where the speed of movement can vary, isokinetic exercises utilize specialized equipment that dynamically adjusts resistance to maintain a predetermined speed throughout the entire exercise.

Imagine performing a leg extension on an isokinetic machine. As you extend your leg (concentric contraction), the machine automatically increases resistance to match the force you're exerting. This ensures your muscles are challenged at a consistent pace, even during the strongest point of the movement. Conversely, during the lowering phase (eccentric contraction), the machine reduces resistance to maintain the same speed as you control the movement.

The key feature of isokinetic training lies in the controlled speed element:

- **Constant Speed, Varied Resistance:** Isokinetic machines regulate resistance based on your force output, ensuring you maintain a consistent speed throughout the entire range of motion. This can be particularly beneficial for targeting specific weaknesses within a movement pattern.
- **Reduced Risk of Injury:** The controlled nature of isokinetic training, with its emphasis on constant speed, might help minimize the risk of injuries associated with sudden changes in tempo or uncontrolled movements.
- **Potential for Rehabilitation:** The ability to isolate muscle groups and precisely control movement speed makes isokinetic training a valuable tool in rehabilitation programs for recovering from injuries.

Here are some additional points to consider:

- **Limited Specificity:** While isokinetic training offers unique benefits, it may not translate directly to activities that require variations in speed and power output.
- **Accessibility:** Isokinetic equipment is not as widely available as traditional free weights or machines.

Overall, isokinetic contractions offer a specialized training method at a constant speed with potential benefits for strength development, injury prevention, and rehabilitation.

4.4.24 Measurement of Muscular Strength

Assessing muscular strength is a cornerstone for understanding human capabilities and limitations. This understanding is crucial in various aspects of our lives, from designing effective tools and products to optimizing performance in recreational activities and even evaluating safety risks.

When designing a method to measure muscular strength, several factors require careful consideration. The primary objective of the assessment plays a key role. Are we aiming to evaluate a person's suitability for a specific task? Perhaps we're designing a product that needs to be comfortable and functional for users with varying strength levels. Or maybe the goal is to identify potential risks associated with a particular activity. Each objective necessitates a tailored approach to strength testing.

The specific muscles involved in the task at hand also influence the testing methodology. Different muscle groups have varying strengths and functions. Identifying the key muscles used in the activity ensures the chosen test accurately reflects the demands placed on the user.

Cost is another important consideration. Complex testing equipment can be expensive, and simpler, more affordable methods might be equally effective depending on the objective. Additionally, the nature of the task itself shapes the testing approach. Is it a static exertion of force, a dynamic movement, or a combination of both?

In the realm of ergonomics, which focuses on designing workplaces and products to fit the human body, three broad categories of strength testing methodologies are commonly employed:

- **Isometric Testing:** This method measures strength exerted against a fixed object, with no movement occurring at the joint.
- **Dynamic Testing:** This category encompasses two main types:
  - **Isoinertial Testing:** Here, the weight or resistance remains constant throughout the movement.
  - **Isokinetic Testing:** This method involves maintaining a constant speed of movement, with the resistance automatically adjusting to match the user's exerted force.

- **Psychophysical Testing:** This approach relies on subjective feedback from the user regarding their perceived exertion or level of effort required to perform the task.

4.4.24.1 Isometric Strength Testing

Isometric strength is the ability of muscles to generate force or torque while maintaining a fixed length. Unlike dynamic movements, isometric contractions involve no change in joint angles. To assess this static strength in various muscle groups, a diverse range of isometric strength testing equipment exists, highlighting the lack of standardization in this widely studied area.

The popularity of isometric testing stems from its relative simplicity. Compared to other methods, it's easier to conduct, measure, and interpret the results. Additionally, isometric testing offers the benefits of being user-friendly and minimizing the risk of injury.

However, a key criticism of this method lies in its limited applicability to real-world tasks. Most jobs, particularly those involving material handling, are inherently dynamic. Static isometric tests may not accurately reflect the demands placed on the body during such activities.

Despite this limitation, isometric testing remains a widely accepted approach in ergonomics research and application. The historical success of applying isometric test results to task evaluation, injury prediction, and product design has made it an invaluable tool. Even though real-world tasks are often dynamic, the insights gleaned from isometric testing have proven effective in informing ergonomic practices.

4.4.24.2 Dynamic Strength Testing

Dynamic strength testing stands in contrast to isometric testing, which deals primarily with static contractions. For dynamic, the focus shifts to assessing muscular strength during movement. This category encompasses a wide range of methodologies, each offering unique insights into human capabilities.

One type of dynamic strength assessment is isoinertial testing. Imagine lifting a weight a set distance. In isoinertial testing, the weight (or resistance) itself remains constant throughout the movement. This method provides valuable information about how effectively muscles generate force against a fixed mass.

Another approach within the dynamic strength testing realm is isokinetic testing. Here, the focus is on maintaining a constant speed of movement throughout a predetermined range of motion. The key difference from isoinertial testing lies in the resistance. In isokinetic testing, the resistance automatically adjusts to match the force exerted by the user. This allows for a more controlled and standardized assessment, as the user's speed dictates the level of difficulty.

4.4.24.3 Psychophysical Testing

The realm of ergonomics utilizes testing methods that not only consider physical capabilities of humans but also their perception of exertion. Psychophysical assessment for determining maximum dynamic lifting capacities offers a unique approach in this regard. This method, used for many years in physiological research, leverages subjective feedback from the participant.

"Psychophysical" refers to a testing methodology where individuals perform lifts to identify their personal limits of safe exertion. Essentially, participants gauge their own acceptable level of risk based on how much weight they feel comfortable lifting. This approach relies heavily on several key factors:

- **Participant Cooperation:** The success of the test hinges on the participant's willingness to openly communicate their perceived exertion levels. Honest feedback ensures a more accurate picture of their lifting capacity.
- **Evaluator Expertise:** The role of the evaluator is crucial. They must possess the knowledge to guide the participant safely through increasing weight loads while ensuring proper lifting technique. Moreover, the evaluator needs to interpret the participant's feedback accurately, distinguishing genuine discomfort from hesitation or apprehension.
- **Shared Judgment:** Ultimately, both the participant and evaluator share responsibility for a safe and accurate assessment. The participant must be vocal about their discomfort, while the evaluator needs to judiciously adjust weights and terminate the test if safety concerns arise.

This method's strength lies in its ability to capture an individual's unique perception of acceptable exertion. However, its heavy reliance on subjective feedback introduces an element of variability. Despite this limitation, when conducted with expertise and sound judgment, psychophysical assessment can offer valuable insights into a person's maximum dynamic lifting capacities. This knowledge can then be used to inform ergonomic interventions aimed at optimizing work practices and preventing injuries.

4.4.24.4 Procedure for Strength Testing

Regardless of the chosen methodology—static (isometric) or dynamic (isoinertial, isokinetic, or psychophysical)—designing a robust strength testing process requires careful consideration of several key factors outlined by Gallagher et al. (1998). Here is a breakdown of these essential elements:

1. **Equipment:** The type of strength test dictates the equipment needed. Isometric testing might utilize hand dynamometers, while dynamic assessments might involve weightlifting equipment or specialized machines. Availability and suitability of the equipment are crucial considerations.
2. **Subject Instructions:** Clear and concise instructions are paramount for ensuring consistent test execution and reliable data collection. Participants need to understand the task, proper technique, and safety protocols.
3. **Measurement Duration:** The length of time a participant exerts force during the test needs to be carefully determined. It should be long enough to capture maximal effort but not so long as to lead to fatigue or compromise in form.

4. **Rest Intervals:** Adequate rest periods between trials are essential for allowing muscles to recover and ensuring subsequent measurements reflect true strength capacity. The optimal rest duration depends on factors like the intensity and duration of the exertion phase.
5. **Body Posture:** Standardized body positioning during the test is crucial for isolating the target muscle groups and minimizing the influence of other muscle actions on the results. Clear instructions and visual aids can help participants maintain proper posture.
6. **Number of Trials:** Performing multiple trials can help account for variability in performance and provide a more reliable picture of an individual's strength. However, an excessive number of trials can lead to fatigue, potentially skewing the results.
7. **Subject's Physical Condition:** Factors like fatigue, recent injuries, or underlying health conditions can significantly impact strength test results. Accounting for these factors through pre-screening or questionnaires helps ensure the validity of the data.
8. **Postural Control:** The level of postural control required during the test can influence the results. For instance, isometric tests might involve a fully supported posture, while dynamic assessments might require maintaining balance during movement. Deciding on the appropriate level of postural control depends on the specific goals of the test.
9. **Environment:** Environmental factors like temperature, humidity, and noise levels can subtly influence participant performance. Maintaining a consistent and controlled testing environment helps minimize these external influences.
10. **Subject Motivation:** A participant's level of motivation can significantly impact the effort they put into the test. Strategies to encourage and motivate participants can help ensure they exert maximal force during the trials, leading to more accurate results.

4.5 DESIGNING A STRENGTH TESTING PROTOCOL

The initial protocol for this strength testing method, established by Caldwell et al. in 1974, has undergone refinements over time based on research advancements. While originally designed for static testing, this protocol can be adapted for dynamic assessments as well. The following provides the details of the Caldwell protocol, as described by Plog and Quinlan (2002), specifically for measuring static strength:

4.5.1 Test Procedure

1. **Measurement Duration:** Assess static strength exertion for a steady 4 seconds.
2. **Data Analysis:** Disregard the initial and final 1-second periods of the exertion phase. The average value recorded during the remaining 3 seconds represents the strength measurement.

4.5.2 Subject Instructions

1. **Informed Consent:** Briefly explain the purpose of the test, the equipment involved, and the testing procedures.
2. **Clear Instructions:** Provide detailed and objective instructions, avoiding emotional cues or coaching.
3. **Effort Instructions:** Instruct the subject to gradually increase their effort to maximum level within 1 second and then maintain that exertion for the remaining 4 seconds.
4. **Special Considerations:** Acknowledge that specific conditions, like finger strength testing, might necessitate adjustments to the timing and procedures.
5. **Positive Feedback:** Offer qualitative, non-comparative, and encouraging feedback throughout the test. Avoid immediate feedback during the exertion phase.
6. **Motivation Management:** Refrain from using incentives like rewards, competition, or external pressures (fear, spectators, noise, heat) that could influence the subject's motivation.

4.5.3 Rest Periods

- Allow a minimum of 2 minutes rest between exertion phases. Extend this period if the subject exhibits signs of fatigue.

4.5.4 Test Description

Prior to the test, provide a clear description of the following aspects:

- **Muscles Involved:** Specify the body parts and primary muscles targeted during the test.
- **Body Posture:** Describe the required body posture and any permitted movements.
- **Support and Resistance:** Explain the type of body support or reaction force provided during the test.
- **Equipment Coupling:** Detail how the subject will be connected to the measuring device.
- **Strength Measurement Device:** Specify the instrument used to measure and record strength data.

4.5.5 Subject Information

- **Selection Criteria:** Describe the population and sample selection process, including sample size.
- **Health:** Recommend a health survey or medical examination to assess the general health of participants.

- **Demographics:** Collect data on gender and relevant anthropometric measurements.
- **Training and Occupation:** Inquire about any training or occupational factors related to strength.

4.5.6 Data Reporting

Following data collection and analysis, the results are used to draw inferences about the subject population. The following provides a list of the statistical data to be reported:

- **Central Tendency:** Measures like mean, median, and mode.
- **Variability:** Measures like standard deviation, variance, and range.
- **Outliers:** Identification of any data points that deviate significantly from the norm.
- **Data Skewness:** Assessment of whether the data distribution leans toward one extreme.
- **Data Visualization:** Presentation of data through graphical representations like histograms.
- **Percentiles:** Calculation of percentiles within the data set.

4.6 SUMMARY

The muscular system is the foundation of motion, converting chemical energy from food and drinks into mechanical energy, allowing us to perform tasks and navigate the world. Ergonomics, the science of designing workplaces and tools to fit the human body, hinges on a deep understanding of this intricate dance between muscles and the nervous system. The application of ergonomic principles is intended to create optimal interactions between the human body and its environment. By understanding how muscles work, ergonomists can design workplaces, equipment, and processes that are compatible with our musculoskeletal system. This reduces the risk of injuries, promotes efficiency, and ultimately, fosters a safer and more comfortable work environment.

4.7 CASE STUDY

Introduction

In ergonomics, understanding how the human body functions during movement is critical for designing safe and efficient work environments. This case study examines the integrated roles of the muscular and nervous systems, explaining how they collaborate to control and execute movement. The nervous system provides signals to muscles, which respond by contracting or relaxing to produce the required action. This coordination is crucial in both manual labor and repetitive tasks to avoid strain and injury. The case study will explore these functions within the context of a telemarketing workstation, where workers frequently interact with keyboards, phones, and monitors.

Problem Overview

The key ergonomic challenge in this scenario involves understanding how improper workstation design can lead to discomfort and even injury, particularly in the neck, wrists, and fingers. Prolonged sitting and repetitive hand movements, combined with awkward postures, increase the risk of cumulative trauma disorders (CTDs). This case study will focus on how the muscular and nervous systems are affected by such tasks and highlight ergonomic risks that may lead to injury.

Analysis of the Muscular and Nervous Systems in Task Execution

- **Muscular System:**

In a typical telemarketing scenario, the muscles responsible for fine motor movements, such as the flexor muscles in the fingers and wrists, are constantly engaged during typing. Larger postural muscles, such as those in the neck, back, and shoulders, maintain stability as workers remain seated for long periods.

- **Nervous System:**

The nervous system plays a critical role by sending motor signals from the brain to muscles. When a worker types, for example, the brain communicates with the muscles in the hands and fingers through the peripheral nervous system. This constant flow of information allows for quick and precise movements. Feedback from the muscles, known as proprioception, informs the nervous system about body positioning, which adjusts accordingly to maintain balance and posture.

Ergonomic Risks in Telemarketing

- **Neck Flexion:**

Workers must continuously look down at forms and then back up at their monitors. This repetitive action stresses the neck muscles and, over time, could lead to muscular imbalances or injury. The nervous system compensates by altering posture, but without breaks or ergonomic interventions, strain builds up.

- **Wrist Pressure:**

Constant typing without appropriate wrist support causes mechanical pressure on the wrist muscles. The nervous system detects this discomfort, but without proper ergonomics, the repetitive nature of the work exacerbates the stress on the hand and wrist muscles, increasing the risk of CTDs such as carpal tunnel syndrome.

Ergonomic Intervention

- **Workstation Adjustments:**

To address the musculoskeletal and nervous system interaction in such tasks, ergonomic solutions must focus on reducing strain. For instance, providing adjustable chairs, monitors, and desks allows workers to maintain neutral postures, reducing unnecessary muscle tension.

- **Task Variability:**

Introducing task variability or incorporating frequent breaks allows muscles to rest and recover. Regular breaks give the nervous system time to reset, decreasing the likelihood of overstimulation and reducing the risk of musculoskeletal fatigue (see Table 4.2).

Conclusion

The interplay between the muscular and nervous systems is vital in movement execution, particularly in repetitive tasks like those performed in telemarketing operations. Understanding how these systems function together can inform ergonomic solutions that prevent musculoskeletal injuries. This case study highlights the importance of incorporating ergonomic design in workplace settings to optimize movement efficiency and minimize health risks.

TABLE 4.2
Workstation Height Summary

Workstation Height (in)	Number
26.50	11
26.75	154
27.00	71
28.00	90
28.25	18
30.25	3
30.50	59
30.75	1
31.00	4
Total	411

Exercises

1. **Define muscular work.**

- Use scholarly databases (e.g., PubMed, ScienceDirect) to review definitions and context for muscular work in ergonomics and biomechanics literature.
- Use AI tools (like ChatGPT or Google Scholar summaries) to compare definitions across fields such as sports science, occupational health, and physiology.
- Interview a physical therapist or occupational health professional to gather applied examples of muscular work in job roles.

2. **Explain how energy (work) can be exerted without actually performing work as defined by laws of physics.**

- Research the physiological vs. physical definitions of work using academic resources and review articles.
- Use an AI tool to generate examples of static muscular effort (e.g., holding a tool overhead).
- Create a short case study involving an ergonomist or athletic trainer describing real-life occupational or sports tasks where such discrepancies occur.

3. **Explain the muscle contraction process.**

- Use diagrams from trusted educational sources online (e.g., Khan Academy, NIOSH, NIH) to map out the physiological muscle contraction process.
- Compare your findings with a description generated by an AI assistant and verify accuracy with scholarly sources.
- Interview a kinesiology or anatomy professor to validate your explanation and gather additional insights.

4. **Explain the chemical processes that take place in aerobic and anaerobic contraction.**

- Conduct a literature search on the metabolic pathways involved in both aerobic and anaerobic muscle activity (e.g., glycolysis, Krebs cycle, lactic acid formation).
- Use AI tools to summarize key differences and relate them to occupational tasks (e.g., long-distance walking vs. heavy lifting).
- Find real-world examples of these processes by interviewing a sports physiologist or a firefighter about energy demands in their work.

5. Define the central nervous system and the peripheral nervous system.

- Create a visual diagram using online anatomy resources to map the CNS and PNS.
- Use AI to provide a side-by-side summary of the two systems and their functions.
- Interview a neurologist or physical rehabilitation specialist to understand the functional implications of these systems in injury or disease.

6. What is the difference between a reflex and a muscular movement such as kicking a ball?

- Review academic sources explaining spinal reflexes and voluntary motor control.
- Use AI to generate a comparison table between reflex arcs and intentional motor responses.
- Interview a physiotherapist or neurologist to learn how these responses differ in rehabilitation or sports training contexts.

7. Explain why muscles fatigue more quickly in static loading. Identify an occupational task where static loading is required.

- Search the literature on muscle fatigue in ergonomics (e.g., posture-related fatigue, EMG studies).
- Use AI to summarize the physiological explanation and occupational relevance of static loading.
- Interview someone working in construction, welding, or dentistry to describe their experience with static loading and fatigue.

8. Discuss the different types of muscle contractions.

- Define isometric, concentric, and eccentric contractions using anatomy and exercise science literature.
- Use AI to find occupational tasks or sports examples for each contraction type.
- Interview a strength coach, physical therapist, or ergonomist to gather real-world examples.

9. Describe a process for dynamic strength testing.

- Research standardized methods (e.g., isokinetic dynamometry) using scholarly databases.
- Use AI to summarize protocols from occupational or clinical testing environments.

- Interview a certified ergonomist or occupational therapist to learn how dynamic strength is assessed in practice, especially in job placement or injury recovery.

10. Develop a process for collecting static strength data.

- Search ISO or NIOSH guidelines for standardized methods in static strength assessment.
- Use AI to generate a draft testing protocol, then validate or refine it using academic sources or textbooks.
- If possible, observe a professional performing strength assessment or conduct a mock interview with a professional who uses such tests in the field.

REFERENCES

Alexander, R. M. (2014). Tendon elasticity and muscle function. *Journal of Experimental Biology*, *217*(16), 2895–2901.

Armstrong, T. J., Buckle, P., Fine, L. J., Hagberg, M., Jonsson, B., Kilbom, Å., ... Silverstein, B. A. (2014). A conceptual model for work-related neck and upper-limb musculoskeletal disorders. *Scandinavian Journal of Work, Environment & Health*, *40*(1), 2–19.

Astrand, P.-O., & Rodahl, K. (1970). *Textbook of work physiology*. McGraw-Hill.

Balsom, P. D., Gaitanos, G. C., Soderlund, K., & Ekblom, B. (2014). High-intensity exercise and muscle glycogen availability in humans. *Journal of Applied Physiology*, *66*(1), 329–336.

Bernard, B. P. (2014). *Musculoskeletal disorders and workplace factors* (Vol. 97). CRC Press.

Blomstrand, E., & Saltin, B. (2015). The role of glycogen in muscle fatigue during prolonged exercise. *Scandinavian Journal of Medicine & Science in Sports*, *25*(2), 19–28.

Bloswick, D. S., & Miller, J. M. (2016). Ergonomics in the workplace: Muscle and joint physiology. *Journal of Occupational Safety and Ergonomics*, *22*(4), 321–330.

Bridger, R. S. (2018). *Introduction to ergonomics* (4th ed.). CRC Press.

Brooks, G. A., & Mercier, J. (2015). Balance of carbohydrate and lipid utilization during exercise: The "crossover" concept. *Journal of Applied Physiology*, *76*(6), 2253–2261.

Burgess, T. L. (2014). Physical exertion, fatigue, and recovery: A systematic review of the literature. *Journal of Occupational Safety and Ergonomics*, *20*(3), 15–24.

Chaffin, D. B., Andersson, G. B. J., & Martin, B. J. (2019). *Occupational biomechanics* (4th ed.). Wiley.

Cooke, R. (2015). The sliding filament model: 50 years of research. *Journal of Muscle Research and Cell Motility*, *36*(3), 247–252.

Cowan, R. (2017). *Galen and the Greek tradition*. Routledge.

David, G. (2005). Ergonomic methods for assessing exposure to risk factors for work-related musculoskeletal disorders. *Occupational Medicine*, *55*(3), 190–199.

Gallagher, S., & Heberger, J. R. (2015). Examining the interaction of force and repetition on musculoskeletal disorder risk: A systematic literature review. *Human Factors*, *55*(1), 108–124.

Gallagher, S., Heberger, J. R., & Punnett, L. (2014). Musculoskeletal disorders. *Occupational Ergonomics: Principles of Work Design*, *23*(4), 257–266.

Gallagher, S., Marras, W. S., Litsky, A. S., & Burr, D. (1998). Load-sharing between muscles and passive tissues during lifting. *Journal of Biomechanics*, *31*(2), 159–168.

Gastin, P. B. (2016). Energy system interaction and relative contribution during maximal exercise. *Sports Medicine, 31*(10), 725–741.

Gladden, L. B. (2016). Lactate metabolism: A new paradigm for the third millennium. *Journal of Physiology, 558*(1), 5–30.

Gnaiger, E. (2020). Mitochondrial pathways and respiratory control: Bioenergetics, the TCA cycle, and the electron transport chain. *Journal of Bioenergetics and Biomembranes, 52*(3), 227–242.

Gollnick, P. D., Piehl, K., & Saltin, B. (2015). Selective glycogen depletion pattern in human muscle fibres after exercise of varying intensity and at varying pedalling rates. *The Journal of Physiology, 241*(1), 45–57.

Gordon, A. M., Homsher, E., & Regnier, M. (2015). Regulation of contraction in striated muscle. *Physiological Reviews, 80*(2), 853–924.

Gordon, A. M., Huxley, A. F., & Julian, F. J. (2016). The variation in isometric tension with sarcomere length in vertebrate muscle fibres. *The Journal of Physiology, 184*(1), 170–192.

Grandjean, A. C., & Vuori, I. (2014). The importance of hydration in labor-intensive industries. *International Journal of Occupational and Environmental Health, 20*(1), 45–56.

Green, H. J., & Jones, S. (2018). Muscle fatigue: Its impact and management in the workplace. *Journal of Occupational Health, 60*(4), 290–300.

Hägg, G. M. (2017). Static work loads and occupational musculoskeletal disorders. *Ergonomics in Action, 41*(2), 188–202.

Hedge, A. (2016). *Ergonomics for improved productivity*. CRC Press.

Helander, M. G. (1995). *A Guide to the Ergonomics of Manufacturing*. Taylor & Francis, Bristol, PA.

Huxley, A. F. (1957). Muscle structure and theories of contraction. *Progress in Biophysics and Biophysical Chemistry, 7*, 255–318.

Jeukendrup, A. E., & Gleeson, M. (2018). *Sports nutrition: From lab to kitchen* (2nd ed.). Bloomsbury Publishing.

Kapoor, S., Sareen, R., & Gupta, M. (2018). The historical contributions of Albrecht von Haller and Luigi Galvani in understanding muscle contraction. *Journal of Clinical and Diagnostic Research, 12*(3), YE01–YE04.

Karwowski, W. (2018). The discipline of ergonomics and human factors: An evolutionary perspective. *Applied Ergonomics, 69*, 1–10.

Keating, A., & O'Brien, T. (2016). Current status of stem cell therapy for spinal cord injury and degenerative diseases. *Stem Cells Translational Medicine, 5*(3), 281–289.

Kim, J., Lee, J., & Park, J. (2015). Application of electromyography in ergonomics: A review. *International Journal of Industrial Ergonomics, 48*, 36–44.

Kovács, M., Tóth, J., Hetényi, C., Málnási-Csizmadia, A., & Sellers, J. R. (2015). Mechanism of blebbistatin inhibition of myosin II. *Journal of Biological Chemistry, 280*(41), 35557–35563.

Kroemer, K. H., Kroemer, H. B., & Kroemer-Elbert, K. E. (1994). *Ergonomics: How to design for ease and efficiency*. Prentice Hall.

Kroemer, K. H. (2017). Fitting the task to the human: A textbook of occupational ergonomics. *Ergonomics, 61*(4), 534–541.

Kumar, S., & Kumar, S. (2020). Work-rest scheduling in repetitive tasks: Ergonomics and worker efficiency. *International Journal of Industrial Ergonomics, 72*, 102875.

Martyn, D. A., & Gordon, A. M. (2017). Muscle structure and function: From molecular motors to biomechanics. *Annual Review of Physiology, 79*(1), 71–94.

McArdle, W. D., Katch, F. I., & Katch, V. L. (2015). *Exercise physiology: Nutrition, energy, and human performance*. Lippincott Williams & Wilkins.

Middlesworth, M. (2017). The impact of rest breaks on worker productivity: An ergonomic perspective. *Journal of Occupational Health Psychology, 12*(3), 257–264.

Nishikawa, K. C., Monroy, J. A., & Uyeno, T. A. (2016). Muscle function: The roles of calcium and ATP in muscle contractions. *Advances in Muscle Physiology, 8*(5), 101–112.

Perry, C. G., Kane, D. A., & Lanza, I. R. (2016). Bioenergetics of exercise and cellular metabolism. *Exercise and Sports Sciences Reviews, 44*(3), 120–126.

Phillips, S. M. (2014). Protein requirements and supplementation in strength sports. *Nutrition and Athletic Performance, 16*(3), 203–209.

Plog, B. A., & Quinlan, P. J. (2002). *Fundamentals of Industrial Hygiene* (5th ed.). National Safety Council.

Powers, S. K., & Howley, E. T. (2017). *Exercise physiology: Theory and application to fitness and performance*. McGraw-Hill Education.

Punnett, L., & Wegman, D. H. (2014). Work-related musculoskeletal disorders: The epidemiologic evidence and the debate. *Journal of Electromyography and Kinesiology, 24*(6), 815–824.

Rassier, D. E., & Herzog, W. (2014). The sliding filament theory: 50 years later. *Journal of Applied Physiology, 117*(1), 4–5.

Rizzolatti, G., & Sinigaglia, C. (2016). *The mirror mechanism: A basic principle of brain function*. Oxford University Press.

Sawka, M. N., Burke, L. M., & Eichner, E. R. (2015). Exercise and fluid replacement. *Journal of Medicine & Science in Sports & Exercise, 47*(2), 99–110.

Siebert, T., Till, O., & Blickhan, R. (2019). Muscle force generation and economy during locomotion: Insights from simulations and experiments. *Journal of Biomechanics, 92*(5), 112–120.

Sparrow, W. A., Begg, R. K., & Parker, S. (2021). Energetics of work and movement: Impact on fatigue and injury. *Journal of Applied Physiology, 120*(6), 1003–1015.

Tipton, K. D., & Wolfe, R. R. (2018). Protein and amino acids for athletes. In *Nutritional strategies to enhance performance* (pp. 191–207). Routledge.

Waersted, M. (2014). Biomechanical exposure in relation to musculoskeletal disorders in occupational settings. *Occupational Medicine, 55*(3), 156–157.

5 Anthropometry

5.1 LEARNING GOALS

The objective of this chapter is to introduce the study of anthropometrics and teach principles that can be used to support anthropometric design. The process for data collection, calculation of statistics and identification of appropriate data sources will be covered.

5.2 KEY TOPICS

- History of anthropometry
- Anthropometric design principles
- Principles of universal design
- The myth of the "average" human
- Anthropometric terminology
- Anthropometric measurement
- Static and dynamic dimensions
- Variability
- Anthropometric data
- Anthropometric tools
- Anthropometric data sources
- Design using anthropometric data
- Anthropometrics and biomechanics
- Anthropometric application

5.3 INTRODUCTION AND BACKGROUND

Ergonomics, as a discipline, has significantly evolved from its early focus on fitting individuals to tasks through training and selection procedures. Historically, pre-World War II ergonomics prioritized the adjustment of workers to their environments, largely by selecting individuals who were most suited to the demands of specific jobs (McCormick, 2014). However, the complexities introduced during wartime efforts, especially in operating intricate technologies, underscored the limitations of this approach. Despite comprehensive training, even the most proficient personnel struggled with the effective operation of highly complex systems. This led to the fundamental shift in ergonomics that we recognize today: the design of work tasks, tools, and environments should be aligned to fit the human user, not the other way around (Wilson, 2017).

DOI: 10.1201/9780429342806-5

This shift was critical in the post-war era, often referred to as the "knob and dial" era, which spurred extensive research into human-machine interaction, particularly in optimizing control designs and visual displays. Researchers began to rigorously study human capabilities and limitations, acknowledging that successful task performance is inherently constrained by human physical, cognitive, and sensory factors (Dul & Weerdmeester, 2018).

5.3.1 Anthropometry and Its Role in Ergonomics

Anthropometry, the scientific study of the measurements and proportions of the human body, is central to modern ergonomics. Its application ensures that equipment, workspaces, and tasks are designed to accommodate a diverse range of users. The term "anthropometry" is derived from the Greek words *anthropos* (human) and *metron* (measure), and its roots can be traced back to the late 19th century, when it was first used by Alphonse Bertillon in criminal identification (Pheasant & Haslegrave, 2016). While anthropometry's origins were in criminal justice, its transition into occupational ergonomics began in earnest during the 20th century, as designers and engineers recognized its importance in creating human-centered designs.

Today, anthropometry informs ergonomic practices across a wide array of industries. In occupational settings, anthropometric data is used to optimize the design of tools, workstations, and personal protective equipment (PPE). This ensures that these items not only enhance productivity but also reduce the risk of injury by fitting the natural physical dimensions of the workers. Studies indicate that poorly designed work environments, which fail to account for human variability in body dimensions, can lead to increased musculoskeletal disorders (MSDs) (Jung et al., 2015). For instance, improperly designed seating or tools that do not consider the range of hand sizes within a population can cause discomfort, reduced efficiency, and even long-term health issues.

5.3.2 The Importance of Inclusive Design

One of the key benefits of anthropometric data is its ability to support inclusive design. In ergonomics, inclusive design refers to the approach of creating products, environments, and systems that are accessible, usable, and beneficial to as many people as possible, regardless of age, ability, gender, or other individual differences. This approach emphasizes designing with the full range of human diversity in mind, aiming to reduce barriers and enhance usability across a spectrum of physical, cognitive, and sensory abilities.

The variation in human body dimensions across populations due to factors such as gender, ethnicity, and age means that a "one-size-fits-all" approach is no longer viable in modern occupational design. Ergonomics must account for this diversity to create workspaces that are accessible and efficient for all users (Stewart et al., 2014). For instance, the implementation of adjustable workstations allows individuals of different heights and body types to work comfortably and efficiently, reducing the physical strain that might otherwise result from poorly fitted environments.

Anthropometry also plays a pivotal role in designing for populations with specific needs, such as those with disabilities. In fields such as prosthetics and orthotics, for example, customized designs based on individual anthropometric data are essential for both the functionality and comfort of users (Garneau & Igarashi, 2020). These personalized solutions are made more accessible through technological advancements like 3D printing, which allows for the rapid and cost-effective production of custom products tailored to individual body measurements.

5.3.3 Technological Advancements in Anthropometry

In recent years, technological innovations have revolutionized how anthropometric data is collected and applied. Traditionally, anthropometry involved labor-intensive methods, requiring extensive physical measurements of human subjects. However, the development of digital and virtual tools has transformed this process. For example, Virtual Reality (VR) technologies now allow for the creation of digital models that represent a wide range of anthropometric characteristics. This capability enables designers to simulate how different body types will interact with various environments and products, allowing for more inclusive and user-centered design processes (Duffy et al., 2015).

Similarly, 3D scanning technology has enhanced the precision and efficiency of anthropometric data collection. With the ability to quickly and accurately capture detailed body measurements, 3D scanning is particularly valuable in the design of ergonomic equipment, personal protective equipment, and workstations. It ensures that products can be tailored to meet the specific dimensions of individual users or population groups. When combined with 3D printing technology, this data allows for the rapid prototyping and production of ergonomically optimized tools and equipment, enhancing both comfort and performance (Guerra et al., 2017).

5.3.4 The Future of Ergonomics and Anthropometry

The integration of anthropometry with emerging technologies is shaping the future of ergonomic design. As workplaces continue to evolve in response to new demands—such as the increase in remote work, aging workforces, and the growing emphasis on workplace health and safety—ergonomists will rely on anthropometric data more than ever to create environments that prioritize worker well-being and productivity (Chaffin et al., 2019). The future of occupational ergonomics will not only involve the refinement of current practices but will also leverage new technologies to further enhance the inclusivity and adaptability of workplace designs.

5.4 WHAT IS ANTHROPOMETRY?

Anthropometry is the scientific study of human body measurements, encompassing not only height and weight but also more detailed physical dimensions such as size, shape, mass, depth, and range of motion. This field is integral to ergonomics, which seeks to create products and environments that accommodate human needs and promote health, safety, and productivity in the workplace (McCormick, 2014). By aligning design with the physical characteristics of users, anthropometry helps prevent fatigue, discomfort, and injury, which are common outcomes of poorly designed workstations and tools (Dul & Weerdmeester, 2018).

5.4.1 Anthropometry: Static and Dynamic Measurements

Anthropometry is generally divided into two main branches: static and dynamic. Static anthropometry focuses on measurements of the body in a stationary position, such as the distance between the shoulders, the length of the forearms, or the circumference of the head. These measurements are critical for designing furniture, tools, and personal protective equipment, ensuring they fit users comfortably when they are not in motion (Cakir et al., 2014).

Dynamic anthropometry, on the other hand, examines the body's movements and functional ranges. This includes data on how far an individual can reach, how much they can bend at the waist, and the range of motion of their limbs. Dynamic anthropometry is particularly important in the design of workstations and tools that require physical activity, such as assembly lines or manufacturing environments. By understanding the limits of human movement, designers can create tools and workspaces that optimize efficiency while minimizing the risk of injury (Stewart et al., 2014).

For example, in the design of industrial workstations, understanding both the static dimensions and dynamic movements of workers allows for the creation of environments that reduce strain on the body. Research has shown that workstations designed with anthropometric data can reduce the risk of musculoskeletal disorders, a leading cause of workplace injuries (Jung et al., 2015).

5.4.2 Applications in Workplace Design and Safety

The application of anthropometry in ergonomics extends beyond comfort; it has significant implications for workplace safety, efficiency, and productivity. For instance, the design of factory workstations based on anthropometric data can reduce the likelihood of injury by ensuring that tools are placed within easy reach, reducing the need for awkward postures or repetitive movements (Wilson, 2017). By designing workspaces that align with natural body movements, the risk of cumulative trauma disorders (CTDs) can be minimized, resulting in lower absenteeism and improved worker performance (Chaffin et al., 2019).

In the office environment, anthropometric data informs the design of ergonomic chairs, desks, and computer setups. Adjustable chairs that cater to different body types, desks with customizable heights, and computer monitors placed at optimal viewing angles are examples of anthropometric principles in action. These designs help maintain good posture, reduce eye strain, and prevent repetitive strain injuries (RSIs), ultimately boosting productivity and job satisfaction (McCormick, 2014).

The US military extensively utilizes anthropometric data to design equipment, clothing, and vehicles that accommodate soldiers of various body shapes and sizes.

Ensuring that gear fits properly is critical for both safety and performance, as ill-fitting equipment can impede movement and cause fatigue (Gordon et al., 2016).

In healthcare, anthropometry plays a critical role in the design of medical devices, including prosthetics and orthotics. Prosthetics tailored to an individual's unique body measurements not only improve comfort but also enhance functionality and mobility, leading to better patient outcomes (Garneau & Igarashi, 2020). In hospitals, anthropometry is used to design patient beds, examination tables, and surgical tools that optimize both patient comfort and clinician effectiveness.

5.4.3 The Importance of Inclusive Design

Inclusive design, a concept closely tied to anthropometry, aims to create products and environments that accommodate the full range of human diversity. This includes variations in body size, shape, age, gender, and physical abilities. Designing for inclusivity ensures that products are accessible to as many users as possible, reducing the risk of discomfort or injury for groups that may not conform to the "average" body type (Pheasant & Haslegrave, 2016). For example, the use of adjustable desks and chairs in office environments ensures that both shorter and taller individuals can work comfortably and efficiently.

Without thc application of anthropometric data, the consequences can be severe. Poorly designed workstations can lead to musculoskeletal issues, increased fatigue, and a decline in productivity. Research has shown that ergonomic designs based on accurate anthropometric data can significantly reduce the prevalence of workplace injuries and improve overall worker well-being (Dul & Weerdmeester, 2018).

5.4.4 Anthropometric Design Principles

There are three key principles to consider when using anthropometric data in design. These principles can be combined in a single project, and the importance of each will vary depending on factors such as who will be using the design, the potential consequences of not considering diverse body sizes, and the overall cost of the design.

5.4.5 General Guidelines

- Design for Adjustable Range
    - Provide a wide range of adjustments, particularly for health and safety-critical applications (e.g., driving a car, working in an office).
    - Increase the adjustability to prevent excluding sections of the user population due to design limitations.
- Design for Extremes
    - Design for the maximum value of a feature (e.g., height of doors, size of buttons).
    - Design for the minimum value of a feature (e.g., distance from controls).
    - Traditionally addresses extremes by designing for:

    - 5th percentile female
    - 95th percentile male
- Design for Average User
    - Use 50th percentile figures for relevant dimensions.
    - Suitable when focusing on one dimension and health and safety are not significant issues (e.g., auditorium seats).
    - Legislation, such as the Americans with Disabilities Act, requires additional accommodations in specific environments.

5.4.6 Principles of Universal Design

Universal design involves creating products, buildings, and environments that are accessible and usable by the widest range of people, without the need for adaptation or specialized design. Originating from a collaborative effort by architects, designers, engineers, and researchers, universal design aims to integrate inclusivity into every aspect of environmental, product, and communication design. The Center for Universal Design at North Carolina State University outlines seven key principles that guide this approach, useful not only in the creation of new designs but also in evaluating existing ones and developing educational materials.

5.4.6.1 Principle 1: Equitable Use

Designs should be appealing and usable to people with diverse abilities.

- Ensure consistent user experience where feasible.
- Prevent segregation or stigmatization of any users.
- Equal access to privacy, security, and safety.
- Attractive design for all users.

5.4.6.2 Principle 2: Flexibility in Use

Designs accommodate a range of preferences and abilities.

- Offer choices in methods of use.
- Accessible to right or left-handed individuals.
- Support user accuracy and adjust to their pace.

5.4.6.3 Principle 3: Simple and Intuitive Use

Ease of use should be evident, regardless of the user's skills or concentration level.

- Eliminate unnecessary complexity.
- Align with user expectations and intuitive behaviors.
- Support varying literacy and language skills.
- Organize information logically with effective prompts.

5.4.6.4 Principle 4: Perceptible Information

Communication of information must be effective under all conditions.

- Multiple modes (visual, verbal, tactile) for essential information.
- High contrast for essential versus supplementary information.
- Ensure information is legible and distinguishable.
- Compatible with assistive devices used by people with sensory limitations.

5.4.6.5 Principle 5: Tolerance for Error

Designs minimize risks and adverse outcomes of unintended actions.

- Optimize elements to reduce hazards.
- Provide clear warnings and fail-safe features.
- Design tasks requiring alertness to prevent inadvertent actions.

5.4.6.6 Principle 6: Low Physical Effort

Use of the design should be efficient, comfortable, and fatigue-minimizing.

- Maintain a neutral body posture.
- Reasonable operating forces and minimal repetitive actions.
- Reduce sustained physical effort.

5.4.6.7 Principle 7: Size and Space for Approach and Use

Ensure accessible size and space for everyone, regardless of physical stature or mobility.

- Clear visibility and comfortable reach for all users.
- Accommodate varying hand sizes and grip strengths.
- Ample space for assistive devices or personal assistance.

5.4.7 Myth of the Average Human

In ergonomics, the concept of an "average user" is often referenced; however, this notion is fundamentally flawed. No individual embodies average dimensions across all body parts, as human body proportions do not conform to a linear model. For example, a person with long arms may not necessarily have a similarly proportioned neck or legs. Consequently, rather than designing for an idealized "average person," ergonomists should consider a broader range of body sizes and proportions. Employing percentiles—such as the 5th through the 95th percentile—enables designers to create products and environments that accommodate the majority of the population, ensuring usability and comfort across diverse users.

5.5 ANTHROPOMETRIC TERMINOLOGY

Before exploring the methods used to obtain anthropometric measurements, it is essential to clarify several key terms. A solid understanding of these terms will provide the foundation needed to comprehend the traditional methods and definitions associated with anthropometric analysis.

5.5.1 Definitions of Measurements

- **Height** and **breadth** are straight-line measurements in opposite directions. Height goes up and down, while breadth goes side-to-side.
- **Depth** is similar to breadth but goes front-to-back.
- **Distance** is a general term for any straight-line measurement between two points on the body.
- **Curvature** follows a bend or curve on the body, not necessarily a perfect circle.
- **Circumference** is the total distance around a curved part of the body.
- **Reach** is the farthest distance an arm or leg can extend when stretched out.

5.5.2 Anthropometric Planes

In anthropometry, we use three main imaginary planes to measure the human body: sagittal, coronal, and transverse.

- The sagittal plane splits the body into left and right sides. A specific midline sagittal plane, called the median plane, divides the body perfectly in half. We use terms like medial and lateral when referring to this plane. Medial means closer to the midline, while lateral means further away.
- The coronal plane, also known as the frontal plane, cuts the body into front (anterior) and back (posterior) sections. These sections can be equal or unequal.
- The transverse plane, also called the cross-sectional plane, is a horizontal plane that divides the body into upper (cranial) and lower (caudal) portions. These sections are typically unequal.

Human body movements are complex and highly coordinated, and understanding these movements is essential to analyzing human functionality. The following provides a detailed overview of primary movements within various anatomical planes, along with additional relevant information.

Sagittal Plane (divides the body into left and right sides):

- **Flexion:** This bending motion decreases the angle between two body parts, like bending your elbow to touch your shoulder.
- **Extension:** The opposite of flexion, extension increases the angle between body parts, like straightening your arm from a bent elbow.
- **Dorsiflexion:** This movement brings the top of your foot closer to your shin, like pulling your toes up toward your knee.
- **Plantarflexion:** Pointing your toes down is plantarflexion, which moves the top of your foot away from your shin.

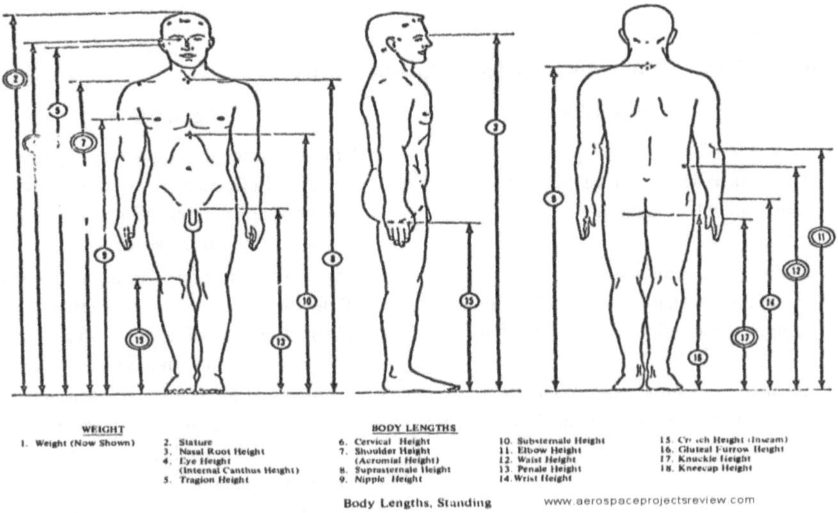

FIGURE 5.1 Anthropometric diagram of body lengths in standing position.

Coronal Plane (divides the body into front and back sides):

- **Adduction:** Imagine bringing your arms to your sides – that's adduction, movement toward your body's midline.
- **Abduction:** The opposite of adduction, abduction is moving a limb away from your midline, like raising your arm out to the side.
- **Inversion:** This movement lifts the inner edge of your foot toward your midline, like turning your foot inwards.
- **Eversion:** Eversion lifts the outer edge of your foot away from your midline, like turning your foot outwards.
- **Elevation:** This refers to the upward movement of your shoulder blade, like shrugging your shoulders.
- **Depression:** The opposite of elevation, depression moves your shoulder blade down, like sinking your shoulders.

Transverse Plane (divides the body into upper and lower sections):

- **Rotation:** This twisting motion turns a body part around a fixed point, like turning your head to look over your shoulder.
- **Pronation**: This rotation turns your forearm so your palm faces down, like turning a screwdriver.
- **Supination:** Supination rotates your forearm so your palm faces up, like the motion of unscrewing a lid.

TABLE 5.1
Common Body Measures and Their Applications

	Dimensions	Applications
1	*Stature* The vertical distance from the floor to the top of the head, when standing	A main measure for comparing population samples. Reference for the minimal height of overhead obstructions Add height for more clearance, hat, shoes, stride
2	*Eye height, standing* The vertical distance from the floor to the outer comer of the right eye, when standing	Origin of the visual field of a standing person. Reference for the location of visual obstructions and of targets such as displays; consider slump and motion
3	*Shoulder height (acromion), standing* The vertical distance from the floor to the tip (acromion) of the shoulder, when standing [2]	Starting point for arm length measurements; near the center of rotation of the upper arm Reference point for hand reaches; consider slump and motion
4	*Elbow height, standing* The vertical distance from the floor to the lowest point of the right elbow, when standing, with the elbow flexed at 90°	Reference for height and distance of the work area of the hand and the location of controls and fixtures; consider slump and motion
5	*Hip height (trochanter), standing* The vertical distance from the floor to the trochanter landmark on the upper side of the right thigh, when standing	Traditional anthropometric measure, indicator of leg length and the height of the hip joint. Used for comparing population samples
6	*Knuckle height, standing* The vertical distance from the floor to the knuckle (metacarpal bone) of the middle finger of the right hand, when standing	Reference for low locations of controls, handles, and handrails; consider slump and motion of the standing person
7	*Fingertip height, standing* The vertical distance from the floor to the tip of the extended index finger of the right hand, when standing	Reference for the lowest location of controls, handles, and handrails; consider slump and motion of the standing person
8	*Sitting height* The vertical distance from the sitting surface to the top of the head, when sitting	Reference for the minimal height of overhead obstructions Add height for more clearance, hat, trunk motion of the seated person
9	*Sitting eye height* The vertical distance from the sitting surface to the outer comer of the right eye, when sitting	Origin of the visual field of a seated person Reference point for the location of visual obstructions and of targets such as displays; consider slump and motion
10	*Sitting shoulder height (acromion)* The vertical distance from the sitting surface to the tip (acromion) of the shoulder, when sitting	Starting point for arm length measurements; near the center of rotation of the upper arm Reference for hand reaches; consider slump and motion

(*Continued*)

TABLE 5.1 (CONTINUED)
Common Body Measures and Their Applications

	Dimensions	Applications
11	*Sitting elbow height* The vertical distance from the sitting surface to the lowest point of the right elbow, when sitting, with the elbow flexed at 90°	Reference for the height of an armrest, of the work area of the hand, and of keyboard and controls; consider slump and motion of the seated person
12	*Sitting thigh height (clearance)* The vertical distance from the sitting surface to the highest point on the top of the horizontal right thigh, with the knee flexed at 90°	Reference for the minimal clearance needed between seat pan and the underside of a structure, such as a table or desk; add clearance for clothing and motions
13	*Sitting knee height* The vertical distance from the floor to the top of the right kneecap, when sitting, with the knees flexed at 90°	Traditional anthropometric measure for lower leg length Reference for the minimal clearance needed below the underside of a structure, such as a table or desk; add height for shoe
14	*Sitting popliteal height* The vertical distance from the floor to the underside of the thigh directly behind the right knee; when sitting, with the knees flexed at 90°	Reference for the height of a seat; add height for shoe
15	*Shoulder–elbow length* The vertical distance from the underside of the right elbow to the right acromion, with the elbow flexed at 90° and the upper arm hanging vertically	Traditional anthropometric measure for comparing population samples
16	*Elbow–fingertip length* The distance from the back of the right elbow to the tip of the extended middle finger, with the elbow flexed at 90°	Traditional anthropometric measure Reference for fingertip reach when moving the forearm in the elbow
17	*Overhead grip reach, sitting* The vertical distance from the sitting surface to the center of a cylindrical rod firmly held in the palm of the right hand	Reference for the height of overhead controls operated by a seated person. Consider ease of motion, reach, and finger/hand/arm strength
18	*Overhead grip reach, standing* The vertical distance from the standing surface to the center of a cylindrical rod firmly held in the palm of the right hand	Reference for the height of overhead controls operated by a standing person. Add shoe height Consider ease of motion, reach, and finger/hand/arm strength
19	*Forward grip reach* The horizontal distance from the back of the right shoulder blade to the center of a cylindrical rod firmly held in the palm of the right hand	Reference for forward reach distance Consider ease of motion, reach, and finger/hand/arm strength
20	*Arm length, vertical* The vertical distance from the tip of the right middle finger to the right acromion, with the arm hanging vertically	A traditional measure for comparing population samples Reference for the location of controls very low on the side of the operator Consider ease of motion, reach, and finger/hand/arm strength

(*Continued*)

TABLE 5.1 (CONTINUED)
Common Body Measures and Their Applications

	Dimensions	Applications
21	*Downward grip reach* The vertical distance from the right acromion to the center of a cylindrical rod firmly held in the palm of the right hand, with the arm hanging vertically	Reference for the location of controls low on the side of the operator Consider ease of motion, reach, and finger/hand/arm strength
22	*Chest depth* The horizontal distance from the back to the right nipple	A traditional measure for comparing population samples. Reference for the clearance between seat backrest and the location of obstructions in front of the trunk
23	*Abdominal depth, sitting* The horizontal distance from the back to the most protruding point on the abdomen	A traditional measure for comparing population samples Reference for the clearance between seat backrest and the location of obstructions in front of the trunk
24	*Buttock–knee depth, sitting* The horizontal distance from the back of the buttocks to the most protruding point on the right knee, when sitting with the knees flexed at 90°	Reference for the clearance between seat backrest and the location of obstructions in front of the knees
25	*Buttock–popliteal depth, sitting* The horizontal distance from the back of the buttocks to back of the right knee just below the thigh, when sitting with the knees flexed at 90°	Reference for the depth of a seat
26	*Shoulder breadth (biacromial)* The distance between the right and left acromion	A traditional measure for comparing population samples. Indicator of the distance between the centers of rotation of the two upper arms
27	*Shoulder breadth (bideltoid)* The maximal horizontal breadth across the shoulders between the lateral margins of the right and left deltoid muscles	Reference for the lateral clearance required at shoulder level Add space for ease of motion and tool use
28	*Hip breadth, sitting* The maximal horizontal breadth across the hips or thighs, whatever is greater, when sitting	Reference for seat width Add space for clothing and ease of motion
29	*Span* The distance between the tips of the middle fingers of the horizontally outstretched arms and hands	A traditional measure for comparing population samples Reference for sideway reach
30	*Elbow span* The distance between the tips of the elbows of the horizontally outstretched upper arms when the elbows are flexed so that the fingertips of the hands meet in front of the trunk	Reference for the lateral space needed at upper body level for ease of motion and tool use

(*Continued*)

TABLE 5.1 (CONTINUED)
Common Body Measures and Their Applications

	Dimensions	Applications
31	*Head length* The distance from the glabella (between the browridges) to the most rearward protrusion (the occiput) on the back, in the middle of the skull	A traditional measure for comparing population samples Reference for headgear size
32	*Head breadth* The maximal horizontal breadth of the head above the attachment of the ears	A traditional measure for comparing population samples. Reference for headgear size
33	*Hand length* The length of the right hand between the crease of the wrist and the tip of the middle finger, with the hand flat	A traditional measure for comparing population samples Reference for hand tool and gear size. Consider manipulations, gloves, tool use
34	*Hand breadth* The breadth of the right hand across the knuckles of the four fingers	A traditional measure for comparing population samples Reference for hand tool and gear size, and for the opening through which a hand may fit Consider manipulations, gloves, tool use
35	*Foot length* The maximal length of the right foot, when standing	A traditional measure for comparing population samples Reference for shoe and pedal size
36	*Foot breadth* The maximal breadth of the right foot, at right angle to the long axis of the foot, when standing	A traditional measure for comparing population samples Reference for shoe size, spacing of pedals
37	*Weight (kg)* Nude body weight taken to the nearest tenth of a kilogram	A traditional measure for comparing population samples Reference for body size, clothing, strength, health, etc. Add weight for clothing and equipment worn on the body

Source: Adapted from Kroemer, K. H. E., *The Occupational Ergonomics Handbook*, Taylor & Francis, Boca Raton, FL, 2006.

Note: Descriptions of dimensions from Gordon et al. (2016) [26] with their reference numbers in brackets.

5.5.3 Anthropometric Measurement

As described, anthropometry involves obtaining various measurements of the human body to analyze physical characteristics and functional capabilities. These measurements extend beyond standard indicators such as height and weight and can encompass dimensions such as:

- **Size and Shape:** This encompasses volume and density of body parts.
- **Flexibility:** This refers to the range of motion in joints.

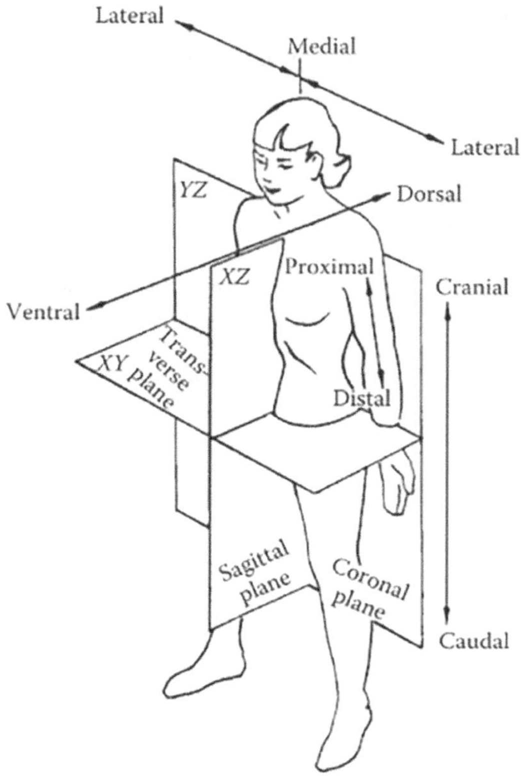

FIGURE 5.2 Provides an example of stature and seated measurement for American male.

- **Strength:** This involves measuring how much force a person can exert.
- **Body Layout:** This includes distances between different anatomical landmarks on the body (like the distance between your shoulder and elbow).

Interestingly, anthropometry can also extend to how well we sense the world around us, including our hearing, vision, and touch.

5.5.3.1 Stature Measurements

Accurate measurement of stature is fundamental across various fields, including healthcare, ergonomics, anthropology, and sports science. However, obtaining precise measurements requires more than simply using a measuring tape; the posture of the individual being measured significantly influences the results, as even minor variations can lead to discrepancies. This section examines the intricacies of stature measurement, emphasizing the impact of posture, timing, and other critical factors essential for achieving optimal accuracy.

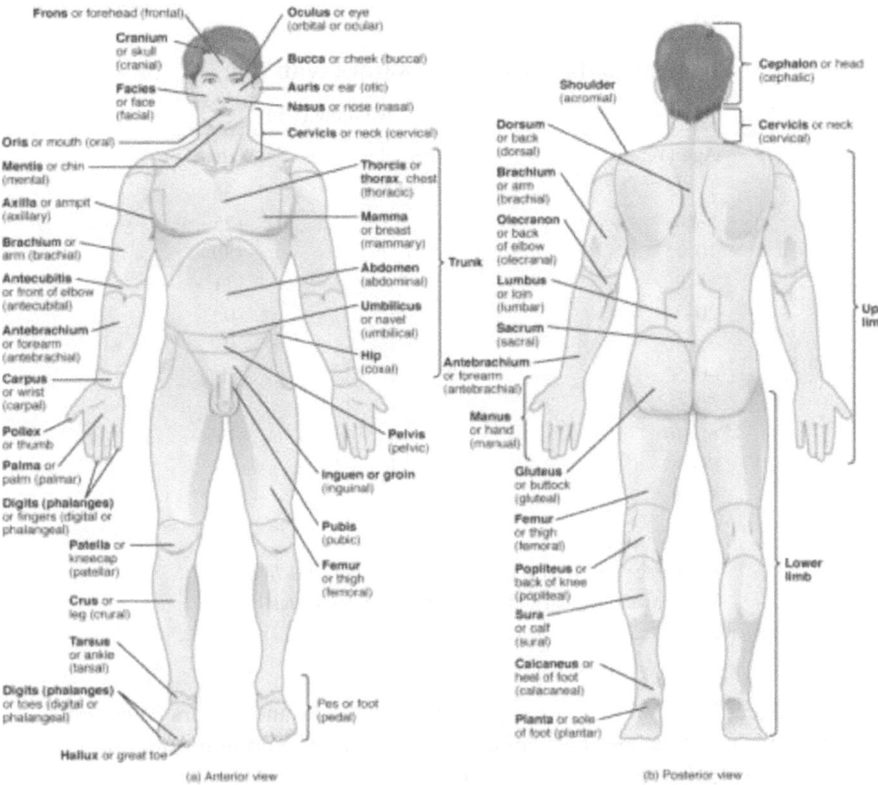

FIGURE 5.3 Illustrates anatomical reference points for accurate stature measurement from both anterior and posterior views.

5.5.3.2 The Postural Play

The human body is not a rigid structure; daily activities and even the act of standing can affect posture. The following provides an overview of common measurement positions and their respective implications for accuracy and consistency in anthropometric assessments.

- **Natural Standing:** This reflects our everyday posture, but it might underestimate height slightly due to the natural inward curve of the spine (kyphosis) in the lower back.
- **Erect Standing:** Aiming for a perfectly straight posture can increase readings by up to 2 cm. However, achieving consistent "perfect" posture across individuals can be challenging, introducing variability into the measurements.
- **Wall-Supported Standing:** This method promotes a straighter posture by encouraging the subject to stand with their shoulder blades, buttocks, and back of the head firmly against a wall. To further enhance consistency, a

TABLE 5.2
Approximate Changes in Stature with Age

	Change (cm)	
Age (Years)	**Females**	**Males**
1–5[a]	+36	+36
5–10	+27	+25
10–15	+22	+30
15–20	+7	+11
20–35[b]	–1	0
35–40	–1	0
40–50	–1	–1
50–60	–1	–1
60–70	–1	–1
70–80	–1	–1
80–90	–1	–1

Source: Kroemer et al., 1994.

[a] Average stature at age 1: females 72 cm, males 74 cm.

[b] Average maximal stature: females 163 cm, males 176 cm.

flat object like a book can be placed on the head and a mark made on the wall where it touches. This creates a clear reference point for measurement, minimizing discrepancies between subjects.

- **Supine Position:** Lying flat on one's back yields the tallest stature measurement. Gravity's influence on spinal compression is minimized in this position. However, it's not a practical posture for everyday comparisons and isn't representative of our typical height.

5.5.3.3 Seated Measurements

Accurate measurement of an individual in a seated position requires attention to detail to ensure that the recorded data accurately reflects their true body dimensions. The following is a comprehensive guide to achieving a standardized seated posture, facilitating optimal measurement accuracy and consistency.

5.5.4 Creating a Stable Foundation

- **The Chair:** Select a firm chair with a flat seat surface for the subject to sit on. Avoid chairs with cushions that may compress or rounded seats that can cause the subject to roll forward. Ideally, the chair should be adjustable to allow for minor height modifications if needed.
- **Seat Level:** Ensure the chair is positioned so the seat is parallel to the floor. This creates a crucial 90° angle between the subject's torso and thighs.

Imagine a straight line running along the subject's back and thighs—in an ideal seated position, this line should be perpendicular to the floor.

5.5.5 Optimizing Leg Posture

- **Knee Angle:** Ask the subject to bend their knees at a 90° angle. A simple way to achieve this is to have them sit with their heels comfortably touching the floor. If their feet dangle because the chair is too high, provide a sturdy box or footrest to ensure their feet are flat and supported. This prevents unnecessary strain on the hamstrings and ensures consistent leg positioning for measurement purposes.
- **Lower Leg Alignment:** With the feet flat on the floor, the subject's lower legs (from the knee down) should be vertical. This position helps maintain proper blood circulation and prevents ankle strain during the measurement process.

5.5.6 Maintaining Proper Body Alignment

- **Posture:** Instruct the subject to relax their shoulders and maintain a straight, upright posture throughout the measurement process. Avoid slouching or leaning forward, as this can significantly alter the measurements, particularly for the torso and hip width. Imagine a straight line running along the subject's earlobe, shoulder, and hip—in proper posture, these points should be roughly aligned.
- **Armrests:** If the chair has armrests, advise the subject to keep their arms relaxed at their sides or gently resting on the armrests without pushing their shoulders up. This helps maintain a natural shoulder position for accurate upper body measurements.

5.5.7 Measurements of Physical Properties of Body Segments

Biomechanical analysis treats the human body as a complex machine. Static and dynamic anthropometric measurements provide key data for this analysis, by measuring and quantifying various physical aspects. These measurements include:

- Lengths of limbs and segments
- Body volume and composition
- Weight distribution and center of mass
- Inertia of different body parts

This detailed anthropometric data is crucial not only used in biomechanics, but also in ergonomic design. By understanding the body's dimensions and mechanical properties, designers can create tools, furniture, and workplaces that are comfortable, safe, and optimized for human use.

5.5.7.1 Variability

While collecting anthropometric data, some variation is inevitable. However, for subjects from a similar background, the data should generally follow a normal distribution pattern, reflecting consistency with expected anthropometric norms. It's important to not only acknowledge this variability but also understand and explain its sources. Here are three main types of variation to consider:

1. **Individual Fluctuations (Intra-Individual Variability):**
   - This refers to how a single person's measurements change over time due to factors like age, health, and diet.
   - Longitudinal studies, which track the same individuals repeatedly, are ideal for capturing these changes.
2. **Natural Differences Between People (Inter-Individual Variability):**
   - This inherent variation exists between people from the get-go, reflecting their unique biological makeup.
   - Cross-sectional studies, where measurements are taken from different individuals at a single point in time, effectively assess this type of variability.
3. **Generational Shifts (Secular Variability):**
   - This highlights the trend of people becoming larger over time compared to past generations.
   - This is evident when comparing anthropometric data from different eras.

5.5.7.2 Additional Considerations

- **Participant Selection:** In some cases, variability might stem from participants not belonging to the target population.
- **Measurement Error:** Inconsistent measuring techniques or equipment can also introduce variation. To minimize this, taking multiple readings (at least 3) for each measurement is crucial.

5.5.7.3 Data Analysis

When analyzing anthropometric data, it is important to calculate descriptive statistics that summarize the central tendency (average), variation (spread), and identify outliers (extreme values) within the dataset.

5.5.8 Anthropometric Tools

5.5.8.1 Automated Anthropometric Measurement Tools

Anthropometric measurement, a fundamental component of ergonomics, has historically relied on manual methods to collect data on human body dimensions. However, recent technological advancements have facilitated a shift toward automated measurement systems, significantly improving both the precision and efficiency of data collection while reducing human error. These advancements are crucial in occupational ergonomics, where accurate anthropometric data is essential for designing work environments, tools, and protective equipment that accommodate diverse human body dimensions (Dul & Weerdmeester, 2018).

One notable innovation in the field of anthropometry is the Anthropometric Measurement System (AMS) developed by Ergonomic Technologies. The AMS is an electro-mechanical system designed to capture key body dimensions with high accuracy. Using sensors and automated measurement techniques, the AMS electronically transmits the collected data to specialized software, which then processes these measurements to make recommendations, such as appropriate clothing sizes. This system has proven particularly useful in the military, where the proper allocation of uniforms and gear based on individual body measurements is critical for ensuring both comfort and safety (McFarland & Moore, 2015).

The application of the AMS in military contexts exemplifies the importance of automated anthropometric systems in environments where precision is paramount. Ill-fitting uniforms and equipment can lead to a range of problems, from discomfort and distraction to reduced mobility and performance, especially in high-stakes settings like combat (Gordon et al., 2016). By providing real-time, accurate body measurements, the AMS helps optimize the fit and function of garments and equipment, reducing these risks and enhancing operational efficiency.

Beyond the AMS, other automated systems are being developed to further enhance anthropometric data collection. For instance, the Automated Anthropometric Data Measurement System (AADMS), discussed in a report by the Human Systems Integration Information Analysis Center (HSIIAC), is another advanced tool aimed at improving the collection and application of body dimension data. The AADMS integrates sophisticated sensors and data processing algorithms to quickly and accurately measure a range of human body dimensions, reducing the need for manual data entry and minimizing errors (HSIIAC), (2011).

The transition from manual to automated anthropometric measurement systems represents a significant advancement in the field of ergonomics. Manual methods, while valuable, are often time-consuming and prone to human error, which can compromise the reliability of the data. Automated systems, on the other hand, provide faster, more consistent measurements, ensuring that the data used in ergonomic design is both accurate and reproducible (Pheasant & Haslegrave, 2016). This is particularly important in industries where slight variations in measurements can have significant implications for safety and performance, such as aerospace, healthcare, and industrial manufacturing.

5.5.8.2 Applications of Automated Anthropometric Systems in Occupational Ergonomics

The implications of these technologies extend beyond clothing size recommendations. In occupational ergonomics, automated anthropometric systems are being increasingly integrated into the design and optimization of workstations, tools, and personal protective equipment. For example, automated systems can measure a worker's body dimensions and dynamic movement patterns to ensure that workstations are designed to fit their body size and reduce the risk of musculoskeletal disorders (Jung et al., 2015). Automated systems also enable the real-time collection of anthropometric data for specific populations, such as workers in hazardous industries, allowing designers to create PPE that is tailored to the individual's unique body dimensions and needs (Stewart et al., 2014).

In addition to improving workplace design and safety, automated anthropometric systems are contributing to more inclusive ergonomic solutions. By capturing data from diverse populations, including variations in gender, age, and ethnicity, these systems help ensure that ergonomic designs account for a broader range of body types. This is especially important in today's global workforce, where one-size-fits-all approaches to design are increasingly recognized as inadequate (Chaffin et al., 2019). Automated systems allow for the collection of large datasets from various demographic groups, enabling the creation of inclusive designs that enhance safety, comfort, and performance for all workers (Garneau & Igarashi, 2020).

5.5.8.3 Future Directions and Challenges

While the shift toward automated anthropometric systems represents a significant step forward in the field of occupational ergonomics, there are still challenges to be addressed. The cost of implementing these systems can be prohibitive for smaller organizations, and the integration of these technologies into existing workplace practices requires careful planning and training (Guerra et al., 2017). Furthermore, as these systems become more advanced, ensuring data privacy and security will be a growing concern, particularly in industries like healthcare, where sensitive personal information is involved (Duffy et al., 2015).

Looking ahead, the continued development of automated anthropometric systems will likely involve greater integration with other emerging technologies, such as artificial intelligence (AI) and machine learning. These technologies could enable even more sophisticated data analysis, allowing for real-time adjustments to workplace designs and personalized ergonomic solutions based on ongoing anthropometric data collection (Wilson, 2017). Additionally, advancements in virtual and augmented reality may provide new opportunities for simulating and testing ergonomic designs before they are physically implemented, further enhancing the role of anthropometry in occupational ergonomics (Gordon et al., 2016).

5.5.9 Anthropometric Design Aids and Data Sources

Selecting the appropriate anthropometric data set is essential when applying population data to design tasks. Similar to selecting the correct tool for a specific task, it is critical to ensure that the data used accurately reflects the characteristics of the target user population for the final product. In the following sections, we will examine some commonly used anthropometric databases within ergonomics. However, it is important to note that numerous additional resources are available. The primary objective is to identify the data that best aligns with the specific requirements of your design.

5.5.9.1 Civilian American and European Surface Anthropometry Resource (CAESAR)

The Civilian American and European Surface Anthropometry Resource (CAESAR) project is a collaborative effort between government and industry to establish a comprehensive database of human body measurements. This valuable resource was created by scanning over 4,400 people from the United States, Canada, and Europe.

The CAESAR database is unique in its vast scope, encompassing men and women aged 18–65 and ensuring inclusion through diverse sampling methods. It considers factors such as weight, ethnicity, gender, geographic location, and socioeconomic background. The data collection, meticulously conducted between 1998 and 2000, captured three detailed body scans per participant in various postures: standing, full body exposure, and relaxed seating. Standardized procedures were implemented to guarantee the database's future expansion and updates.

The CAESAR project's findings have been leveraged to create the CAESAR product line. This line provides current and reliable anthropometric measurements, catering to the needs of today's generation. This product line is a testament to the project's far-reaching impact, bringing together researchers, designers, academics, and students from a wide range of industries, including apparel, aerospace, and automotive.

5.5.9.2 US Army Anthropometry Survey

The 1988 US Army Anthropometry Survey stands out for its comprehensiveness and data collection methods. It boasts a vast collection of detailed body measurements gathered through a meticulously designed process. While the specific population studied (active duty US Army in 1988) limits generalizability, the data shines in analyzing connections between different body measurements. However, the survey's true strength lies in its applicability to populations resembling the US Army soldiers of that era. This includes groups with similar characteristics like fitness levels, age ranges, and body weight distribution.

5.5.9.3 AnthroKids

Unlike adults, children's bodies have unique features that must be considered during product design. One key difference is their center of gravity, which sits higher due to their proportionally larger heads. Recognizing this distinction, the National Institute of Standards and Technology (NIST) undertook a project to gather anthropometric data specifically on children. This data, collected in two studies conducted in 1975 and 1977, aimed to assist designers in developing safe products for children.

Here's some additional information you might find interesting:

- Not only is the head proportionally larger in children, but their torsos are shorter and their limbs are longer in relation to their total body size compared to adults. This means that child-sized versions of adult products often aren't appropriate.
- In addition to safety, anthropometric data is important for designing children's products that are functional and comfortable. For example, knowing the average grip strength of children at different ages can help designers create tools that are easy for them to hold.

5.5.9.4 International Anthropometric Resource

Understanding human body measurements is crucial in various fields, and anthropometry provides a wealth of data for this purpose. The text mentions two resources that international ergonomists can use to access anthropometric data:

- **The World Engineering Anthropometry Resource (WEAR)** is an international effort to create a global database of anthropometric data specifically for engineering applications.
- **A reference publication by NASA** (published in 1978) contains measurements of 306 different body dimensions from 91 populations around the world. This resource is helpful not just for engineering but also for other fields that can benefit from understanding human body variations across populations.

In addition to ergonomics, anthropometry is also used in:

- **Product design** to ensure products are comfortable and functional for a wide range of users. For instance, knowing the average dimensions of hands helps designers create door handles and tools that are easy to grip for most people.
- **Clothing design** to create clothes that fit well for different body types.
- **Architecture** to design buildings and spaces that are accessible and comfortable for people of all sizes.
- **Biomechanics** to study how the human body moves and how to prevent injuries.
- **Archaeology** to understand the physique of past populations based on skeletal remains.

5.5.9.5 *The International Journal of Clothing Science and Technology*

Anthropometric data plays a crucial role in the development of clothing and footwear sizing systems, as it provides the foundational body measurements needed to design garments that accurately fit the target population. A recent large-scale study conducted in Croatia serves as an illustrative example of how this data is used to create a tailored sizing system for a specific country. This research, published in the *International Journal of Clothing Science and Technology*, represents the first comprehensive and systematic measurement of individuals across all Croatian counties, with the aim of proposing a new sizing system that better reflects the actual body dimensions of the Croatian population (Darko et al., 2006).

Historically, clothing size charts have been developed using anthropometric data. These measurements are crucial in determining the fit and functionality of garments. However, body size and shape can vary significantly between regions and populations, making it essential for countries to have their own specific anthropometric data to ensure the accuracy of their sizing systems (Pheasant & Haslegrave, 2016).

In the Croatian study, researchers measured a wide range of body dimensions, including height, chest circumference, waist circumference, and other key anthropometric variables. This data was then analyzed to develop a sizing system that aligns with the unique body characteristics of the Croatian population. The research highlighted the importance of updating national size charts to reflect the evolving body sizes and shapes within the population, which can be influenced by factors such as nutrition, lifestyle, and socioeconomic changes (Jung et al., 2015).

5.5.10 Global Applications of Anthropometric Data

While the Croatian study is specific to one country, it underscores a broader trend in the global apparel industry: the need for precise anthropometric data in the design and manufacturing of clothing for diverse populations. Companies like Humanics Ergonomics provide comprehensive anthropometric datasets for various countries around the world, offering measurements that include not only adults but also children and disabled individuals. These datasets are invaluable for companies that manufacture clothing for the global market, as they enable designers to create garments that are inclusive and appropriately sized for different populations (McFarland & Moore, 2015).

Accurate anthropometric data ensures a better fit for consumers, which is essential for both comfort and customer satisfaction. Poorly fitting garments can lead to discomfort, reduced mobility, and a negative perception of the brand. By incorporating country-specific anthropometric data, companies can design clothing that better suits the local population, leading to higher customer satisfaction and fewer returns due to sizing issues (Gordon et al., 2016).

5.5.11 The Importance of Population-Specific Sizing Systems

The Croatian anthropometric study highlights the importance of developing population-specific sizing systems. As body size can vary significantly between populations due to genetic, environmental, and lifestyle factors, relying on global or standardized sizing charts can result in poor fit and inefficiency in the clothing manufacturing process (Chaffin et al., 2019). For instance, a standardized European sizing system may not accurately reflect the body dimensions of individuals in different countries, leading to widespread issues with garment fit.

Furthermore, anthropometric data is not static. As populations evolve due to changes in diet, physical activity, and other lifestyle factors, their body sizes and shapes may shift over time. As a result, sizing systems must be regularly updated to reflect these changes. The Croatian study, by systematically measuring the population across all counties, provides a valuable blueprint for how other countries can update their own sizing systems to ensure that clothing fits the current population (Duffy et al., 2015).

5.6 PROCEDURE FOR DESIGN USING ANTHROPOMETRIC DATA

The following section outlines a procedure for utilizing anthropometric data in engineering design, occupational ergonomics or other applications.

5.6.1 Planning and Preparation

1. **Understand the Users:** Clearly define the target population you're designing for. Consider factors like age, gender, and occupation to ensure the anthropometric data is relevant.

2. **Data Availability and Validity:** Check for existing anthropometric data on your target population. If available, assess its validity and usefulness for your design decisions.
3. **Data Acquisition:** If reliable data doesn't exist, gather measurements from a representative sample population. Determine the appropriate sample size to ensure the data accurately reflects the user base.
4. **Percentile Range:** Decide on the percentile range to accommodate in your design. It's common to use the 5th to 95th percentile range for a specific gender (male or female) if the user group is predominantly one sex.
5. **Inclusivity Considerations:** While prioritizing the dominant user group is practical, consider accessibility needs for the other gender. In such cases, you might design for the full range (5th percentile female to 95th percentile male) to ensure inclusivity.

5.6.2 Design and Measurement

1. **Population Characteristics:** Determine the size and characteristics of the target population.
2. **Workspace Design:** Design workspaces and tasks with reachability in mind. The smallest person (5th percentile) should be able to reach all necessary items, and the largest person (95th percentile) should comfortably fit within the designated space.
3. **Reach and Clearance:** Use reach data (5th percentile) to design features within easy reach for everyone. Use clearance data (95th percentile) to ensure there's ample headroom and legroom to prevent people from bumping into things.
4. **Measurement Selection:** Identify the specific anthropometric measurements most relevant to your design goals.
5. **Data Collection:** Collect anthropometric data through direct measurement or by using reliable existing datasets.

5.6.3 Data Analysis and Utilization

1. **Statistical Analysis:** Employ various statistical methods to analyze and interpret the anthropometric data. This includes measures of central tendency (mean, median, mode) and variability (range, standard deviation, skewness, kurtosis). Additionally, calculate measures of relationship (correlation coefficient, regression analysis) to identify potential correlations between anthropometric data and design features.
2. **Data Comparison:** Compare your data with relevant databases to see how your target population fits within the broader anthropometric landscape.
3. **Outcome Measures:** Based on the analysis, determine the applicable anthropometric ranges and measures to be used in your design's final outcomes.

In anthropometry, confidence intervals (CIs) serve as a tool for estimating the range within which a population parameter, such as mean height, is likely to fall. These intervals are typically constructed to capture 90% or 95% of the population with a specified level of confidence. They function as a measure of reliability: a broader interval (representing a higher confidence level) provides greater assurance that the true population parameter is contained within the specified range.

CIs rely on the statistical concept known as the normal distribution, where data tends to cluster around an average value. To further define this concept consider the "empirical rule" which states the how data is described in the normal distribution:

- Roughly 68.76% of data falls within 1 standard deviation of the mean (think of it as a narrow spotlight).
- Around 95.65% falls within 2 standard deviations (a wider beam).
- And a whopping 99.73% falls within 3 standard deviations (almost everything!).

While these intervals give a general sense of population spread, constructing a CI requires the mean, standard deviation, and a desired confidence level. The following formulas demonstrate how to calculate these intervals:

- For a 95% CI (think very confident!), the formula is: 95% CI = mean ± 1.96 * standard deviation. This creates a range where we're 95% sure the true population value resides.
- For a 90% CI (still pretty confident!), the formula is: 90% CI = mean ± 1.64 * standard deviation. This offers a slightly narrower range but with a bit less certainty.

Statistical analysis goes beyond just the average. To truly understand how spread out the data is, we can calculate the coefficient of variation (CV). This statistic tells us how much the data fluctuates relative to the mean, offering valuable insights into whether the data truly follows a normal distribution.

Confidence Intervals and the coefficient of variation together provide a robust set of tools for anthropometric research. CIs indicate the likely range within which the true population value lies, while the CV measures the extent of variability in the data and assesses its conformity to a normal distribution model. Equipped with these metrics, researchers can make more reliable inferences about the population under study.

5.6.4 Anthropometrics and Biomechanics

Biomechanics extends beyond anthropometry by examining the mechanical functioning of the human body in greater depth. It analyzes the body as a system of interconnected segments, comparable to a machine with linked components. Understanding the movement of these segments necessitates accurate measurement of various parameters, including:

- **Link Lengths:** Dimensions like lengths, widths, and circumferences of each body segment.
- **Segment Mass:** Weight and volume of each segment.
- **Center of Mass (COM):** The point where the entire segment's mass is concentrated (often shown as a percentage down from the proximal end in reference materials).
- **Inertia:** A measure of the segment's resistance to changes in motion (affected by mass distribution).

To quantify these parameters, we define key anatomical landmarks. These are easily identifiable points on the body, often joint centers, that act as reference points for measurements. Linear and angular measurement tools are then used to determine the link lengths between these landmarks.

Scientists use various techniques to pinpoint the center of gravity (also called mass center) in different parts of the body. This center of gravity is the imaginary point where all the weight of a body segment seems to be concentrated.

Traditionally, researchers have relied on data from cadavers to estimate these center of gravity locations. This data expresses them as a percentage of the segment's length.

Here are some methods for finding the center of gravity in living subjects:

- **Force Platform:** The person lies on a special platform that measures forces exerted on the body. This information helps calculate the center of gravity for specific limbs.
- **Immersion Technique:** A limb is partially submerged in water. By carefully monitoring water levels as the limb is raised and lowered, scientists can identify the waterline that bisects the center of gravity.

Once the center of gravity is located, its distance from the ends of the limb segment is measured.

To perform biomechanical analysis, the body is modeled as a series of interconnected segments (links). Joint centers, which are the rotation points of these links, are crucial for taking measurements. These measurements are then used in biomechanical modeling software to assess movement and forces acting on the body.

5.6.5 Anthropometry in Application

Anthropometry, the study of human body measurements, is essential for designing various products and environments, from clothing to workplaces. However, it remains underutilized in some industries, such as the airline industry. Although research has not definitively established a link between reduced seat dimensions and safety issues, airlines may be reluctant to adjust seating due to potential financial implications. Passenger discomfort from limited seating space is typically not a primary factor in airline choice, which further discourages airlines from adopting anthropometric principles that could result in more spacious seating arrangements.

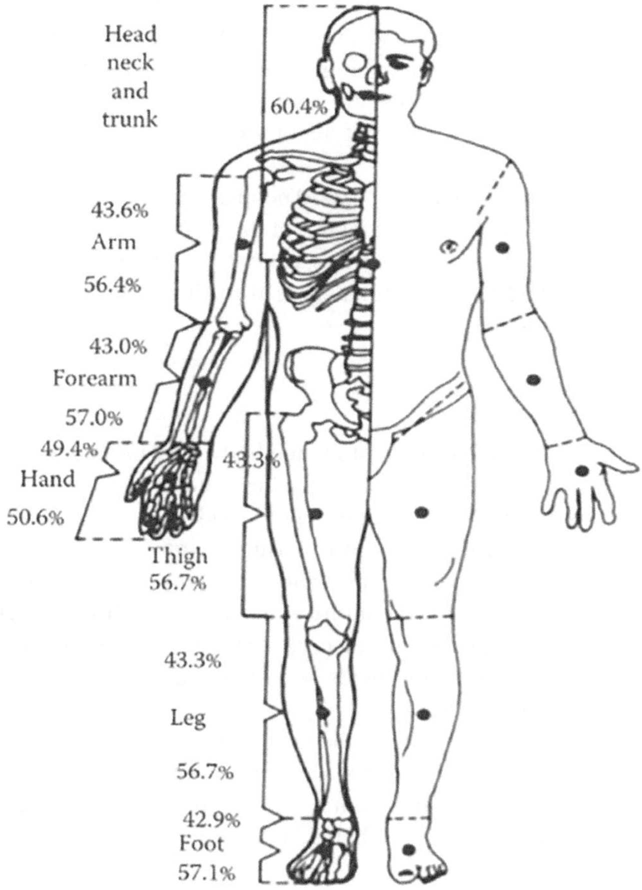

FIGURE 5.4 Provides an example of stature and seated measurement for a 40-year-old Japanese American female.

Despite these constraints, anthropometric considerations are not entirely absent. Measurements such as limb length, torso size, and hip width are integrated into certain aspects of seat design. Additionally, airlines provide options such as seat belt extenders and allow larger passengers to purchase a second seat if needed. Nonetheless, there is considerable potential for enhancing the application of anthropometric data to improve passenger comfort across a broader range of body types, creating a more inclusive flying experience.

One major problem in today's clothing industry is inconsistent sizing. People with very different body shapes can end up wearing the same size! This is because traditional sizing relies on a limited set of measurements, often outdated, that don't account for body shape variations.

The SizeUSA project is a major step forward in creating clothes that fit better. This research project used cutting-edge 3D scanning technology to capture a much more detailed picture of body size and shape in the United States. Imagine a camera system that takes over 200,000 data points from your body in under a minute! This information is then processed to create a highly accurate set of measurements that consider not just height and weight, but also things like shoulder width, waist circumference, and leg length.

The SizeUSA project scanned over 10,000 people, taking into account factors like age, ethnicity, and gender. This vast amount of data can be used by clothing manufacturers to create size charts that are more accurate and reflect the true diversity of body shapes in the US. This will ultimately lead to clothes that fit better and make people feel more confident!

5.7 SUMMARY

Anthropometrics, the scientific study of human body measurements, is fundamental to designing user-centered equipment, processes, and products. Accurate data on dimensions such as height, weight, limb length, and body segment proportions is essential for optimizing product design to ensure comfortable, safe, and effective use, thereby minimizing the risk of injury and enhancing user experience.

Beyond everyday products like furniture and clothing, anthropometrics is crucial for specialized environments. For instance, space station and spacecraft design must precisely accommodate astronauts' needs, optimizing every inch of space to ensure safety and enhance performance.

The role of anthropometrics extends beyond individual products and is a foundational aspect of ergonomics, which seeks to adapt workplaces, recreational facilities, and public spaces to human requirements. This alignment with human dimensions helps to reduce fatigue, promote well-being, and improve functionality.

Despite its benefits, product safety issues—particularly those affecting children—continue to arise, underscoring the necessity for ongoing anthropometric data collection. This need is further amplified by the global rise in obesity, which demands design adaptations to accommodate evolving body sizes.

By consistently gathering and applying anthropometric data, we can develop products and environments that harmonize with the human form, advancing safety, comfort, and functionality across a broad spectrum of applications.

5.8 CASE STUDY

Anthropometrics is the science of obtaining systematic measurements of the human body. These measurements are critical for designing products, workplaces, and environments that fit the needs of people, ensuring comfort, efficiency, and safety. In ergonomics, anthropometric data helps in the design of furniture, tools, and equipment that accommodate a broad range of users. This case study introduces the principles of anthropometric design using an office chair as an example. The process for data collection, calculation of key statistics, and the identification of appropriate data sources will be explored, illustrating how anthropometric principles can support ergonomic design.

Problem Overview

Designing an office chair that fits a wide range of users is essential for maintaining comfort and preventing injuries like musculoskeletal disorders during prolonged use. Many individuals sit for extended periods, and ill-fitting chairs can lead to discomfort and long-term health issues. The primary challenge is designing a chair that is adjustable and adaptable to the variability in human body sizes and postures, ensuring optimal support for people of different heights, weights, and body proportions.

Ergonomic Design Process

1. Data Collection Using Anthropometric Measurements

To create an office chair that fits a broad demographic, anthropometric data is collected. This data includes key body dimensions such as:

- Stature (Height)
- Sitting Height
- Popliteal Height (seat to knee)
- Hip Width
- Thigh Clearance
- Elbow Rest Height
- Weight

Anthropometric data can be sourced from national databases like the *NASA Anthropometric Source Book* or regional databases, as human body sizes can vary across populations.

2. Identifying Percentiles and Range of Users

Using the collected data, statistical analysis is performed to determine the range of user body sizes to accommodate. In ergonomic design, it is common to design for the 5th percentile female to the 95th percentile male to include the majority of users, without compromising the functionality or usability for the extremes.

For example:

- **Seat Height:** Adjustable to accommodate the 5th percentile female (shorter users) to the 95th percentile male (taller users).
- **Backrest Height:** Adjustable for the upper back, considering variations in torso length.

3. Key Anthropometric Design Considerations

Seat Height Adjustability

- The seat height must allow users' feet to rest flat on the ground, with thighs parallel to the floor. Using popliteal height measurements, the chair height can be designed to accommodate a range of 5th percentile female (approximately 38 cm) to 95th percentile male (approximately 55 cm).

Backrest and Lumbar Support

- The backrest should provide support to the lower back (lumbar region), with adjustability to fit users of various torso lengths. Anthropometric data on sitting height informs this adjustment range.

Seat Depth and Width

- Seat depth should be sufficient to support the thighs without causing pressure behind the knees. Anthropometric data on thigh length helps ensure the depth accommodates a variety of leg lengths.

Armrest Height and Width

- Adjustable armrests support the elbows without raising the shoulders or causing slumping. Elbow rest height data provides the necessary range for adjustment.

Calculation of Statistics

Once anthropometric data is collected, the design team calculates the mean, standard deviation, and percentiles to determine the range of body dimensions that need to be accommodated. For example, if designing for a range of users, the team might calculate the 5th, 50th (median), and 95th percentiles for each key measurement (e.g., seat height, seat depth).

Example

For seat height, if the 5th percentile female popliteal height is 38 cm and the 95th percentile male popliteal height is 55 cm, the chair's height adjustability must span this range. Additionally, the mean (average) might be used to set a default midpoint.

Identification of Appropriate Data Sources

Reliable anthropometric data is essential for creating effective ergonomic designs. Commonly used sources for anthropometric data include:

- **National Databases:** The US ANSUR (Army Anthropometric Survey) and CDC's National Health and Nutrition Examination Survey (NHANES) provide comprehensive anthropometric data for various populations.
- ***NASA Anthropometric Source Book***: A globally recognized source for detailed human body measurements.
- **Ergonomics Textbooks and Academic Research Papers:** For specific industry standards or niche population data.

TABLE 5.3
Anthropometric Data (cm)[a]

Measurement	Males 50th Percentile	Males ±1 SD	Females 50th Percentile	Females ±1 SD	Population Percentiles, 50/50 Males/Females 5th	50th	95th
Standing							
1. Forward functional reach							
a. Includes body depth at shoulder	82.6 (79.3)	4.8 (5.6)	74.1 (71.3)	3.9 (4.4)	69.1 (65.5)	77.9 (74.8)	88.8 (86.5)
b. Acromial process to functional pinch	63.8 (62.1)	4.3 (8.9)	62.5 (60.4)	3.4 (6.7)	57.5 (48.5)	65.0 (61.1)	74.5 (74.5)
c. Abdominal extension to functional pinch[b]	23.1	2	20.9	2.1	18.1	22	25.8
2. Abdominal extension depth							
3. Waist height	106.3 (104.8)	5.4 (6.3)	101.7 (98.5)	5.0 (5.5)	94.9 (91.0)	103.9 (101.4)	113.5 (113.0)
4. Tibial height	45.6	2.8	42	2.4	38.8	43.6	49.2
5. Knuckle height	75.5	4.1	71	4	65.7	73.2	80.9
6. Elbow height	110.5 (114.6)	4.5 (6.3)	102.6 (107.1)	4.8 (6.8)	96.4 (98.8)	106.7 (110.7)	116.3 (123.5)
7. Shoulder height	143.7 (146.4)	6.2 (7.8)	132.9 (135.3)	5.5 (6.6)	124.8 (126.6)	137.4 (140.4)	151.7 (156.4)
8. Eye height	164.4	6.1	151.4	5.6	144.2	157.7	172.3
9. Stature	174.5 (177.5)	6.6 (6.7)	162.1 (164.5)	6.0 (7.2)	154.4 (155.1)	168.0 (170.4)	183.0 (188.7)
10. Functional overhead reach	209.6	8.5	199.2	8.6	188	204.5	220.8
Seated							
11. Thigh clearance height	14.7	1.4	12.4	1.2	10.8	13.5	16.5
12. Elbow rest height	24.1	3.2	23.1	3	18.4	23.6	28.9

(Continued)

TABLE 5.3 (CONTINUED)
Anthropometric Data (cm)[a]

	Males		Females		Population Percentiles, 50/50 Males/Females		
Measurement	**50th Percentile**	**±1 SD**	**50th Percentile**	**±1 SD**	**5th**	**50th**	**95th**
13. Midshoulder height	62.4	3.2	58	2.7	54.5	60	66.5
14. Eye height	78.7	3.6	73.7	3.1	69.7	76	83.3
15. Sitting height normal	86.6	3.8	81.8	4	76.6	84.2	91.6
16. Functional overhead reach	128.4	8.5	119.8	6.6	110.6	123.6	139.3
17. Knee height	54	2.7	51	2.6	47.5	52.5	57.7
18. Popliteal height	44.6	2.5	41	1.9	38.6	42.6	47.8
19. Leg length	105.1	4.8	100.7	4.3	94.7	102.8	111.4
20. Upper-leg length	59.4	2.8	57.4	2.6	53.7	58.4	63.3
21. Buttocks-to-popliteal length	49.8	2.5	48	3.2	43.8	49	53.6
22. Elbow-to-fist length	38.5 (37.1)	2.1 (3.0)	34.8 (32.9)	2.3 (3.1)	31.9 (28.9)	36.7 (35.0)	41.1 (41.0)
23. Upper-arm length	36.9 (37.0)	1.9 (2.5)	34.1 (33.8)	2.5 (2.1)	31.0 (28.9)	35.7 (35.0)	39.4 (41.0)
24. Shoulder breadth	45.4	1.9	39	2.1	36.3	42.3	47.8
25. Hip breadth	35.6	2.3	38	2.6	32.4	36.8	41.5
Foot							
26. Foot length	26.8	1.3	24.1	1.1	22.6	25.3	28.4
27. Foot breadth	10	0.6	8.9	0.5	8.2	9.4	10.8
Hand							
28. Hand thickness, metacarpal III	3.3	0.2	2.8	0.2	2.7	3	3.6

(Continued)

TABLE 5.3 (CONTINUED)
Anthropometric Data (cm)[a]

	Males		Females		Population Percentiles, 50/50 Males/Females		
Measurement	**50th Percentile**	**±1 SD**	**50th Percentile**	**±1 SD**	**5th**	**50th**	**95th**
29. Hand length	19	1	18.4	1	17	18.7	20.4
30. Digit two length	7.5	0.7	6.9	0.8	5.8	7.2	8.5
31. Hand breadth	8.7	0.5	7.7	0.5	7	8.2	9.3
32. Digit one length	12.7	1.1	11	1	9.7	11.8	14.2
33. Breadth of digit one interphalangeal joint	2.3	0.1	1.9	0.1	1.8	2.1	2.5
34. Breadth of digit three interphalangeal joint	1.8	0.1	1.5	0.1	1.4	1.7	2
35. Grip breadth, inside diameter	4.9	0.6	4.3	0.3	3.8	4.5	5.7
36. Hand spread, digit one to digit two, first phalangeal	12.4	2.4	9.9	1.7	7.5	10.9	15.5
			joint				
37. Hand spread, digit one to digit two, second phalangeal joint	10.5	1.7	8.1	1.7	5.9	9.3	12.7
			Head				
38. Head breadth	15.3	0.6	14.5	0.6	13.8	14.9	16
39. Interpupillary breadth	6.1	0.4	5.8	0.4	5.2	6	6.7
40. Bi-ocular breadth	9.2	0.5	9	0.5	8.3	9.1	10

(Continued)

TABLE 5.3 (CONTINUED)
Anthropometric Data (cm)[a]

	Males		Females		Population Percentiles, 50/50 Males/Females		
Measurement	50th Percentile	±1 SD	50th Percentile	±1 SD	5th	50th	95th
			Other measurements				
41. Flexion-extension, range of motion of wrist radians (5 7°/rad)	2.33	0.33	2.46	0.26	1.92	2.4	2.8
42. Ulnar-radial range of motion of wrist radians (57°/rad)	1.05	0.23	1.17	0.24	0.81	1.15	1.49
43. Weight, in kilograms	83.2	15.1	66.4	13.9	47.7	74.4	102.9

Source: Adapted from Champney, P.C. (1979); Muller-Borer, B. (1981); Eastman Kodak Company, *Ergonomics Design for People at Work, Vol. 1,* Van Nostrand Reinhold, New York, 1983; NASA RP 1024, *Anthropometric Source Book: Volume 1: Anthropometry for Designers Anthropology* Staff/Webb Associates, NASA, Washington, DC, pp. 7–78, 1978.

Note: The mean, or 50th percentile, values, plus or minus one standard deviation (SD) of the mean, are shown for 43 anthropometric variables (column 1). Variables 1 through 10 are standing heights, clearances, or reaches, and variables 11 through 25 are measurements for the subject seated. Data on American men (columns 2 and 3) and women (columns 4 and 5) are statistically combined to derive the 5th, 50th, and 95th percentile values for a 50/50 mix of these populations (columns 6 through 8). The data are taken primarily from military studies, where several thousand people were studied. The entries shown in parentheses are from industrial studies, where 50–100 women and 100–150 men were studied. The data in the footnotes "a and b" are from a study on 50 women and 100 men in the industry.

[a] These values should be adjusted for clothing and posture.

[b] Add the following for bending forward from hips or waist—Male: waist, 25 ± 7; hips, 42 ± 8. Female: waist, 20 ± 5; hips, 36 ± 9.

TABLE 5.4
International Anthropometry: Adults, Height, and Weight Averages (with Standard Deviations)

	Sample Size	Stature (mm)	Weight (kg)
	Algeria		
Females (1990)	666	1,576 (56)	61 (1)
	Australia		
Females, 77 (8) years old	138	1,521 (70)	61 (13)
Males, 76 (7) years old (2000)	33	1,658 (79)	72 (11)
	Brazil		
Males (1988)	3,076	1,699 (67)	NDA
	China		
Females (Hong Kong)	69	1,607 (54)	NDA
Females (Taiwan) (1994)	300	1,582 (49)	51 (7)
Females (Taiwan) (2000)	About 600	1,572 (53)	52 (7)
Males (Hong Kong) (2000)	286	1,737 (49)	NDA
Males (Canton) (1990)	41	1,720 (63)	60 (6)
Males (Taiwan) (2002)	About 600	1,705 (59)	67 (9)
	Egypt		
Females (1987)	4,960	1,606 (72)	63 (4)
	France		
Female soldiers	328	1,620	58
Male soldiers (1997)	687	1,747	70
Females	5,510	1,625 (71)	62 (12)
Males (IFTH and Goncalves, personal communication, 2006)	3,986	1,756 (77)	77 (13)
	Germany (east)		
Females	123	1,608 (59)	NDA
Males (1986)	30	1,715 (66)	NDA
	India		
Females	251	1,523 (66)	50 (10)
Males (1997)	710	1,650 (70)	57 (11)
East-Central India male farm workers (2002)	300	1,638 (56)	57 (7)
Central India male farm workers (1989)	39	1,620 (50)	49 (6)
South India male workers (1992)	128	1,607 (60)	57 (5)
East India male farm workers (1997)	134	1,621 (58)	54 (67)
	Indonesia		
Females	468	1,516 (54)	NDA
Males (1985)	949	1,613 (56)	NDA
	Iran		
Female students	74	1,597 (58)	56 (10)
Male students (1997)	105	1,725 (58)	66 (10)

(Continued)

TABLE 5.4 (CONTINUED)

International Anthropometry: Adults, Height, and Weight Averages (with Standard Deviations)

	Sample Size	Stature (mm)	Weight (kg)
	Ireland		
Males (1991)	164	1,731 (58)	74 (9)
	Italy		
Females (1991)	753	1,610 (64)	58 (8)
Females (2002)	386	1,611 (62)	58 (9)
Males (1991)	913	1,733 (71)	75 (10)
Males (2002)	410	1,736 (67)	73 (11)
	Jamaica		
Females	123	1,648	61
Males (1991)	30	1,749	68
	Japan		
Females	240	1,584 (50)	54 (6)
Males (1990)	248	1,688 (55)	66 (8)
	Korea (South)		
Female workers (1989)	101	1,580 (57)	54 (7)
	Malaysia		
Females (1988)	32	1,559 (66)	NDA
	The Netherlands		
Females, 20–30 years old (1998)	68	1,686 (66)	67 (10)
Females, 18–65 years old (2002)	691	1,679 (75)	73 (16)
Males, 20–30 years old (1998)	55	1,848 (80)	81 (14)
Males, 18–65 years old (2002)	564	1,813 (90)	84 (16)
	Russia		
Female herders (ethnic Asians)	246	1,588 (55)	NDA
Female students (ethnic Russians)	207	1,637 (57)	61 (8)
Female students (ethnic Uzbeks)	164	1,578 (49)	56 (7)
Female factory workers (ethnic Russians)	205	1,606 (53)	61 (8)
Female factory workers (ethnic Uzbeks)	301	1,580 (54)	58 (9)
Male students (ethnic Russians)	166	1,757 (56)	71 (9)
Male students (ethnic Uzbeks)	150	1,700 (52)	65 (7)
Male factory workers (ethnic Russians)	192	1,736 (61)	72 (10)
Male factory workers (ethnic mix)	150	1,700 (59)	68 (8)
Male farm mechanics (ethnic Asians)	520	1,704 (58)	64 (8)
Male coal miners (ethnic Russians)	150	1,801 (61)	NDA
Male construction workers (ethnic Russians) 1999	150	1,707 (69)	NDA
	Saudi Arabia		
Males (1985)	1,440	1,675 (61)	NDA
	Singapore		
Females (1988)	46	1,598 (58)	NDA

(Continued)

TABLE 5.4 (CONTINUED)
International Anthropometry: Adults, Height, and Weight Averages (with Standard Deviations)

	Sample Size	Stature (mm)	Weight (kg)
Males (pilot trainees) (1995)	832	1,685 (53)	NDA
Sri Lanka			
Females	287	1,523 (59)	774 (22)
Males (1991)	435	1,639 (63)	833 (27)
Sudan			
Males			
Villagers (1981)	37	1,687 (63)	NDA
City dwellers (1982)	16	1,704 (72)	NDA
City dwellers (1982)	48	1,668	NDA
Soldiers (1981)	21	1,735 (71)	NDA
Soldiers (1982)	104	1,728	NDA
Thailand			
Females	250	1,512 (48)	NDA
Females	711	1,540 (50)	817 (27)
Males	250	1,607 (20)	NDA
Males (1991)	1,478	1,654 (59)	872 (32)
Turkey			
Females			
Villagers	47	1,567 (52)	792 (38)
City dwellers	53	1,563 (55)	786 (05)
Male soldiers (1991)	5,108	1,702 (60)	888 (34)
United States			
Females	About 3,800	1,625	NDA
Males (2004)	About 3,800	1,762	NDA
Midwest workers/with shoes and light clothes			
Females	125	1,637 (62)	NDA
Males (1993)	384	1,778 (73)	NDA
US male miners (1993)	105	1,803 (65)	NDA
US Army soldiers			
Females	2,208	1,629 (64)	852 (35)
Males (1989)	1,774	1,756 (67)	914 (36)
North American (Canada and United States)			
Females, 18–26 years old	1,255	1,640 (73)	NDA
Males, 18–65 years old (2002)	1,120	1,778 (79)	NDA
Vietnamese, living in the United States			
Females	30	1,559 (61)	NDA
Males (1993)	41	1,646 (60)	NDA

Source: Kumar, S. (Ed.), *Biomechanics in Ergonomics*, 2nd edn., CRC Press, New York, 2008.

These sources provide critical data needed for accurate design calculations and adjustments.

Outcome

The application of anthropometric principles resulted in an office chair design that fits 90% of the population comfortably, from the 5th percentile female to the 95th percentile male. The chair's adjustability in height, depth, and armrest positioning ensures that it provides ergonomic support for a wide range of body sizes. The chair's design minimizes the risk of musculoskeletal disorders by supporting proper posture and reducing strain during extended periods of sitting.

Conclusion

This case study demonstrates how anthropometric data can be effectively used to support ergonomic design, improving both user comfort and safety. By using reliable data sources, conducting statistical analysis, and applying anthropometric principles, designers can create products that accommodate the natural variation in human body sizes. The office chair case study highlights the critical role of anthropometrics in ensuring ergonomic products meet the needs of a diverse user population.

Exercises

1. **Explain the purpose of universal design and how this might impact the economics, dimensions, and processes of design from an anthropometric perspective.**
   - Use scholarly literature (e.g., Google Scholar, PubMed, ScienceDirect) to define and contextualize *Universal Design (UD)* in ergonomics.
   - Use AI tools to summarize cost-benefit discussions surrounding UD implementation in public and private sectors.
   - Interview an industrial designer or architect to understand real-world impacts of UD on dimensions, user inclusion, and cost.
   - Analyze a product or space designed with UD principles and evaluate its anthropometric inclusiveness.
2. **In the absence of universal design, what ranges should the ergonomist promote design for in order to accommodate the largest portion of the population?**
   - Research percentile-based design strategies using sources like NIOSH, HFES, and ISO 7250.
   - Use AI to generate a summary of common anthropometric guidelines and validate with literature.
   - Interview an ergonomist to discuss how they choose percentiles in product or workplace design.
   - Write a report that explains the design implications of designing for the 5th to 95th percentile range and associated trade-offs.

3. **In emergency situations, how should anthropometric data be used to guide design decisions?**
   - Search for literature on egress design, emergency escape, and safety codes (e.g., NFPA, OSHA) with anthropometric considerations.
   - Use AI tools to generate scenario-based examples (e.g., airplane or submarine escape hatches).
   - Interview professionals in emergency planning, firefighting, or vehicle design for real-world insights.
   - Develop a checklist of anthropometric factors to consider in designing for time-sensitive and life-saving mobility.
4. **Develop a process for collecting anthropometric data in an occupational ergonomic environment.**
   - Research ISO or NIOSH protocols on anthropometric measurements for workplace design.
   - Use AI tools to draft a step-by-step data collection protocol and validate it with published literature.
   - Interview an occupational therapist, physical anthropologist, or workplace health and safety professional to enhance your protocol.
   - Identify potential ethical, cultural, and logistical considerations in data collection.
5. **Contrast and compare the use of manual versus automatic anthropometric measurement tools.**
   - Conduct a literature review comparing manual tools (e.g., calipers, anthropometers) with digital tools (e.g., 3D scanners, photogrammetry).
   - Use AI to generate a comparison table with categories like accuracy, usability, cost, training, and error margins.
   - Interview a lab technician or researcher using both methods to gather practical insights.
   - Visit a virtual lab (e.g., via YouTube or university resources) to observe tool use and report your findings.
6. **Evaluate the use of goniometers for measuring body motions.**
   - Research articles that apply goniometry in ergonomics, physiotherapy, or occupational health.
   - Use AI tools to create a summary of the accuracy, limitations, and ergonomic relevance of goniometers.
   - Interview a physical therapist or biomechanics researcher who uses goniometers regularly.
   - Design a small observational study to practice using a goniometer and report data collection challenges.
7. **Describe how stature is measured and why it is relevant to occupational ergonomics.**
   - Research standardized procedures for measuring standing and sitting height and their significance in ergonomic layout design.
   - Use AI to provide examples of how stature affects workspace planning (e.g., desk height, vehicle cabin design).

- Interview an ergonomics consultant about how they use stature data in workstation evaluations.
- Conduct a brief survey among peers and analyze the variation in stature and its implications for classroom desk design.

8. **Explain why the dimensions of the "average user" should not be the goal of design engineers.**
   - Use AI and scholarly articles to explore the fallacy of designing for the "average" and discuss inclusion challenges.
   - Research anthropometric variability and inclusive design literature (e.g., HFES guidelines).
   - Interview a product designer or ergonomist about inclusive design practices.
   - Identify a product designed for the "average user" and critique its usability across a diverse population.
9. **Describe the effects of age, gender, and size on anthropometrics.**
   - Research changes in anthropometry across the lifespan and between sexes using academic literature and growth charts.
   - Use AI to generate population comparison summaries (e.g., military vs. civilian, pediatric vs. elderly).
   - Interview a pediatrician, occupational therapist, or gerontologist about how these factors impact design.
   - Develop a matrix that shows how design must adapt based on these anthropometric differences in different industries.
10. **Identify three anthropometric data sources and explain how they should be used, the population they are most applicable to, and an industry that will benefit from this data.**
    - Select data sources such as ANSUR II, or the Anthropometric Survey of US Army Personnel II, NHANES, the National Health and Nutrition Examination Survey, and ISO 7250 (international standard).
    - Use scholarly literature to evaluate limitations and scope of each dataset.
    - Interview a professional ergonomist or industrial engineer to understand which dataset they use and why.
    - Write a brief memo recommending appropriate data sources for use in healthcare, transportation, and industrial manufacturing.

REFERENCES

Cakir, A., Robinette, K. M., & Bradtmiller, B. (2014). Anthropometry in design: An overview. *Applied Ergonomics*, *45*(2), 199–208.

Chaffin, D. B., Andersson, G. B., Martin, B. J., & Kumar, S. (2019). *Occupational biomechanics*. Wiley.

Darko, J., Popović, S., & Borisov, D. (2006). A new clothing and footwear sizing system for croatia. *International Journal of Clothing Science and Technology*, *18*(3), 227–235.

Duffy, V. G., Roscoe, R., & Galea, A. (2015). Virtual ergonomic evaluations of workplace designs. *International Journal of Occupational Safety and Ergonomics*, *21*(2), 144–156.

Dul, J., & Weerdmeester, B. (2018). *Ergonomics for beginners: A quick reference guide*. CRC Press.

Garneau, C. J., & Igarashi, T. J. (2020). 3D printing in prosthetics and orthotics: benefits, challenges, and the future of patient-specific devices. *Journal of Prosthetics & Orthotics*, *32*(4), 223–229.

Gordon, C. C., et al. (2016). The US Army Anthropometric Survey (ANSUR II): Methods and summary statistics. *Technical Report*, United States Army.

Guerra, F., Camporeale, P., & Tarquinio, G. (2017). Anthropometry and 3D printing: A revolutionary approach for ergonomics. *Proceedings of the International Conference on Human-Computer Interaction*, 156–167.

Human Systems Integration Information Analysis Center (HSIIAC). (2011). *Automated Anthropometric Data Measurement System Report.*

Jung, H. S., Lee, S. H., & Kim, H. M. (2015). Anthropometric data for ergonomic design in the digital age. *Ergonomics in Design*, *23*(4), 20–24.

McCormick, E. J. (2014). *Human factors in engineering and design*. McGraw-Hill.

McFarland, R. A., & Moore, G. T. (2015). Ergonomics and the evolution of anthropometric measurement tools. *Journal of Occupational Health*, *57*(3), 15–23.

Pheasant, S., & Haslegrave, C. M. (2016). *Bodyspace: anthropometry, ergonomics, and the design of work*. CRC Press.

Stewart, R., Pullen, R., & Newton, J. (2014). Inclusive design: Creating workspaces that suit everyone. *Applied Ergonomics*, *45*(2), 450–457.

Wilson, J. R. (2017). The evolution of ergonomics and human factors: Celebrating the past, designing the future. *Ergonomics*, *60*(1), 1–16.

6 Design of Workplaces and Hand Tools

6.1 LEARNING GOALS

The objective of this chapter is to provide an introduction to basic ergonomic principles of workplace design and the evaluation of hand tools. The student will study the general process for workplace design and be provided with a list of both sample and potential evaluation tools in detail.

6.2 KEY TOPICS

- Workplace design analysis
- Principles of arrangement
- General process for workplace design
- General principles for workplace design
- Designing to fit the moving body
- Designing for the standing operator
- Designing for the sitting operator
- Working in confined and awkward spaces
- Designing for hand and foot operation
- Workplace evaluation tools
- Rapid Upper Limb Assessment (RULA) Tool
- Risk factors and risk prediction
- General guidelines for workplace controls and displays

6.3 INTRODUCTION AND BACKGROUND

The development of workplace design and hand tools is a reflection of the ongoing commitment to enhancing worker safety, efficiency, and comfort through the principles of ergonomics. The study of occupational ergonomics focuses on designing work environments and tools that fit the physical and cognitive capabilities of workers, thereby minimizing the risk of injury and maximizing productivity. This field has evolved significantly over the years, with early observations from pioneers like Bernardino Ramazzini that laid the groundwork for modern ergonomic principles.

DOI: 10.1201/9780429342806-6

6.3.1 Early Observations in Ergonomics

Over three centuries ago, Bernardino Ramazzini, often hailed as the father of occupational medicine, was one of the first to recognize the link between work posture and health. In his landmark work *De Morbis Artificum Diatriba* (Diseases of Workers), Ramazzini observed that prolonged standing could cause significant fatigue due to the continuous muscle tension required to maintain an upright position (Helander, 2020). His observations extended to workers who performed sedentary tasks, noting that poor posture, such as a stooped sitting position, often led to musculoskeletal disorders, including round shoulders, leg numbness, and sciatica. These early insights into the relationship between posture and health laid the foundation for the concept of dynamic work, where movement and posture variation are encouraged to mitigate these negative effects (Bernard, 1997).

6.3.2 Industrial Revolution and Early Ergonomics

The onset of the Industrial Revolution in the late 18th century brought about profound changes in the nature of work and workplace design. Mechanized production processes increased the demand for labor, but early factories prioritized productivity over worker well-being, paying little attention to ergonomics. During this period, workers were often subjected to repetitive, physically demanding tasks without consideration of the long-term health consequences.

A shift in perspective began to emerge in the late 19th century, thanks to the pioneering efforts of Frank and Lillian Gilbreth. The Gilbreths conducted detailed time and motion studies, aiming to increase industrial efficiency by reducing unnecessary motions and improving worker performance. Their work is considered one of the first applications of ergonomics in industrial settings, as it emphasized task simplification and the reduction of worker fatigue through better workplace design and tool use (Gupta et al., 2013). This marked the beginning of applied ergonomics, which continues to be a critical component of occupational safety and health today.

6.3.3 Modern Ergonomics and Workplace Design

In contemporary workplace design, ergonomics has evolved into a comprehensive discipline that addresses the physical, cognitive, and organizational aspects of work environments. Modern ergonomics considers multiple components of the work system, including task demands, the design of equipment, workplace layout, and the interactions between workers and their tools (Hedge, 2016). By integrating ergonomic principles into the design of workplaces and tools, companies can reduce the risk of work-related musculoskeletal disorders (WMSDs), enhance worker productivity, and promote overall well-being.

The design of ergonomic tools has faced significant progress, demonstrating how even simple objects can benefit from ergonomic innovations. For instance, the pencil—initially a Roman metal stylus—evolved over time to include more user-friendly materials like graphite, which replaced lead due to its non-toxic properties. In this instance, encasing the brittle graphite in wood (reflecting the modern pencil seen today) is an example of ergonomic problem-solving aimed at improving the user experience (Petroski, 2010). This example underscores the importance of ergonomic design in enhancing tool usability and reducing physical strain.

6.3.4 Ergonomic Interventions in Office Environments

In today's technology-driven work environments, ergonomics has become especially crucial in office settings, where prolonged sitting, repetitive movements, and poor workstation design can lead to significant health issues. The rise of computer-based work has resulted in a surge of ergonomic interventions aimed at reducing the incidence of musculoskeletal disorders. Adjustable chairs, sit–stand desks, and computer monitors that allow for ergonomic alignment are now common in many workplaces (Robertson et al., 2013). Research shows that these ergonomic improvements not only reduce worker discomfort but also lead to measurable increases in productivity, illustrating the clear benefits of ergonomic interventions in reducing the risks associated with sedentary work.

6.3.5 Ergonomics in Healthcare and Other Industries

Beyond office environments, ergonomics plays a vital role in industries like healthcare, where the physical demands on workers can be intense. In healthcare settings, ergonomics informs the design of medical instruments, the layout of operating rooms, and patient handling techniques. Utilizing ergonomic advancements not only protect healthcare workers from injury but also enhance the quality of patient care by improving efficiency and reducing errors (Nelson & Baptiste, 2020).

In the retail and service sectors, ergonomic principles have contributed to the design of checkout counters, hand-held scanning devices, stock handling equipment, and customer service stations. These ergonomic improvements have reduced the physical demands placed on workers, minimized injury risks, and enhanced overall operational efficiency (David, 2015).

6.3.6 Future Trends: Artificial Intelligence and Personalized Ergonomics

As technology advances, so does the field of ergonomics. The integration of artificial intelligence (AI) and machine learning in workplace design is paving the way for personalized ergonomic solutions. AI systems can collect and analyze large volumes of data on worker movements, postures, and behaviors to provide real-time feedback and recommendations for workstation adjustments (Mathiassen & Winkel, 2020a). This data-driven approach enables the creation of adaptive work environments that are tailored to the individual needs of workers, further enhancing comfort and reducing the risk of injury.

6.4 WORKPLACE DESIGN ANALYSIS

Ergonomics plays a critical role in workplace design, with its primary goal being the optimization of the work environment to improve worker health, safety, and productivity. The Occupational Safety and Health Administration (OSHA) has long recognized the significance of ergonomics in preventing work-related injuries and illnesses. According to OSHA (2013), "The ergonomic approach to workplace design is the most effective and should be the first choice for controlling sources of workplace stress." This statement highlights the importance of ergonomics as a proactive approach to reducing occupational hazards. Ergonomics should prioritize designing tasks, tools, and work environments to fit the worker, rather than focusing on selecting or training workers to adapt to poor designs.

6.4.1 Ergonomics as a Preventive Measure

The ergonomic approach to job design is grounded in the principle of achieving the "best fit" between workers and their tasks, with an emphasis on optimizing work postures, reducing excessive force exertion, minimizing repetitive movements, and limiting prolonged exposure to risk factors associated with musculoskeletal disorders (MSDs) and other occupational injuries (OSHA, 2013). By implementing ergonomic principles in workplace design, organizations can prevent injuries related to poor posture and overexertion, thereby fostering healthier, more efficient, and more productive work environments.

For instance, improving workstations by adjusting the height of surfaces, redesigning tools to reduce grip force, or organizing tasks to encourage natural postures are key examples of ergonomic interventions that address common workplace risks. Such interventions are especially valuable in manual labor jobs, where workers are often exposed to high physical demands, but they are equally important in office settings, where sedentary workers can suffer from repetitive strain injuries (RSIs) related to poor workstation design (Robertson et al., 2013a).

6.4.2 Analysis Techniques for Ergonomic Assessments

A range of analysis techniques are employed in ergonomic assessments to determine the interaction between workers and their workplace environment. These techniques help identify factors contributing to ergonomic issues such as injuries, illnesses, operator discomfort, and decreased productivity. Common methods include worker surveys, interviews, and direct observational analysis of workstations. More advanced approaches involve the use of technology-based tools, such as biomechanical modeling and computer-based task analysis systems, to assess physical demands and ergonomic risks in more detail (Kroemer, 2001).

Surveys and interviews provide qualitative insights into worker discomfort and perceived risks, while observational analysis can reveal specific workplace design issues that may not be immediately apparent through self-report methods. For instance, a detailed observation may reveal that a workstation's design requires a worker to perform a task in an awkward posture, contributing to chronic back pain. Task analysis tools and ergonomic software are often used in combination with these traditional methods to create a more comprehensive picture of the workplace's ergonomic challenges.

6.4.3 Advancements in Ergonomic Tools and Technologies

Recent advancements in ergonomic tools and methodologies have greatly enhanced the ability of ergonomists to design safer, more efficient workplaces. Among these innovations, Digital Human Modeling (DHM) software has become an essential tool for simulating human interactions with workplace environments. DHM allows ergonomists to predict potential ergonomic risks before physical implementation by modeling human postures, movements, and exertions within a virtual environment (Chaffin, 2019). For example, DHM can be used to simulate the physical demands of a task on a variety of body types, helping to optimize the design of tools, workstations, or assembly lines for a diverse workforce.

Another emerging technology with significant implications for workplace ergonomics is wearable technology. Wearable devices equipped with sensors can monitor workers' movements and postures in real-time, providing valuable data on how tasks are performed and where ergonomic risks may exist. This technology enables immediate feedback to workers, allowing them to adjust their posture or movements to reduce strain. In the long term, wearable data can inform more extensive ergonomic interventions by identifying repetitive patterns of risky behavior, such as frequent bending or twisting, which may contribute to injury (Liu & Su, 2020).

6.4.4 Ergonomics and Productivity

Ergonomics is not solely focused on injury prevention; it also plays a crucial role in improving efficiency and productivity. Research consistently shows that ergonomic interventions lead to increases in both worker satisfaction and efficiency. Ergonomic improvements can reduce worker fatigue, enhance comfort, and streamline tasks, all of which contribute to better job performance (Robertson et al., 2013b). For example, the redesign of a workstation to ensure proper alignment of computer screens, chairs, and keyboards can reduce eye strain, back pain, and hand fatigue, allowing workers to maintain productivity for longer periods without discomfort.

6.4.5 Future Directions in Ergonomic Research

As technology continues to advance, the field of ergonomics will likely see even greater integration of data-driven and AI-powered tools. Machine learning algorithms, for example, have the potential to analyze vast datasets collected from wearable devices, allowing for the development of personalized ergonomic interventions. By continuously learning from real-time data, these systems could offer workers individualized suggestions to modify their work habits and adjust their environments to prevent injuries before they occur (Mathiassen & Winkel, 2020b).

With the persistence of these technological advances, the design of human–robot interactions will become an important area of ergonomic research. Ensuring that robots and other automated systems can be operated and maintained without increasing the risk of injury to workers will require a deep understanding of human factors in ergonomics.

6.4.5.1 Principles of Arrangement

In occupational ergonomics, the design of a workspace plays a crucial role in optimizing worker performance, health, and safety. A well-designed workspace not only enhances productivity but also minimizes the risk of musculoskeletal disorders and other occupational injuries (Burdorf & van der Beek, 2016). Given the complexity of modern workplaces, trade-offs between different design considerations are often necessary to achieve a balance between functionality, safety, and user comfort. This balancing act can be managed by adhering to core ergonomic principles, such as the Principles of Arrangement, initially outlined by Sanders and McCormick (2013). These four principles remain foundational for ergonomics professionals and have been further expanded upon in more recent literature (Robertson et al., 2017; Dul & Bruder, 2020).

6.4.5.1.1 Importance Principle

The Importance Principle states that workspace design should prioritize components and activities most critical to the task being performed. By placing essential tools, controls, and information within easy view and reach, the design ensures that operators can access high-priority items without unnecessary strain. This is especially important in safety-critical industries, such as healthcare and aviation, where delays in accessing essential controls could lead to severe consequences (Green & Westgaard, 2019). Prioritizing critical tasks in workspace design not only improves task efficiency and promotes safety, but also reduces cognitive load, as the worker can focus more on decision-making rather than the physical retrieval of tools or materials (Hedge, 2016).

6.4.5.1.2 Frequency of Use Principle

The Frequency of Use Principle suggests that frequently used components should be positioned within the primary reach zone to avoid awkward postures and excessive stretching. According to Kaber and Riley (2015), repetitive motion injuries, such as tendinitis or carpal tunnel syndrome, are common in work environments that neglect this principle. Placing frequently used items in ergonomically optimal locations—such as between shoulder height and waist level—can significantly reduce the risk of these injuries. Additionally, frequent access to tools or controls should require minimal force, thereby reducing muscular fatigue and enhancing operational efficiency (Fernandez & Marley, 2015).

6.4.5.1.3 Functional Grouping Principle

The Functional Grouping Principle recommends that tools, materials, and information should be organized according to the functions they support in the workflow. Categorizing related items together can reduce cognitive and physical effort, making it easier for workers to locate and use them efficiently. For example, in manufacturing settings, tools required for a specific assembly task should be located in close proximity to one another to avoid unnecessary movement and confusion (Dul & Weerdmeester, 2014). This approach not only minimizes errors but also enhances task performance by creating a logical and intuitive workspace arrangement.

6.4.5.1.4 Sequence of Use Principle

The Sequence of Use Principle suggests that workspace layouts should mirror the chronological sequence of operations in a task. This principle is particularly relevant in jobs that involve routine or repetitive sequences, such as assembly line work or laboratory procedures. A sequential arrangement allows workers to perform tasks fluidly, moving from one step to the next with minimal disruption (Wilson, 2014). Poor sequencing, on the other hand, can lead to inefficient workflows, increased task completion times, and elevated risk of injury from excessive movement. By aligning the workspace design with the operational flow, this principle helps streamline processes, reduce unnecessary motions, and enhance productivity (Bridger, 2018).

6.4.5.2 General Process for Workplace Design

Ergonomists require in-depth information about the work environment, task demands, and anticipated workforce demographics to design effective workplaces. There is a five-step approach that streamlines this process:

Step 1: Initial Evaluation

- **Comprehensive Analysis:** Meticulously study the environment, tasks, and any factors impacting task completion. This includes task and process design, workplace layout, equipment, and tools.
- **Worker Capabilities:** Identify any job requirements exceeding worker abilities.
- **Procedure Assessment:** Evaluate written procedures for accessibility, clarity, and how they address situational factors, external influences, scheduling, and stress.

Step 2: Workplace Factor Investigation

- **Tool and Equipment Evaluation:** Identify and assess all tools and equipment used in the workplace.
- **Working Condition Assessment:** Evaluate factors like lighting, noise, temperature, and ventilation.
- **Safety Assessment:** Identify and assess potential safety hazards.

Step 3: Human Performance and Physiological Analysis

- **Performance Measures:** Analyze productivity throughout a shift. Consider metrics like total units completed, output per hour, idle time, distractions, interruptions, and accidents.

- **Output Quality:** Measure quality by recording defects, errors, and frequency of referencing instructions.
- **Physiological and Psychophysical Measures:**
  - **Direct Measurements:** Assess heart rate, blood pressure, respiration rate, muscle activity (i.e., EMG), body temperature, and oxygen consumption.
  - **Subjective Measures:** Consider perceived exertion, difficulty, energy expenditure, and fatigue levels.

Step 4: Data Compilation and Results

- **Summarize Evaluation Outcomes:** Clearly present findings from the analysis.
- **Ranking and Prioritization:** Rank outcomes using a qualitative or quantitative scale (e.g., highly acceptable, acceptable, unacceptable). Prioritize critical factors for design integration.

Step 5: Solution Development

- **Ergonomic Design Strategies:** Leverage collected data to develop ergonomically-informed design solutions.
- **Multi-Tiered Solutions:** Consider various design options with differing levels of complexity, cost, and implementation timelines. This could involve:
  - **Immediate Solutions:** Quick fixes for pressing ergonomic issues.
  - **Interim Solutions:** Mid-term improvements addressing broader ergonomic concerns.
  - **Long-Term Solutions:** Comprehensive redesign plans for optimal ergonomic workplaces.

The steps outlined above provide a framework for conducting a comprehensive workplace analysis. Nonetheless, it is essential to customize this approach based on the specific details of the work environment to meet the unique needs of each workplace. This process is crafted to consider all elements of the workplace environment—including the individual worker, the workstation, the tasks performed, and the tools and equipment used.

Necessary ergonomic adjustments might include reconfiguring the workstation to decrease static muscle fatigue, reduce awkward postures, and enhance visual clarity. Such considerations ensure a holistic improvement of the workstation environment. Additionally, modifying the placement or design of workpieces and tools, such as jigs, clamps, and vises, is critical. These adjustments secure objects in place, reducing the need for workers to adopt uncomfortable postures and decreasing static loading and repetitive stress.

Further ergonomic interventions can include optimizing the layout of components and storage bins to improve accessibility, minimize transportation time and distance, and efficiently use space. Incorporating lift tables is advisable as they diminish the need for manual lifting, bending, and twisting, thus preventing strain injuries. Similarly, footrests can be instrumental in reducing fatigue and back strain, alleviating stress on the legs, back, and neck, enhancing circulation, and promoting correct posture and comfort while seated.

6.5 GENERAL PRINCIPLES OF WORKPLACE DESIGN

The design of workplaces for optimal safety, efficiency, and productivity is critical for maintaining a healthy and thriving workforce. From an occupational ergonomics standpoint, achieving these outcomes is best accomplished during the design phase of the workplace environment, when potential hazards can be mitigated proactively. This approach is known as Design for Safety (DfS), a strategy that integrates safety, ergonomic, and usability considerations early in the design process to minimize risks before workers encounter them (NIOSH, 2019). By doing so, DfS enhances not only employee well-being but also operational efficiency and productivity.

6.5.1 Ergonomic Design and OSHA Toolbox Principles

Ergonomics plays a central role in DfS by focusing on designing workstations, tools, and workflows that align with human capabilities and limitations. OSHA provides valuable resources through its Ergonomic Toolbox, which offers practical guidance on ergonomic principles such as hand tool selection, workstation configuration, and task design (OSHA, 2023). OSHA's Ergonomic Toolbox is a comprehensive resource designed to address common ergonomic challenges in the workplace by applying best practices in design and ergonomics. However, DfS extends beyond ergonomics alone, incorporating a broader spectrum of safety considerations to address hazards comprehensively.

According to the OSHA Ergonomic Toolbox, several principles should be adhered to when designing safe and ergonomically sound workspaces:

1. **Neutral Postures:** Ensure work tasks are designed to allow for neutral postures of the spine, neck, and limbs, minimizing awkward positions that can lead to musculoskeletal disorders (MSDs).
2. **Repetitive Motion Reduction:** Design workflows to limit repetitive motion, incorporating automation where possible to reduce the physical strain on workers.
3. **Force Minimization:** Tools and equipment should be selected or designed to minimize the force required to complete tasks, thereby reducing the risk of overexertion injuries.
4. **Workstation Adjustability:** Promote the use of adjustable workstations to accommodate workers of different body sizes and allow for both standing and seated work.

These principles align with broader occupational ergonomics theories, which focus on the reduction of work-related injuries through better design of tools, workspaces, and job tasks (Karwowski & Marras, 2018).

6.5.2 Designing for Safety, Efficiency, and Sustainability

While ergonomics is fundamental to workplace design, DfS encompasses additional elements essential for creating safe, efficient, and sustainable workplaces. These include maintainability, safety, usability, comfort, and reliability.

6.5.2.1 Maintainability

Maintainability involves designing workplaces and equipment in a manner that ensures their longevity and ease of maintenance. This is essential for reducing downtime and improving operational efficiency.

- **Product Lifespan:** High-quality materials and modular designs enhance durability and allow for easy replacement of parts, reducing the need for full equipment replacement (Baines et al., 2017).
- **Clear Instructions:** Providing intuitive user manuals and maintenance training ensures that personnel can perform repairs efficiently and safely (Singh et al., 2015).
- **Accessibility:** Workspaces should be designed with adequate space for maintenance personnel to perform their tasks comfortably and safely, adhering to ergonomic principles to minimize awkward postures during maintenance work (ISO 9241-11, 2018).
- **Hazard Identification:** The proactive identification of hazards during the design phase, such as pinch points or sharp edges, allows for the integration of warning systems or physical barriers to reduce the risk of injury (Cheng et al., 2019).

6.5.2.2 Safety

In occupational ergonomics, safety must be integrated from the initial design phase. Safety is an essential element of ergonomics and should always be a priority in workplace design decisions. This involves identifying potential hazards and mitigating them through design features such as machine guards, fail-safe mechanisms, and proper signage.

- **Safety as a Priority:** Effective workplace design integrates safety features such as emergency stop buttons, machine guards, and proper ventilation (HSE, 2018).
- **Balancing Safety with Productivity:** Safety should not come at the expense of productivity. Controls and interfaces should be intuitive, minimizing the potential for human error while allowing for efficient task completion (Wilson, 2014).

6.5.2.3 Usability

Usability is a critical component in ensuring that equipment and tools are easy to operate and aligned with the physical and cognitive abilities of workers.

- **Human-Centered Design:** Equipment and workstations must account for anthropometric data, ensuring that design accommodates a wide range of body sizes and shapes (Pheasant & Haslegrave, 2016) for the worker population. For example, adjustable equipment and workstations should be used to help to accommodate users with different stature and reach.
- **Clear Instructions:** Documentation and training should be designed to be straightforward and accessible, ensuring that all workers understand how to operate equipment safely and efficiently (Marras et al., 2009). The language should also be written to the expected reading level of the workers as well as the use of symbols to further illustrate key aspects of the instructions.

6.5.2.4 Comfort

Comfort plays a critical role in reducing worker fatigue and improving productivity. Ergonomic workstation design focuses on ensuring that employees can maintain good posture, experience minimal strain, and perform tasks with optimal efficiency.

- **Postural Support:** Workstations should be designed to allow for natural body postures and minimal exertion, particularly for tasks that require prolonged standing or repetitive motions (Van der Beek et al., 2014).
- **Environmental Factors:** Attention must also be paid to factors such as lighting, noise levels, and temperature, as they contribute significantly to worker comfort and overall job satisfaction (Hedge, 2016).

6.5.2.5 Reliability and Sustainability

The final component of DfS is the reliability and sustainability of equipment and the work environment. Reliable tools and systems reduce downtime and repair costs, while sustainability ensures that the workplace has a minimal environmental footprint.

- **Design for Durability:** Equipment should be designed for long-term use, taking into account the expected operational environment and patterns of use (BS EN ISO 6385, 2016).
- **Sustainability:** Sustainable design considerations, such as energy-efficient systems and the use of recyclable materials, contribute to an environmentally friendly workplace (Luttropp & Lagerstedt, 2014).

6.5.3 Designing to Fit the Moving Body

Human beings are inherently dynamic, and their performance, comfort, and well-being can be significantly enhanced when they are allowed to vary their posture, workload, and work position throughout the workday. As research by Gallagher et al. (2013) and more recent studies (Holtermann et al., 2019) have shown, static postures and repetitive motions are key contributors to fatigue and musculoskeletal disorders. Thus, modern ergonomic design principles aim to support a variety of postures and task-specific motions, allowing workers to transition seamlessly between positions, reducing physical strain, and optimizing overall performance (National Institute for Occupational Safety and Health [NIOSH], 2020).

6.5.3.1 Evolution of Postural Ergonomics

Traditionally, ergonomics emphasized three primary postures: sitting, standing, and lying down. These foundational postures were deemed sufficient for most industrial tasks of the 20th century. However, with the rise of diversified work environments in manufacturing, healthcare, office settings, and service industries, a broader array of work postures is now considered essential. Modern ergonomic interventions must accommodate not only sitting and standing but also more dynamic and complex positions, such as kneeling (either on one or both knees), squatting, and leaning against supports or perches (American College of Occupational and Environmental Medicine [ACOEM], 2020). These additional postures allow for more flexibility and comfort, reducing the strain associated with prolonged static positions (Bishop & Demeter, 2017).

6.5.3.2 Assessing the Impact of Postures in Ergonomics

To properly assess the effects of different work postures on human health and performance, ergonomists analyze both independent and dependent variables. Independent variables encompass aspects such as body movements, postural changes, and the biomechanical and physiological metrics used to quantify these variations. These include measures such as the range of motion, the frequency of posture changes, and the maximum displacement of joints or body segments during a task (Gallagher et al., 2013).

On the other hand, dependent variables reflect the outcomes and effects of these postural changes and include metrics across several disciplines:

1. **Physiological Measures:** These include oxygen consumption, heart rate, blood pressure, and electromyographic (EMG) activity. For instance, EMG is used to assess muscle activation patterns in relation to task demands, helping to determine whether certain postures cause excessive muscle fatigue (Rohmert et al., 1975).
2. **Medical Outcomes:** This primarily focuses on the risk and prevalence of musculoskeletal disorders and cumulative trauma disorders (CTDs). The relationship between prolonged static postures and the development of MSDs has been extensively documented in ergonomic literature (Holtermann et al., 2019).
3. **Biomechanical Parameters:** Biomechanical analysis investigates forces on the body, such as spine disc pressure and joint loadings. Studies utilizing biomechanical modeling have demonstrated how specific postures

can increase disc pressure in the lumbar spine, contributing to lower back pain in workers engaged in prolonged sitting or heavy lifting (Cholewicki & McGill, 1996).

4. **Engineering Metrics:** These involve evaluating productivity and mechanical forces exerted on workplace equipment. For example, ergonomists might assess the force needed to operate a machine while in various postures, aiming to minimize physical demands and enhance operational efficiency (Gallagher et al., 2013).
5. **Psychophysical Assessments:** Subjective discomfort ratings are essential in understanding workers' perceptions of physical strain. These ratings, often collected through standardized surveys, have shown significant correlations with physiological measures such as heart rate and perceived exertion, providing a reliable tool for ergonomic evaluations (NIOSH, 2020; Rohmert et al., 1975).

6.5.3.3 Importance of Subjective Discomfort Ratings

Subjective discomfort ratings remain a critical metric in ergonomic assessments, particularly because they offer insight into workers' personal experiences of physical strain and discomfort. Research has demonstrated strong correlations between these subjective ratings and objective physiological measures, such as heart rate and muscle activity (Holtermann et al., 2019). Well-designed surveys, such as the Borg CR10 scale or visual analog scales (VAS), can provide valuable information for tailoring ergonomic interventions and improving workplace comfort (Rohmert et al., 1975).

6.5.3.4 Role of Biomechanics in Dynamic Task Analysis

Biomechanical principles are central to understanding the complex interactions between the human body and the work environment. By examining body movements, joint loadings, and muscle forces, biomechanical analysis allows for a detailed understanding of how different postures and motions affect the risk of injury and overall performance (Gallagher et al., 2013). Occupational ergonomists use various techniques to evaluate dynamic tasks:

- **Modeling Tools:** These tools, such as the 3D Static Strength Prediction Program (3DSSPP), simulate workplace environments and predict the impact of different postures and tasks on musculoskeletal loading (Cholewicki & McGill, 1996).
- **Video Analysis:** By capturing real-time movements, video analysis allows ergonomists to break down complex tasks into individual movements, helping identify areas of high risk for injury (Holtermann et al., 2019).
- **Mathematical Models:** These models, such as the NIOSH lifting equation, predict forces and stresses on the body during lifting or other manual tasks, offering a quantitative assessment of risk (NIOSH, 2020).
- **Measurement Tools:** Instruments such as goniometers and EMG sensors assess joint angles, muscle activity, and force production during task execution, providing empirical data to refine workplace designs (Rohmert et al., 1975).

In modern occupational ergonomics, designing workplaces that accommodate a range of postures and task demands is crucial for enhancing both worker comfort and productivity. By considering the physiological, biomechanical, psychophysical, and medical outcomes of postural variation, ergonomists can create environments that reduce the risk of MSDs and improve overall health and well-being. As workplaces continue to evolve, so too must ergonomic practices, ensuring that workers remain at the center of design efforts, with their dynamic needs met by thoughtful, evidence-based interventions (Holtermann et al., 2019).

6.5.4 Designing for the Standing Operator

In occupational ergonomics, the design of standing workstations plays a crucial role in promoting worker comfort, productivity, and health. Although sitting remains the dominant posture in many work environments, standing postures can offer significant ergonomic advantages, particularly when tasks require enhanced mobility, larger work area coverage, or exertion of greater force. To optimize standing workstations, attention must be given to several ergonomic principles, including the height of the workstation, body posture, and environmental factors that influence worker performance and well-being.

6.5.5 When Standing is Ergonomically Appropriate

Standing postures are preferable in certain work contexts where specific task requirements align better with the physiological capabilities of the body in an upright position. Key scenarios where standing may be ergonomically advantageous include:

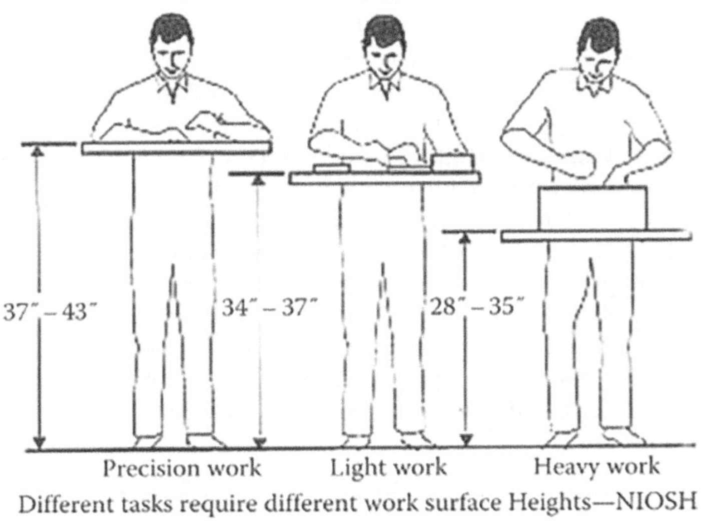

FIGURE 6.1 Standing workstation design dimensions.

- **Tasks Requiring Large Force Exertion:** When performing tasks that demand significant manual force, such as those involving hand tools or heavy objects, standing provides better leverage and allows for greater use of body weight to assist in the task (Wulff et al., 2014). This reduces the risk of overloading smaller muscle groups, such as those in the upper extremities.
- **Mobility Requirements:** In tasks that involve frequent movement or changing positions, standing allows workers to move more freely and respond quickly to dynamic work conditions. For example, tasks in assembly lines, healthcare settings, or retail often require continuous movement, which is facilitated by standing (Sullivan et al., 2019).
- **Large Workspace Coverage:** When workers need to reach across wide work areas, standing enables broader reach and easier access to tools, materials, or equipment positioned in different locations (Waters et al., 2016). This is common in jobs such as inspection, packaging, or large-scale assembly.

6.5.6 Importance of Proper Working Height in Standing Workstations

A key element in the ergonomic design of standing workstations is the height of the work surface relative to the worker's body. Incorrect working height can lead to musculoskeletal discomfort and long-term injury, particularly in the back, shoulders, and upper extremities. For optimal ergonomic design, the elbow height serves as a crucial benchmark for determining the correct working height, as it offers an ideal balance between strength and ease of movement (Kroemer et al., 2015).

- **Elbow Height as the Benchmark:** The work surface should be positioned near elbow height, which allows the worker to maintain a neutral posture without excessive reaching or bending. The most efficient working range for the arms falls slightly below the elbow level, reducing strain on the shoulders and upper arms (Gallagher et al., 2016).
- **Working Height Too High:** When the work surface is set above shoulder height, workers are forced to elevate their shoulders repeatedly, which can lead to discomfort and conditions such as shoulder impingement or tendinitis (Ciriello et al., 2017). Elevated shoulder positions are biomechanically disadvantageous, as they overburden the muscles of the neck and shoulders.
- **Working Height Too Low:** Conversely, workstations that are too low require workers to stoop or bend at the waist, placing undue stress on the lower back. Prolonged stooping can contribute to lower back pain and may increase the risk of disc herniation (Roffey et al., 2014).

6.5.7 Recommended Working Heights Based on Task Type

Research on anthropometric data provides clear guidelines on recommended working heights for various task types in standing workstations. The average elbow height varies slightly between genders, with an average height of 1,070 mm for men and 1,000 mm for women (Jürgens et al., 2014). To reduce musculoskeletal risk, the working height should be adjusted according to the task:

- **Delicate Tasks (e.g., drawing, fine assembly):** For tasks requiring precision and minimal exertion, the work surface should be positioned 50 to 100 mm above elbow height. This height facilitates close visual inspection while supporting the arms to minimize strain on the upper body (Kroemer & Grandjean, 2015).
- **Manual Tasks (e.g., light assembly, general tasks):** For tasks involving light manual labor, such as tool use or material handling, the recommended work surface height is 100 to 150 mm below the elbow level. This positioning allows for efficient use of tools while maintaining neutral postures (Bridger, 2018).
- **Heavy Tasks (e.g., woodworking, heavy assembly):** When tasks require significant manual exertion, such as lifting or manipulating heavy objects, the work surface should be 150 to 400 mm below elbow height. This configuration provides workers with more mechanical advantage by allowing them to use their upper body strength effectively (Punnett & Wegman, 2016).

6.5.8 Additional Ergonomic Considerations for Standing Workstations

Beyond the height of the workstation, several additional factors must be considered in the design of standing work environments to optimize worker comfort and minimize fatigue.

- **Foot Space and Clearance:** Adequate foot space, including sufficient toe and knee clearance, is essential for allowing workers to adjust their posture and shift their weight during tasks. This helps to distribute load across different muscle groups and reduces the static strain associated with prolonged standing (Sundstrup et al., 2016).
- **Floor Surface and Anti-Fatigue Measures:** The floor surface should be smooth and obstacle-free to reduce tripping hazards. Anti-fatigue mats, soft-soled shoes, or elastic flooring can significantly reduce the discomfort associated with standing for long periods, especially by minimizing pressure on the feet and lower extremities (King et al., 2014). Studies have shown that anti-fatigue mats reduce perceived discomfort and can contribute to lower levels of musculoskeletal strain (Cham & Redfern, 2020).
- **Sit–Stand Workstations:** To mitigate the risks associated with static standing postures, sit–stand workstations or semi-standing supports can be introduced. These designs allow workers to alternate between sitting and standing, reducing static load on the lower limbs and providing rest breaks for muscles (Parry & Straker, 2014). However, it is critical to ensure that such supports do not limit leg movement or cause compression in the lower limbs, which could lead to ischemia (Lin et al., 2017).

6.5.9 Limitations of Sit–Stand Supports

While sit–stand supports offer ergonomic benefits, they also have some limitations:

- **Stability Concerns:** Certain sit–stand supports may lack sufficient stability, particularly for workers with balance impairments or when performing tasks requiring precise movement.
- **Restricted Movement:** Some sit–stand configurations can restrict the range of motion needed for specific tasks, particularly in jobs requiring frequent position changes or reaching.
- **Ischemia Risk:** Inadequately designed sit–stand devices can result in lower limb ischemia if they prevent full leg extension or place pressure on the thighs, reducing blood flow and leading to discomfort or numbness over time (Waters et al., 2016).

In conclusion, the design of standing workstations should focus on promoting healthy postures, reducing musculoskeletal strain, and optimizing task efficiency. By carefully considering factors such as working height, task demands, and environmental conditions, ergonomic principles can be effectively applied to improve the well-being of workers and enhance overall productivity.

Table 6.1 provides more detailed dimensions for designing standing workstations for men and women.

Sufficient room for the feet of the operator must also be provided, including toe and knee space in a standing workplace. This space allows the worker to move closer to the workspace. The floor at the workstation should be flat and free of obstacles to reduce the risk of tripping. Likewise, sufficient friction between the soles of shoes and floor surfaces is necessary to reduce the risk of slips and falls. Anti-fatigue floor mats may provide some relief from foot, leg, and back discomfort during prolonged standing, although more research is needed to determine their long-term effectiveness.

To manage static loading, employers will sometimes provide stand-sit or semi-stand chairs. These offer partial support between standing and sitting, promoting comfort for specific tasks without compromising performance. However, it's crucial to choose well-designed options that ensure proper leg extension to prevent ischemia (reduced blood flow) in the legs while providing a reliable and safe option to reduce the load of excessive standing in task performance. Stability and range of motion should also be considered to avoid discomfort and limitations in task execution.

6.6 DESIGNING FOR THE SITTING OPERATOR

Humans are not biomechanically adapted for long-term static postures, whether it is standing or sitting. Therefore, sitting for extended periods of time can lead to lower back pain, foot swelling, and an increased risk of chronic diseases. While sitting requires less muscle activity than standing, recent research suggests it's important to change postures throughout the shift to reduce spinal compression and muscle fatigue.

TABLE 6.1
NIOSH Ergonomic Toolbox: General Workstation Design Principles

Tray 9-A. General Workstation Design Principles

1. Make the workstation adjustable, enabling both large and small persons to fit comfortably and reach materials easily.
2. Locate all materials and tools in front of the worker to reduce twisting motions. Provide sufficient work space for the whole body to turn.
3. Avoid static loads, fixed work postures, and job requirements in which operators must frequently hold static postures for long periods.
   a. Lean to the front or the side.
   b. Hold a limb in a bent or extended position.
   c. Tilt the head forward more than 15°.
   d. Support the body's weight with one leg.
4. Set the work surface above elbow height for tasks involving fine visual details and below elbow height for tasks requiring downward forces and heavy physical effort.
5. Provide adjustable, properly designed chairs with the following features:
   a. Adjustable seat height.
   b. Adjustable up and down back rest, including a lumbar (lower back) support.
   c. Padding that will not compress more than an inch under the weight of a seated individual
   d. Chair that is stable to floor at all times (five-leg base).
6. Allow the workers, at their discretion, to alternate between sitting and standing. Provide floor mats or padded surfaces for prolonged standing.
7. Support the limbs: provide elbow, wrist, arm, foot, and back rests as needed and feasible.
8. Use gravity to move materials.
9. Design the workstation so that arm movements are continuous and curved. Avoid straight-line, jerking arm motions.
10. Design so arm movements pivot about the elbow rather than around the shoulder to avoid stress on shoulder, neck, and upper back.
11. Design the primary work area so that arm movements or extensions of more than 15 in are minimized.
12. Provide dials and displays that are simple, logical, and easy to read, reach, and operate.
13. Eliminate or minimize the effects of undesirable environmental conditions such as excessive noise, heat, humidity, cold, and poor illumination.

Tray 9-B. Design Principles for Repetitive Hand and Wrist Tasks

1. Reduce the number of repetitions per shift. Where possible, substitute full or semiautomated systems.
2. Maintain neutral (handshake) wrist positions:
   a. Design jobs and select tools to reduce extreme flexion or deviation of the wrist.
   b. Avoid inward and outward rotation of the forearm when the wrist is bent to minimize elbow disorders (i.e., tennis elbow).
3. Reduce the force or pressure on the wrists and hands:
   a. Wherever possible, reduce the weight and size of objects that must be handled repeatedly.
   b. Avoid tools that create pressure on the base of the palm that can obstruct blood flow and nerve function.
   c. Avoid repeated pounding with the base of the palm d Avoid repetitive, forceful pressing with the finger tips.

(Continued)

TABLE 6.1 (CONTINUED)
NIOSH Ergonomic Toolbox: General Workstation Design Principles

4. Design tasks so that a power rather than a finger pinch grip can be used to grasp materials. Note that a pinch grip is five times more stressful than a power grip.
5. Avoid reaching more than 15 inches in front of the body for materials:
   a. Avoid reaching above shoulder height, below waist level, or behind the body to minimize shoulder disorders.
   b. Avoid repetitive work that requires full arm extension (i.e., the elbow held straight and the arm extended).
6. Provide support devices where awkward body postures (elevated hands or elbows and extended arms) must be maintained. Use fixtures to relieve stressful hand/arm positions.
7. Select power tools and equipment with features designed to control or limit vibration transmissions to the hands, or alternatively design work methods to reduce time or need to hold vibrating tools.
8. Provide for protection of the hands if working in a cold environment. Furnish a selection of glove sizes and sensitize users to problems of forceful overgripping when worn.
9. Select and use properly designed hand tools (e.g., grip size of tool handles should accommodate majority of workers).

Tray 9-C. Hand Tool Use and Selection Principles

1. Maintain straight wrists. Avoid bending or rotating the wrists. Remember, bend the tool, not the wrist. A variety of bent-handle tools are commercially available.
2. Avoid static muscle loading. Reduce both the weight and size of the tool. Do not raise or extend elbows when working with heavy tools Provide counter-balanced support devices for larger, heavier tools.
3. Avoid stress on soft tissues. Stress concentrations result from poorly designed tools that exert pressure on the palms or fingers. Examples include short-handled pliers and tools with finger grooves that do not fit the worker's hand.
4. Reduce grip force requirements. The greater the effort to maintain control of a hand tool, the higher the potential for injury. A compressible gripping surface rather than hard plastic may alleviate this problem.
5. Whenever possible, select tools that use a full-hand power grip rather than a precision finger grip.
6. Maintain optimal grip span. Optimum grip spans for pliers, scissors, or tongs, measured from the fingers to the base of the thumb, range from 6–9 cm The recommended handle diameters for circular-handle tools such as screwdrivers are 3–5 cm when a power grip is required, and 0.75–1.5 cm when a precision finger grip is needed.
7. Avoid sharp edges and pinch points Select tools that will not cut or pinch the hands even when gloves are not worn.
8. Avoid repetitive trigger-finger actions Select tools with large switches that can be operated with all four fingers. Proximity switches are the most desirable triggering mechanism.
9. Isolate hands from heat, cold, and vibration. Heat and cold can cause loss of manual dexterity and increased grip strength requirements excessive vibration can cause reduced blood circulation in the hands causing a painful condition known as white-finger syndrome.
10. Wear gloves that fit. Gloves reduce both strength and dexterity. Tight-fitting gloves can put pressure on the hands, while loose-fitting gloves reduce grip strength and pose other safety hazards (e.g., snagging).

(*Continued*)

TABLE 6.1 (CONTINUED)
NIOSH Ergonomic Toolbox: General Workstation Design Principles

Tray 9-D. Design Principles for Lifting and Lowering Tasks

1. Optimize material flow through the workplace by
   a. Reducing manual lifting of materials to a minimum
   b. Establishing adequate receiving, storage, and shipping facilities
   c. Maintaining adequate clearances in aisle and access areas
2. Eliminate the need to lift or lower manually by
   a. Increasing the weight to a point where it must be mechanically handled
   b. Palletizing handling of raw materials and products
   c. Using unit load concept (bulk handling in large bins or containers)
3. Reduce the weight of the object by
   a. Reducing the weight and capacity of the container
   b. Reducing the load in the container
   c. Limiting the quantity per container to suppliers
4. Reduce the hand distance from the body by
   a. Changing the shape of the object or container so that it can be held closer to the body
   b. Providing grips or handles for enabling the load to be held closer to the body
5. Convert load lifting, carrying, and lowering movements to a push or pull by providing
   a. Conveyors
   b. Ball caster tables
   c. Hand trucks
   d. Four-wheel carts

Tray 9-E. Design Principles for Pushing and Pulling Tasks

1. Eliminate the need to push or pull by using the following mechanical aids, when applicable:
   a. Conveyors (powered and non-powered)
   b. Powered trucks
   c. Lift tables
   d. Slides or chutes
2. Reduce the force required to push or pull by
   a. Reducing side and/or weight of load
   b. Using four-wheel trucks or dollies
   c. Using non-powered conveyors
   d. Requiring that wheels and casters on hand trucks or dollies have (i) periodic lubrication of bearings, (ii) adequate maintenance, and (iii) proper sizing (provide larger diameter wheels and casters)
   e. Maintaining the floors to eliminate holes and bumps
   f. Requiring surface treatment of floors to reduce friction
3. Reduce the distance of the push or pull by
   a. Moving receiving, storage, production, or shipping areas closer to work production areas
   b. Improving the production process to eliminate unnecessary materials handling steps
4. Optimize the technique of the push or pull by
   a. Providing variable-height handles so that both short and tall employees can maintain an elbow bend of 80°–100°
   b. Replacing a pull with a push whenever possible
   c. Using ramps with a slope of less than 10%

(Continued)

TABLE 6.1 (CONTINUED)
NIOSH Ergonomic Toolbox: General Workstation Design Principles

Tray 9-F. Design Principles for Carrying Tasks

1. Eliminate the need to carry by rearranging the workplace to eliminate unnecessary materials movement and using the following mechanical handling aids, when applicable:
   a. Conveyors (all kinds)
   b. Lift trucks and hand trucks
   c. Tables or slides between workstations
   d. Four-wheel carts or dollies
   e. Air or gravity press ejection systems
2. Reduce the weight that is carried by
   a. Reducing the weight of the object
   b. Reducing the weight of the container
   c. Reducing the load in the container
   d. Reducing the quantity per container to suppliers
3. Reduce the bulk of the materials that are carried by
   a. Reducing the size or shape of the object or container
   b. Providing handles or hand-grips that allow materials to be held close to the body
   c. Assigning the job to two or more persons
4. Reduce the carrying distance by
   a. Moving receiving, storage, or shipping areas closer to production areas
   b. Using powered and non-powered conveyors
5. Convert carry to push or pull by
   a. Using non-powered conveyors
   b. Using hand trucks and push carts

Source: Adapted from design checklists developed by Ridyard, D., *Appl. Ergon. Technol.*, pp. 19046–3129; DHHS (NIOSH) publication number 97–117, http://www.cdc.gov/hiosh/docs/97-117/pdfs/97-117. pdf website, accessed April 18, 2011.

When designing a workstation for a sitting operator, it is important to consider:

- **Adjustable Features:** Adjustable workstations that allow for both sitting and standing postures are becoming increasingly popular. Studies have shown that sit–stand workstations can improve employee comfort, productivity, and overall well-being.
- **Free Leg and Foot Movement:** The free space required by the legs and feet for movement in all directions should be considered.
- **Ergonomic Chair Design:** An adjustable chair which meets ergonomic design criteria should be considered. This includes features like lumbar support to promote good posture and minimal trunk flexion.

- **Working Area Height:** The preferred working area is right in front of the body at about elbow height, with the upper arms hanging. Proper monitor placement is also crucial to prevent neck strain.
- **Seat Height and Comfort:** Seat heights should range from 35 to 50 cm, and the comfort of the seat is determined by the design of the seat pan, backrest, material that touches the body and adjustability.

6.6.1 Elements of an Ergonomic Chair

The relationship between sedentary work and musculoskeletal disorders, particularly back pain, is well-documented in ergonomic and occupational health research (Lin et al., 2020). Proper chair design plays a crucial role in reducing the risk of such disorders by supporting the body's natural posture, promoting movement, and reducing strain on various body parts. This section explores the key ergonomic principles and technical specifications to consider when selecting chairs for occupational settings, emphasizing the need for customization and adjustability based on individual needs.

6.6.2 Key Ergonomic Features of a Chair

1. **Lumbar Support:** Lumbar support is one of the most critical features of an ergonomic chair, as it supports the natural inward curve of the lower spine, reducing the risk of lumbar strain and promoting proper spinal alignment (Katzman et al., 2019). A chair with adjustable lumbar support allows users to position the support in accordance with their unique spinal curvature, which is essential for maintaining optimal posture throughout the workday. Insufficient lumbar support can result in the flattening of the spine's natural curve, which contributes to discomfort and long-term musculoskeletal problems (Zemp et al., 2018).
2. **Waterfall Seat Edge:** A waterfall seat edge, where the front edge of the seat slopes gently downward, prevents excessive pressure from building up behind the thighs. This design feature is crucial for maintaining proper blood circulation to the legs, which can otherwise be restricted when the edge of the seat presses against the lower limbs, leading to numbness or discomfort (Office Ergonomics Research Committee, 2018). Poor circulation is a contributing factor to fatigue and may lead to more serious circulatory problems in long-term seated work.
3. **Adjustable Armrests:** Proper armrest design and adjustability are vital for upper body support and can significantly impact shoulder and neck posture. Adjustable armrests allow users to maintain a neutral shoulder position, reducing strain on the trapezius muscles and preventing over-elevation or depression of the shoulders (Bernard, 2019). Armrests should ideally allow for adjustment in both height and width to accommodate

various tasks, particularly those involving repetitive movements like typing or using a mouse.

4. **Seat Pan and Backrest Adjustability:** The ability to adjust seat height, depth, and backrest angle ensures that the chair can be tailored to fit the user's body size and workstation configuration. An adjustable seat pan allows users to maintain a seated position where their feet are flat on the floor or on a footrest, with knees positioned at roughly a 90° angle, reducing pressure on the thighs and lower back (Shirazi et al., 2021). Similarly, backrest angle adjustability supports varying recline preferences, which is essential for reducing spinal compression over long periods.
5. **Breathability and Material Considerations:** The chair's material can also influence user comfort, particularly in climates where heat retention may be a concern. Mesh materials, for example, enhance airflow around the body, preventing heat buildup and allowing for greater comfort during prolonged use (Coenen et al., 2017). Chairs made from less breathable materials, such as certain types of foam or leather, may cause discomfort in warmer environments due to increased perspiration.
6. **Headrest (Optional):** While not always necessary, a headrest can provide additional support during tasks that involve leaning back, such as reading or attending virtual meetings. For workers who spend a significant amount of time reclining in their chairs, an adjustable headrest can help support the cervical spine, reducing strain on the neck muscles (Kroemer, 2001).

6.7 DYNAMIC POSTURE AND MOVEMENT: THE FOUNDATION OF ERGONOMIC DESIGN

Traditional ergonomics promoted the idea of a single "correct" posture for seated work, often emphasizing an upright, static posture. However, recent ergonomic research has shifted toward a more dynamic approach, emphasizing the importance of movement and frequent posture changes during seated work (NIOSH, 2020). Workers are now encouraged to avoid remaining in one position for extended periods and instead adopt a variety of postures throughout the workday to prevent fatigue and musculoskeletal strain. This approach is supported by findings that prolonged static postures increase the risk of musculoskeletal disorders by placing continuous pressure on certain body parts without allowing time for recovery (Straker et al., 2019).

The modern ergonomic chair should facilitate this dynamic approach by providing easy adjustability and supporting a range of postures. For instance, a chair that allows for recline encourages users to periodically change their position and take pressure off the spine. Furthermore, promoting micro-movements—such as tilting, shifting weight, or adjusting sitting angles—helps prevent the cumulative effects of static loading on the musculoskeletal system.

6.7.1 Avoiding Problematic Chair Designs

Not all design features commonly found in office chairs are ergonomically beneficial. For example, forward-declining seats, sometimes seen in minimalist or "active sitting" designs, have been shown to increase discomfort and fatigue over long periods by placing excessive pressure on the shins and encouraging slouched postures. Such designs may also promote slipping forward in the chair, reducing overall stability and increasing the risk of poor posture. Chairs that lack adequate lumbar support or are too hard and flat may also lead to long-term discomfort and contribute to the development of work-related musculoskeletal disorders (Punnett & Wegman, 2016).

6.7.2 Conclusion: Selecting the Right Chair for Occupational Health

Choosing an ergonomic chair is a fundamental aspect of designing a workstation that prioritizes musculoskeletal health. Key features such as lumbar support, adjustable armrests, and seat adjustability must be carefully considered to match the needs of the individual worker. Moreover, promoting movement and posture variation throughout the day is crucial in preventing work-related musculoskeletal disorders. By integrating ergonomic principles into chair design and use, workplaces can reduce the risk of musculoskeletal injuries and improve overall worker comfort and productivity.

6.8 WORKING IN CONFINED AND AWKWARD SPACES

Confined space operations present significant hazards across various industries, leading to substantial risks of injury and fatality. These operations are commonly encountered in sectors such as construction, mining, utilities, space/aerospace industry, and wastewater management. Confined spaces are defined by their limited entry and exit points, restricted air circulation, and the potential for hazardous conditions, making them inherently dangerous. According to OSHA, confined spaces include tanks, vessels, silos, storage bins, hoppers, vaults, and pits (OSHA, 2019). Despite regulatory improvements, confined spaces continue to pose substantial risks, necessitating ongoing research and the implementation of ergonomic solutions to minimize worker harm.

6.8.1 Confined Space Incidents: Prevalence and Data

Recent data from the Bureau of Labor Statistics (BLS) underscores the continued danger of confined space operations. In 2016, confined spaces were identified as a leading contributor to workplace fatalities, particularly in industries such as construction and manufacturing (BLS, 2018). While comprehensive, real-time data on confined space incidents are not always readily available, the BLS report highlighted that atmospheric hazards and physical entrapment were key factors in these fatalities. This indicates a pressing need for further study and intervention strategies to address confined space-related injuries and deaths.

6.8.2 Key Ergonomic Risk Factors in Confined Space Operations

Research into confined space hazards has consistently identified several key risk factors that workers may encounter. Rekus (2012) and subsequent studies have highlighted the following ergonomic and environmental risks:

1. **Atmospheric Hazards:** These include oxygen deficiency, accumulation of toxic gases (e.g., hydrogen sulfide, carbon monoxide), and flammable vapors, which may lead to asphyxiation, poisoning, or explosion (Zhao et al., 2019).
2. **Physical Hazards:** Confined spaces may involve risks such as crushing from collapsing structures, falling objects, electrical hazards, and physical entrapment due to narrow openings and constrained work areas (Rekus, 2012).
3. **Manual Handling and Physical Strain:** Workers in confined spaces are often required to perform tasks in awkward postures, increasing the risk of musculoskeletal injuries (McAtamney & Corlett, 1993). Tasks such as lifting, pulling, or working overhead in tight spaces place substantial biomechanical stress on the body, particularly the lower back, shoulders, and knees (Putz-Anderson et al., 2017).
4. **Noise and Vibrations:** Confined spaces often amplify noise levels, leading to auditory damage or distraction, which increases the risk of accidents. Additionally, vibrations from heavy machinery may exacerbate physical strain (Burdorf et al., 2015).
5. **Psychological Stressors:** Confined space operations can also trigger psychological challenges, including claustrophobia, anxiety, and fatigue, which may compromise focus and situational awareness (Chen et al., 2014). High cognitive load, combined with the physical demands of the task, can reduce worker performance and increase the likelihood of accidents.

6.8.3 Ergonomic and Safety Interventions

Addressing ergonomic concerns in confined space operations is essential for reducing the biomechanical and psychological strain experienced in confined spaces. Several strategies can be employed to mitigate the risks:

1. **Task Design and Workstation Layout:** Ensuring that tools and equipment are designed for easy use in confined spaces is paramount. This includes providing ergonomic tools that reduce awkward postures and using adjustable work platforms to minimize the need for bending, stooping, or overreaching (Corlett, 2016).
2. **Personal Protective Equipment (PPE):** PPE, such as respiratory protection and fall harnesses, must be carefully designed to ensure comfort and usability in confined spaces. For example, lightweight, flexible materials

should be prioritized to prevent additional physical strain (Burdorf et al., 2015).

3. **Work–Rest Cycles:** The physical and cognitive demands of confined space work necessitate carefully planned work–rest cycles to prevent overexertion. Studies have shown that frequent breaks can reduce fatigue and improve overall performance in physically demanding environments.
4. **Automation and Robotics:** Whenever possible, the use of robotics and automation to perform high-risk tasks can reduce worker exposure to hazardous conditions (Beniwal et al., 2017). This is particularly important in tasks involving hazardous materials or extreme environmental conditions.

6.8.4 Psychological Considerations

Confined space work also presents psychological challenges, such as maintaining focus and situational awareness in a high-stress environment. Claustrophobia and anxiety are common, as the restricted movement and isolation can contribute to mental strain. Ergonomic principles, such as task rotation and cognitive training, can help alleviate some of these psychological burdens. Training programs should incorporate techniques to enhance workers' ability to manage stress, stay focused, and maintain situational awareness in confined environments (Chen et al., 2014).

6.8.5 Regulatory Framework and Ergonomics Integration

The primary regulatory standard for confined space operations in the United States is OSHA's 29 CFR 1910.146, "Permit-Required Confined Spaces" (OSHA, 2019). This regulation mandates a comprehensive confined space entry program that includes identification, permitting, atmospheric testing, worker training, emergency response, and rescue procedures. Ergonomic considerations should be integrated into these regulatory requirements to enhance worker safety and well-being. For example, including ergonomic risk assessments as part of the permitting process can help identify tasks or environments that pose heightened biomechanical or cognitive risks.

6.8.6 Future Research Directions

Despite existing research, there remains a need for further investigation into the ergonomic challenges associated with confined space work. Future studies should focus on:

1. **Developing a Comprehensive Risk Factor Matrix:** specific to various industries that frequently encounter confined spaces. Such a matrix would facilitate targeted ergonomic interventions.
2. **Biomechanical and Physiological Studies:** that examine how workers' bodies respond to the physical demands of confined space work. This

includes assessing the impact of posture, force exertion, and repetitive motions.
3. **Cognitive Workload Analysis:** to better understand the psychological demands of confined space tasks, particularly in high-stress, time-sensitive operations.

6.9 DESIGNING FOR HAND USE

The human hand is an intricate and highly functional part of the human anatomy. It consists of over 27 bones, 29 major joints, 48 nerves, and 123 ligaments. This complex network enables a wide range of postures, movements, and dexterity, making the hand incredibly versatile. In various occupational environments, the hand is extensively used for physical tasks, capable of performing a large variety of activities that can be broadly categorized as follows:

1. **Fine Manipulation of Objects:** These activities require minimal displacement and force. Examples include writing, adjusting controls, and assembling small parts.
2. **Rapid Movements Toward an Object:** These movements demand moderate accuracy and small exertions of force. Examples include reaching for a switch and operating it.
3. **Frequent Movements Between Targets:** These tasks require high accuracy with little force, such as in assembly tasks involving picking parts from bins and assembling them.
4. **Forceful Activities with Little or Moderate Displacement:** These activities include tasks like turning a hand tool against resistance.
5. **Forceful Activities with Large Displacement:** Examples include using a hammer to drive a nail into a wall.

Advancements in ergonomic research over the past decade have further emphasized the importance of healthy hand function in occupational health. Understanding the biomechanics and functional capacities of the hand is crucial in designing tools and workspaces that optimize human performance and reduce injury risks.

Three key considerations when designing tasks for hand use are accuracy, strength, and displacement. Fitts' Law (Fitts, 1954) remains a foundational source for guiding efficient and accurate hand movements. It predicts the time required to move to a target on a surface and can be used to optimize task design for accuracy.

6.9.1 Fitts' Law: A Balancing Act

The core principle of Fitts' Law is that movement time is a function of task difficulty. This difficulty is directly related to the distance to the target and inversely related to the size of the target. Larger, closer targets are easier to reach than smaller, farther ones.

The mathematical formula for Fitts' Law captures this relationship:

$$MT = a + b \log_2(2A/W)$$

Where:

- MT = Movement Time
- a & b = Empirically determined constants
- A = Distance from the starting point to the center of the target
- W = Width of the target

6.9.2 Recent Developments in Fitts' Law

While the core principle remains strong, research suggests Fitts' Law might benefit from incorporating both target dimensions. A 2019 study by Gillan (Gillan, 2019] found that Fitts' Law may need to account for the width of the target perpendicular to the movement direction as well. This suggests that a wider target in general (not just in the movement direction) might be easier to reach.

These refinements highlight the ongoing exploration of Fitts' Law and its role in optimizing user experience design.

The term $\log_2(2A/W)$, known as the Index of Difficulty, measures the difficulty of a motor task based on distance and target size. Fitts' Law is a cornerstone for designing tasks involving hand movements, particularly for precision. Interestingly, Fitts' Law can also be applied to foot motion.

Drury's modification of Fitts' Law (Drury, 1975) predicts movement time for operating foot pedals. It defines the Index of Difficulty as:

$$ID = \log_2 (A/W+S) +.05$$

Where:

- ID = Index of Difficulty
- A = Movement Amplitude (distance) from start to centerline of target
- W = Width of the target
- S = Shoe sole width

This formula considers shoe sole width, an important factor in foot control design. By incorporating Fitts' Law principles, designers can optimize foot controls, pedals, and other foot-activated devices for ease of use and efficiency, ultimately improving user experience.

6.9.3 Hand Tool Design

Designing ergonomic hand tools is a multidimensional and complex process that requires the integration of several factors to ensure both user safety and efficiency. A well-designed tool should reduce the risk of injury while enhancing task performance by optimizing human interaction with the tool (Dul & Weerdmeester, 2014). Below, key considerations for ergonomic hand tool design are discussed from a scientific and technical perspective.

6.9.4 Fit and Anthropometry

One of the fundamental principles of ergonomic hand tool design is ensuring that the tool fits the user's hand size. Anthropometric data plays a crucial role in this, as hand tools must be adaptable to a wide range of hand sizes and shapes to minimize excessive force during use. A study by Wimer et al. (2017) demonstrated that tools poorly matched to hand dimensions can significantly increase grip force, thereby raising the risk of muscle fatigue and injury. Properly fitting tools allow for a comfortable grip, reducing the likelihood of muscular strain during prolonged use.

6.9.5 Strength and Energy Efficiency

Hand tools should be designed to take advantage of the user's natural strength while minimizing energy expenditure. This concept relates to the mechanical advantage offered by the tool, which can either enhance or hinder user performance. Tools with optimized mechanical leverage, such as ergonomic pliers with spring-assist mechanisms, allow users to perform tasks with less force, conserving energy and reducing muscle fatigue (Hedge, 2016). Tools that fail to account for user strength levels can lead to excessive force application, increasing the risk of musculoskeletal disorders.

6.10 POSTURE AND BIOMECHANICS

Neutral wrist and arm postures are essential in reducing biomechanical stress during tool use. Tools that promote awkward postures—such as wrist deviation or arm elevation—place excessive strain on the musculoskeletal system, particularly the upper limbs. A neutral posture, where the wrist is aligned with the forearm, minimizes joint and tendon stress. Studies show that tools designed with inappropriate handles or angles can force users into non-neutral positions, significantly increasing the risk of conditions like carpal tunnel syndrome and tendinitis (Mendell et al., 2015). Tools should, therefore, be designed to encourage proper postural alignment, enhancing biomechanical efficiency.

6.10.1 Consequences of Poor Ergonomic Design

Ignoring ergonomic principles in hand tool design has substantial health implications, particularly in occupational settings where tools are used repetitively or for extended periods. Poorly designed tools can contribute to a wide range of musculoskeletal and vascular disorders in the hands and upper extremities.

TABLE 6.2
Example of an MTM Table of Motion Times

		Methods-Time Data Table I-Reach			Leveled Time TMUs			
		Le						
Case	Description	Distance moved (in)	A STD.	A Hand in Mot	A with C, D or B	B Hand in Mot	C or D	E
A	Reach to object in	1			2 1		3 6	
	fixed location or to	2			4 3		5 9	
	object in other hand	3			5 9		7 3	
	or on which other hand rests	4	6 1	4 9	7 1	4 3	8 4	6 8
B	Reach to single object	5	6 5	5 3	7 8	5 0	9 4	7 4
	in location that may	6	7 0	5 7	8 6	5 7	10 1	8 0
	vary slightly from	7	7 4	6 1	9 3	6 5	10 8	8 7
	cycle to cycle	8	7 9	6 5	10 1	7 2	11 5	9 3
C	Reach to object in	9	8 3	6 9	10 8	7 9	12 2	9 9
	group	10	8 7	7 3	11 5	8 6	12 9	10 5
		12	9 6	8 1	12 9	10 1	14 2	11 8
		14	10 5	8 9	14 4	11 5	15 6	13 0
D	Reach to very small	16	11 4	9 7	15 8	12 9	17 0	14 2
	object or where	18	12 3	10 5	17 2	14 4	18 4	15 5
	accurate grasp is	20	13 1	11 3	18 6		19 8	16 7
	required	22	14 0	12 1	20 1		21 2	18 0
E	Reach to indefinite	24	14 9	12 9	21 5		22 5	19 2
	location to get hand	26	15 8	13 7	22 9		23 9	20 4
	in position for body	28	16 7	14 5	24 4		26 3	21 7
	balance or next motion or out of way	30	17 5	15 3	25 8		26 7	22 9

Source: Maynard et al. (1948).

6.10.2 Acute Musculoskeletal Disorders

Acute injuries resulting from improper hand tool design can manifest as pain, discomfort, or inflammation in muscles, tendons, and joints. Such conditions are typically short-term but can become chronic if ergonomic factors are continually ignored. In a systematic review, Davis and Kotowski (2015) found a significant relationship between hand tool use and the onset of acute musculoskeletal symptoms, such as hand and wrist strain, often observed in industries like manufacturing and construction.

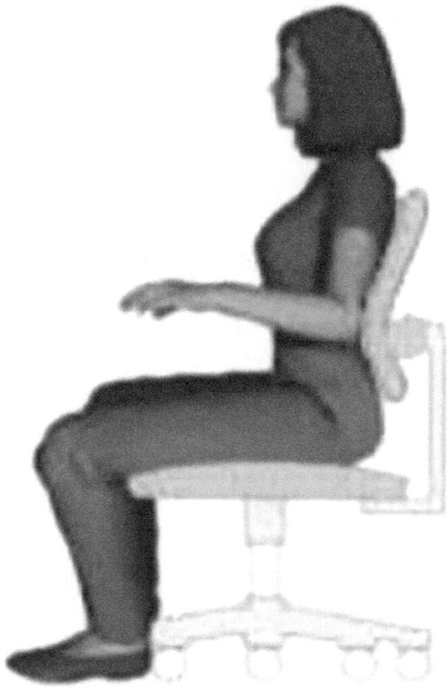

FIGURE 6.2 Seated worker.

TABLE 6.3
Nonfatal Occupational Injuries Due to Hand Tools in Major US Industry Divisions (2002)

Industry	Total Number	Due to Hand Tools
All private industry	1,436,194	66,588
Agriculture, forestry, and fishing	31,520	2,185
Mining	11,355	758
Construction	163,641	14,439
Manufacturing	280,005	16,032
Transportation and public utilities	168,632	3,116
Wholesale and retail trade	372,192	19,179
Finance, insurance, and real estate	36,689	1,399
Services	372,159	9,443

Source: Kumar, S. (Ed.), *Biomechanics in Ergonomics*, 2nd edn., CRC Press, New York, 2008.

TABLE 6.4
Incidence of Nonfatal Occupational Injuries in the United States (2002) Due to Different Types of Non-Powered Hand Tools Classified by Major US Industry Divisions

Tool Type	Total Number	Agriculture, Forestry, and Fishing	Mining	Construction	Manufacturing	Transportation and Public Utilities	Wholesale Trade	Retail Trade	Finance, Insurance, and Real Estate	Services
A	162.6	208.6	198.8	276.8	174.5	270.6	175.0	153.6	53.5	133.8
B	10.6	16.2	15.6	30.6	12.0	12.3	8.0	10.8	3.0	7.0
C	5.1	9.4	11.9	13.5	5.8	3.6	3.9	7.9	1.6	2.3
D	0.1	—	0.2	0.3	0.1	—	0.1	—	—	0.0
E	2.7	5.2	0.5	5.1	2.3	0.7	2.1	6.3	0.8	1.2
F	0.5	2.0	—	3.0	0.4	0.7	0.2	0.1	0.4	0.3
G	0.1	—	3.0	0.2	0.3	0.2	—	—	—	0.1
H	0.1	—	—	—	0.1	0.0	—	0.1	—	0.1
I	0.5	0.6	2.3	2.6	0.9	0.5	0.5	0.1	0.2	0.1
J	0.1	—	—	0.2	0.1	—	—	—	—	0.0
K	0.4	0.6	2.8	0.9	0.9	0.7	0.6	0.2	0.1	0.1
L	0.6	0.8	2.9	1.1	0.6	0.7	0.3	1.0	0.1	0.4

Source: Kumar, S. (Ed.), Biomechanics in Ergonomics, 2nd edn., CRC Press, New York, 2008.

Note: A, total in all private industries; B, tools, instruments, and equipment; C, non-powered hand tools; D, non-powered boring hand tools; E, non-powered cutting hand tools; F, non-powered digging hand tools; G, non-powered gripping hand tools; H, non-powered measuring hand tools; I, non-powered striking and nailing hand tools; J, non-powered surfacing hand tools; K, non-powered turning hand tools; L, non-powered hand tools such as crowbars, pitch-forks, etc.

6.10.3 Chronic Musculoskeletal Disorders

Long-term exposure to poorly designed hand tools increases the risk of chronic musculoskeletal disorders, such as carpal tunnel syndrome, tendinitis, and tenosynovitis. Carpal tunnel syndrome, in particular, is linked to repetitive hand tool use with non-neutral wrist postures and high force exertion (Rempel et al., 2021). These chronic conditions can severely impact worker productivity and quality of life, leading to prolonged medical treatments and even surgical intervention.

6.10.4 Vascular Disorders

Hand tools that generate significant vibration levels—such as pneumatic drills or grinders—can cause vascular issues in the hands. Vibration-induced white finger (VWF), also known as hand-arm vibration syndrome (HAVS), is a common condition among workers exposed to prolonged vibration. This disorder is characterized by reduced blood flow to the fingers, leading to numbness, tingling, and, in severe cases, tissue damage. Research by Griffin (2014) indicates that hand tools exceeding vibration exposure limits dramatically increase the risk of developing VWF, particularly in jobs requiring high-frequency tool use.

6.10.5 Vibration-Induced Disorders

Prolonged exposure to hand tools that generate high vibration levels can lead to vibration-induced disorders such as VWF. Vibration-induced white finger is caused by restricted blood flow due to mechanical vibration, leading to numbness, tingling, and loss of dexterity in the fingers. According to a study by Niu et al. (2016), the risk of developing VWF increases with the duration and intensity of tool vibration, making it essential to design tools that minimize vibration exposure as well as offering personal protective equipment to reduce the impact and exposure of the vibration.

Ergonomic hand tool design is a critical component in reducing occupational injuries and enhancing worker performance. By considering anthropometry, biomechanics, and minimizing vibration and force, tool designers can significantly reduce the incidence of musculoskeletal and vascular disorders. Failure to account for these ergonomic factors can result in acute and chronic injuries, leading to lost productivity, increased medical costs, and long-term health impacts on workers.

Hand grip on a tool, also known as tool coupling, is crucial for determining both the amount of force transferred and the precision of that transfer. According to Kroemer (2001), ten generally recognized tool couplings include:

1. **Digit Touch:** A single finger makes contact with the object.
2. **Palm Touch:** A section of the palm or hand touches the object.
3. **Finger Palmar Grip (Hook Grip):** One or more fingers hook onto a ridge or handle, often used when thumb opposition isn't required.
4. **Thumb–Fingertip Grip (Tip Pinch):** The tip of the thumb opposes the tip of a single finger.
5. **Thumb–Finger Palmar Grip (Pad Pinch or Plier Grip):** The pad of the thumb opposes the pad(s) of one or more fingers near their tips.

6. **Thumb–Forefinger Side Grip (Lateral Grip or Side Pinch):** The thumb opposes the side of the forefinger.
7. **Thumb–Two-Finger Grip (Writing Grip):** The thumb and two fingers, typically the forefinger and middle finger, oppose each other near their tips.
8. **Thumb–Fingering Enclosure (Disk Grip):** The pads of the thumb and three or four fingers oppose each other near the tips, without the object touching the palm.
9. **Finger–Palm Enclosure (Collet Enclosure):** Most or all of the inner hand surface makes contact with and encloses the object.
10. **Power Grasp:** The entire inner surface of the hand grasps a cylindrical handle positioned parallel to the knuckles.

Hand tools are devices used for various tasks, requiring the user to generate all the force for operation. This allows for precise control over the applied energy, except in the case of some hammering motions. Here's a breakdown of common hand tool actions and examples:

- **Percussive (Striking):** These tools transfer force through a swinging or forceful push/pull on the handle. Examples include hammers, mallets, and chisels.
- **Scraping:** Scraping tools remove material by pushing, pulling, or holding the handle against the workpiece. Examples include putty knives, paint scrapers, and files.
- **Rotation or Boring:** These tools create holes or drive fasteners by turning, pushing, pulling, or holding the handle. Examples include hand drills, screwdrivers (including manual ratcheting screwdrivers), and awls.
- **Squeezing:** Squeezing tools apply concentrated force through a squeezing action on the handle. Examples include pliers, clamps, and locking pliers.
- **Cutting:** Cutting tools sever materials by pulling or pushing a sharp blade while holding the handle. Examples include scissors, knives, hacksaws, and coping saws.

When operating power tools, the user controls the tool's position and movement, while an external power source generates the force. Sudden resistance during use can cause injuries, and vibrations can lead to overuse disorders (OUDs) or cumulative trauma disorders (CTDs). Here are some key points for safe and comfortable use of both power tools and hand tools:

- **Hand Posture:** Proper hand and wrist position are crucial. The wrist should remain straight, avoiding strain on muscles, tissues, nerves, and blood vessels. Ideally, the hand's grasp centerline should be around 70° to the forearm axis. It's better to bend the tool at the working joint then bend your wrist.
- **Handle Design:** Tool handles should be designed for a comfortable and secure grip, following the natural form of the hand. While some bulge or texture can improve grip, it shouldn't compromise comfort during extended use. Friction surfaces can also enhance grip and reduce the risk of dropping tools.

6.11 HAND TOOL DESIGN PRINCIPLES (ADAPTED FROM KROEMER, 2001)

- **Force Direction:** Whenever possible, push or pull the tool in the same direction as your forearm, with the handle positioned directly in front of it.
- **Wrist Position:** Maintain a straight wrist for optimal control and comfort.
- **Grip and Handle Interaction:** The handle shape and surface texture should provide a secure and comfortable grip.
- **Pressure Point Avoidance:** Design the handle to avoid creating pressure points or pinch points on the hand.
- **Vibration Reduction:** Minimize vibrations transmitted to the hand through the tool design or appropriate damping materials.
- **Automation for Repetitive Tasks:** When feasible, consider using automated tools for repetitive or forceful tasks to reduce strain on the operator's hands and arms.

The ergonomic design of hand tools significantly impacts operator comfort and safety. A study by Dababneh et al. (2004) evaluated the ergonomic impact of non-powered hand tools commonly used in construction and carpentry. The researchers developed a 16-item checklist to assess various ergonomic factors, including handle size, shape, grip strength requirements, and vibration levels.

6.11.1 Power Hand Tools

Musculoskeletal disorders are a significant concern in workplaces that rely heavily on power tools. Recognizing this issue, the National Institute for Occupational Safety and Health (NIOSH) conducted a study titled "Ergonomic Study of Power Hand Tools in the Automotive Manufacturing Industry" in 1994. This research established a framework for evaluating the ergonomic design of power tools, focusing on key factors that can influence operator comfort and safety.

6.11.2 Key Ergonomic Evaluation Factors

- **Tool Weight and Balance:** A lighter tool with balanced weight distribution reduces fatigue and improves control during operation.
- **Grip Design:** Handle size, shape, and texture significantly impact grip comfort and stability. An ergonomic grip should fit the natural curve of the hand and minimize pressure points.
- **Torque Reaction:** Power tools can generate kickback or twisting forces during use. Ergonomic design should minimize these forces to prevent injuries.
- **Vibration:** Exposure to excessive vibration can lead to hand-arm vibration syndrome (HAVS) and other health issues. Tools with low vibration levels are preferred.

First thing first: The tool will do the job with the desired quality and will last as expected:
Yes: continue with the checklist;
No: reject the tool.

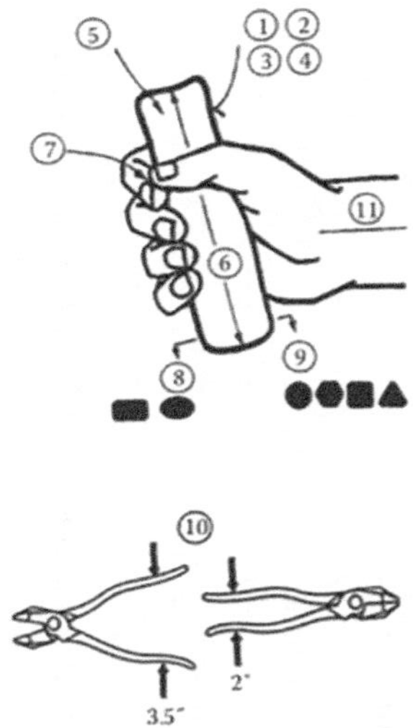

Checklist for the Ergonomic Evaluation of Hand Tools

Considering the job the tool is designed for and the work environment, respond to each item on the checklist by "Yes," "No," or "NA" (not applicable). Place the score that corresponds to your response in the "score" column. Add the scores of all items to get the total score of the tool. Maximum score is 100.
(*Items 7, 8, 9, and 10 are not applicable for all tools.)

Item	Ergonomic Feature	Yes	NA	No	Score
1	Grip surface is non-slippery.	+10		0	
2	Grip surface does not have sharp edges, undercuts, deep ribs, and/or finger grooves.	+10		0	
3	Grip surface is electrically insulated; tool handle is either made of wood or coated with rubber or soft plastic.	+10		0	
4	Grip surface is thermally insulated; it will not get hot or cold quickly when working in a hot or cold environment.	+2		0	
5	Handle is made of wood, or grip surface is coated with semi-pliable material; not too hard and not too soft, similar to the rubber used in the soles of sport shoes.	+10		0	
6	Grip length is 4 – 6"; handle does not end inside the palm of the hand.	+10		0	
7	For one-handle tools: Size of handle cross section is not too small or too large. The index finger and the thumb are allowed to overlap by 3/8" when gripping (for hammers and hammer-like tools, overlap of 1" is acceptable).	+8*	0	0	
8	For one-handle tools other than screwdrivers: Shape of handle cross section is oval or rounded-edge rectangular.	+2*	0	0	
9	For screwdrivers: The basic shape of handle cross section is circular, hexagonal, square, or triangular.	+2*	0	0	
10	For two-handle tools (plier-like): Grip span is greater or equal than 2" when fully closed and less than or equal to 3.5" when fully open.	+10*	0	0	
11	Angle of the handle is formed so that the work can be done keeping a straight wrist.	+10		0	
12	The tool weight is less than 5 lb.	+10		0	
13	The tool can be used with either hand.	+2		0	
14	The tool can be used with the worker's dominant hand.	+10		0	
15	The tool will allow a two-handed operation (using both hands at the same time).	+4		0	
16	The tool and accessories are clearly marked and/or color coded so they are easy to identify; colors are bright and tool contrasts with the surroundings of the work area.	+2		0	
	Total score of the tool (100 points possible)				

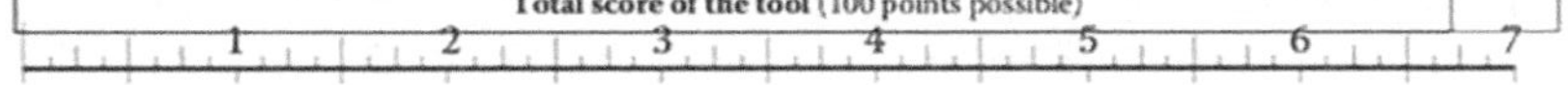

Measurement scale (inches)

FIGURE 6.3 Checklist non-powered hand tools.

6.11.3 Beyond the 1994 Study

While the 1994 study provides a valuable foundation, the ergonomic evaluation of power tools has evolved in recent years. Here are some additional factors considered in more contemporary evaluations:

- **Noise Level:** Excessive noise can contribute to hearing loss and fatigue.
- **Ease of Use:** Intuitive controls and clear user interfaces can reduce operator strain.
- **Tool Accessibility:** Easy access to controls and comfortable working postures are crucial for preventing awkward reaching or bending.

Ergonomic evaluation plays a vital role in selecting and using power tools safely and comfortably. By considering the factors mentioned above, workplaces can minimize the risk of MSDs and promote operator well-being.

Vibration is a major concern for workers who use power tools regularly. These tools can generate vibrations at low, moderate, and high frequencies. In severe cases, high vibration levels can damage blood vessels and nerves in the hands, leading to long-term health problems.

6.11.4 Hand–Arm Vibration Syndrome (HAVS)

The risk of vibration-related injuries increases with frequent or prolonged use of vibrating power tools. HAVS is a musculoskeletal condition caused by repeated exposure to vibration from pneumatic, electric, hydraulic, or gasoline-powered hand tools. This condition can be permanent and debilitating.

6.11.5 Common Power Tools that Generate Vibration

Power hand tools come in various forms, including:

- Electric drills
- Staple guns
- Pneumatic torque wrenches
- Nail guns
- Angle grinders
- Floor sanders

These are just a few examples; it is important to be aware of the potential vibration risks associated with any power tool being used.

6.11.6 Additional Considerations

- Look for tools with lower vibration ratings when possible.

- Implement work practices that minimize vibration exposure, such as limiting daily use time and taking regular breaks.
- Employers should provide proper personal protective equipment (PPE) such as vibration-dampening gloves.
- Workers experiencing symptoms of HAVS should seek medical attention promptly.

While power tools offer convenience and efficiency, they also present various safety hazards. Operators can be exposed to:

- **Falling or Flying Objects:** Debris ejected during operation can cause injuries.
- **Abrasive or Splashing Materials:** Sparks from grinding tools or hot liquids from power washers can cause burns.
- **Harmful Airborne Contaminants:** Dust from sanding, fumes from painting, and vapors from solvents pose respiratory risks.

6.11.7 Five Basic Rules for Safe Power Tool Use

Following these fundamental safety practices from OSHA (2011) can significantly reduce the risk of injuries when using power tools:

1. **Maintain Your Tools:** Regularly inspect and maintain tools to ensure they are in proper working order. Dull blades, damaged cords, or malfunctioning parts can increase the risk of accidents.
2. **Match the Tool to the Task:** Always use the right tool for the job. Using the wrong tool can be ineffective and potentially dangerous.
3. **Inspect Before Use:** Before each use, thoroughly examine the tool for any damage, loose parts, or worn components. Do not operate a faulty tool.
4. **Follow the Manual:** Read and understand the manufacturer's instructions for the safe operation and maintenance of the specific tool you're using.
5. **Protect Yourself:** Wear appropriate personal protective equipment (PPE) based on the potential hazards associated with the tool and the task. This may include safety glasses, gloves, respirators, hearing protection, or face shields.

6.12 DESIGNING FOR FOOT OPERATION

Foot controls offer a distinct advantage: they can generate significant force. However, compared to hand movements, foot operation is generally slower, less precise, and requires more energy expenditure.

6.12.1 Foot Controls and Posture

Posture plays a crucial role in effective and comfortable foot control use. Here's a breakdown of considerations for seated and standing operators:

- **Seated Operators:** For tasks requiring repeated foot actuation, a seated position is preferred. The seat provides support, allowing for easier foot movement.
    - **Light Downward Forces:** Ideal knee angles are between 105–110°.
    - **Strong Forward Forces:** Aim for knee angles between 135–155°, ideally with good back support from the seat.
- **Standing Operators:** Controls and workstations for standing operators should allow for comfortable knee bending. Avoid tasks that overstress the foot through excessive pedal use. Ensure the foot controls are positioned within a comfortable reach zone.

6.12.2 Foot Control Design Principles (adapted from Helander, 2020)

- **Seated Operation Preferred for Frequent Use:** Limit repeated foot control use to seated positions for better ergonomics.
- **Pushing Direction:** Design foot controls for pushing motions that align with the lower leg for optimal force application.
- **Force and control:**
    - **Small Forces:** Utilize ankle tilting for applying small amounts of force.
    - **Large Forces:** Engage the entire leg for exerting larger forces, ideally with proper back support when seated.
- **Prioritize Control Needs:** Avoid designs requiring simultaneous fine control, continuous operation, and quick foot movements.

6.13 WORK ENVELOPES

The primary workstation, also known as the work envelope, should be designed to accommodate the natural reach of both men and women. This space defines the optimal zone for placing tasks, equipment, and tools to minimize excessive reaching, awkward postures, and discomfort.

6.13.1 Key Considerations for Standing Work Envelopes

- **Minimize Torso Bending:** Standing workers should avoid excessive forward bending. The functional workspace should primarily rely on arm and hand extension.
- **Accommodate the 5th Percentile Female:** The workstation design should ensure comfortable reach for workers of smaller stature (typically the 5th percentile female).
- **Frequent Tasks within Elbow Reach:** Frequently performed tasks should be positioned within easy reach, ideally with elbows at the sides, reducing overall reach distance.
- **Tailored to the Workforce:** Workspace dimensions should be based on the specific anthropometry of the employee population and consider all equipment and tools used during tasks.

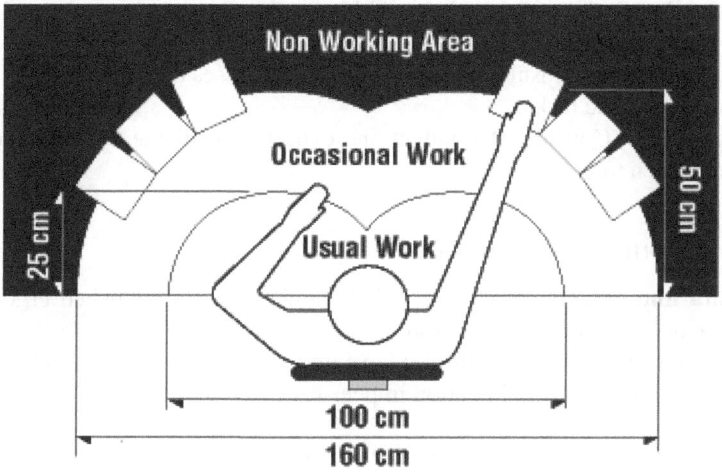

FIGURE 6.4 Ergonomic chair range of dimensions. Source: Canadian Centre for Occupational Health and Safety https://www.ccohs.ca/oshanswers/ergonomics/standing/standing_basic.html

6.14 WORKPLACE EVALUATION TOOLS

Ergonomic analysis is crucial for creating safe and comfortable workplaces. Selecting the appropriate tools is essential for a thorough and effective evaluation.

6.14.1 Helpful Resources

- ***The Occupational Ergonomics Handbook* (Marras et al., 2009):** This comprehensive resource provides guidance on various ergonomic analysis tools and techniques.
- **Dr. Thomas Bernard's *Ergonomic Analysis Tools Summary* (University of South Florida):** Dr. Bernard's work offers a valuable summary of ergonomic analysis tools. Consider searching for the latest update available on the University of South Florida website.

6.14.2 Choosing the Right Tool

The selection of an ergonomic analysis tool depends on several factors:

- **Work Environment:** Consider the specific work setting being evaluated (office, factory, etc.)
- **Risk Level:** Identify the potential ergonomic risks associated with the tasks performed in the workplace.
- **Risk Types:** Determine whether MSDs, fatigue, or other ergonomic issues are the types of risks associated with the chosen tool.

- **User Population:** The tool should be suitable for the physical capabilities of the workforce.
- **Compatibility:** Ensure the tool effectively assesses the relevant risk factors for the specific tasks.
- **Resources:** Consider the budget and expertise available for implementing the chosen tool.

6.14.3 Additional Considerations

- **Data and Norms:** The chosen tool should utilize established ergonomic data and norms for comparison.
- **Evaluation:** Look for tools that have undergone some form of independent evaluation, such as publication in peer-reviewed journals.

JACK Ergonomic Software is a powerful human modeling and simulation tool developed by the University of Pennsylvania (Siemens, 2024). This software empowers researchers, designers, and product developers with a virtual environment to evaluate various design and system parameters critical for ergonomics, biomechanics, and human physiology.

6.14.4 Benefits of JACK Software

- **Improved Design and Workplaces:** JACK allows users to position biomechanically accurate digital humans of various sizes within virtual environments. This enables the evaluation of product designs and workplace tasks for potential ergonomic issues before physical prototypes are built or changes are implemented in the real world.
- **Enhanced Ergonomics and Biomechanics Analysis:** Digital humans in JACK can be assigned tasks and their movements analyzed. This virtual simulation provides realistic conditions for in-depth ergonomic assessments, helping identify potential risks associated with musculoskeletal strain, discomfort, or injury.
- **Streamlined Development Process:** By identifying and addressing ergonomic issues early in the design phase, JACK software can help streamline the development process, potentially saving time and resources.

Ergonomic software like JACK empowers early identification of potential ergonomic risks associated with product designs and workplace layouts. This foresight enables adjustments to be made during the design phase, preventing costly rework and potential worker injuries later.

6.15 RAPID UPPER LIMB ASSESSMENT TOOL (RULA)

RULA is a quick and effective method for assessing the risk of musculoskeletal disorders associated with upper body postures during work tasks. It functions as a

TABLE 6.5
Contents of the AET

Part A—Work System Analysis

1. Work objects
   1.1 Material work objects (physical condition, special properties of the material, quality of surfaces, manipulation delicacy, form, size, weight, dangerousness)
   1.2 Energy as work object
   1.3 Information as work object
   1.4 Man, animals, plants as work objects
2. Equipment
   2.1 Working equipment
      2.1.1 Equipment, tools, machinery to change the properties of work objects
      2.1.2 Means of transport
      2.1.3 Controls
   2.2 Other equipment
      2.2.1 Displays, measuring instruments
      2.2.2 Technical aids to support human sense organs
      2.2.3 Work chair, table, room
3. Work environment
   3.1 Physical environment
      3.1.1 Environmental influences
      3.1.2 Dangerousness of work and risk of occupational diseases
   3.2 Organizational and social environment
      3.2.1 Temporal organization of work
      3.2.2 Position in the organization of work sequence
      3.2.3 Hierarchical position in the organization
      3.2.4 Position in the communication system
   3.3 Principles and methods of remuneration
      3.3.1 Principles of remuneration
      3.3.2 Methods of remuneration

Part B—Task Analysis

1. Tasks relating to material work objects
2. Tasks relating to abstract work objects
3. Man-related tasks
4. Number and repetitiveness of tasks

Part C—Job Demand Analysis

1. Demands on perception
   1.1 Mode of perception
      1.1.1 Visual
      1.1.2 Auditory
      1.1.3 Tactile
      1.1.4 Olfactory
      1.1.5 Proprioceptive

(*Continued*)

TABLE 6.5 (CONTINUED)
Contents of the AET

1.2 Absolute/relative evaluation of perceived information
1.3 Accuracy of perception
Part C—Job Demand Analysis
2. Demands for decision
2.1 Complexity of decision
2.2 Pressure of time
2.3 Required knowledge
3. Demands for response/activity
3.1 Body postures
3.2 Static work
3.3 Heavy muscular work
3.4 Light muscular work, active light work
3.5 Strenuousness and frequency of movements

screening tool, evaluating the biomechanical and postural load on the neck, trunk, and upper limbs.

6.15.1 Key Features of RULA

- **Postural Snapshot:** RULA focuses on a specific posture within the work cycle, typically the one with the highest risk or longest duration.
- **Laterality:** The assessment can target either the right, left, or both sides of the body simultaneously.
- **Repetitive Tasks:** For tasks with extended cycles, RULA allows for posture evaluation at regular intervals.
- **Risk Scoring:** RULA scores indicate the level of intervention needed to reduce the risk of MSDs.

6.15.2 Steps in RULA Assessment

1. **Select the Work Cycle Stage:** Choose the specific stage of the task to be evaluated.
2. **Laterality:** Determine if the assessment will focus on the right, left, or both arms.
3. **Posture Scoring:** Use the RULA scoring sheet or software like JACK to assign scores to each limb based on its posture.
4. **Review Accuracy:** Double-check the assigned scores for accuracy.
5. **Calculate RULA Score:** Utilize tables or software to calculate the final RULA score, which indicates the risk level.

Once you have the final RULA score (grand score), consult the RULA action level list to determine the associated risk level:

- **Score 1 or 2:** These scores indicate a *low risk* of MSDs. No immediate action is required, but consider monitoring the task for future changes.
- **Score 3 or 4:** These scores represent a *medium risk*. Investigate ways to improve the ergonomics of the task, such as providing adjustable workstations or implementing job rotation.
- **Score 5 to 6:** These scores signify a *high risk*. Urgent action is necessary to reduce the risk of MSDs. This might involve redesigning the work process, using assistive equipment, or providing additional training for workers.
- **Score 7:** This score indicates a *very high risk*. The task should be redesigned or alternative methods implemented immediately to eliminate the risk of MSDs.

6.15.3 Beyond the Grand Score

While the grand score provides a quick risk assessment, it's also important to review the individual body segment scores. These scores can reveal specific postural deviations that contribute to the overall risk. Look for:

- **High Scores in Neck, Trunk, or Upper Limb Sections:** These indicate areas requiring particular ergonomic attention.

6.15.4 Taking Action for Improvement

- **Address Undesirable Postures:** Based on the body segment scores, identify and address any problematic postures through adjustments to workstations, tools, or work methods.
- **Ergonomic Intervention Strategies:** Explore various ergonomic interventions to improve worker posture, such as providing proper training, implementing rest breaks, or incorporating adjustable equipment.

6.16 RISK FACTORS AND RISK PREDICTION

While many ergonomic screening methods rely on qualitative measures, the McCauley-Bell and Badiru (1993) model offers a unique approach. It quantifies the risks associated with MSDs by considering three key factors:

- **Task Factors:** This category includes elements like repetitive motions, awkward postures, and force requirements of the job.
- **Personal Factors:** Individual characteristics such as age, fitness level, and pre-existing injuries are taken into account.
- **Organizational Factors:** This category considers aspects like training provided, work schedules, and recovery time allowed.

6.16.1 Calculating Risk Levels

The model utilizes fuzzy linear equations to calculate a risk level for each of these three categories. These individual risk scores are then combined to generate a summative overall risk level.

6.17 GENERAL GUIDELINES FOR WORKPLACES WITH CONTROLS AND DISPLAYS

The information age has brought a surge in workstations with complex control panels and displays. Effective design is crucial for user comfort and productivity. Here are key principles adapted from Van Cott and Kinkade (1972) and Sanders and McCormick (1987).

6.17.1 Prioritization of Elements

- **Primary Visual Tasks (Priority 1):** The most important visual information should be positioned for optimal viewing, with clear and unobstructed lines of sight.
- **Primary Controls (Priority 2):** Controls that directly interact with the primary visual tasks should be readily accessible and easy to operate.
- **Control–Display Relationships (Priority 3):** Controls should be situated close to their corresponding displays. Additionally, ensure a logical and consistent relationship between control movements and display changes (e.g., turning a knob clockwise increases a displayed value).

6.17.2 Workflow and User Convenience

- **Sequential Use (Priority 4):** Elements used in a specific sequence should be arranged in a logical flow to minimize unnecessary reaching or hand movements.
- **Frequent Use (Priority 5):** Controls and displays used frequently should be positioned for convenient access, minimizing reaching or awkward postures.

6.17.3 Consistency and Standardization

- **Consistency Within a System (Priority 6):** The layout of controls and displays should be consistent throughout a particular system to avoid confusion for users familiar with similar setups.
- **Consistency Across Systems (Optional):** Where possible, consider adopting layouts that are consistent with industry standards or common across different systems to enhance user learnability.

These core principles provide a solid foundation for designing user-friendly and efficient workstations.

6.18 SUMMARY

Effective workplace design goes beyond just functionality. It should consider the needs of a diverse workforce with varying body sizes and builds. This inclusivity ensures everyone can work comfortably, maintain proper postures to avoid injury, and efficiently access the tools they need.

Key ergonomic approaches to promote efficiency include providing adjustable furniture like chairs and desks, offering flexible lighting options for individual needs, maintaining proper ventilation for temperature and air quality, and strategically placing tools within easy reach. These elements can significantly benefit businesses or organizations and help reduce the risk of musculoskeletal disorders by minimizing awkward postures and strain. Comfortable workstations can also lead to a more positive work environment, potentially boosting employee morale and satisfaction. Ultimately, ergonomics can enhance productivity by reducing discomfort and fatigue, leading to a more focused and efficient workforce.

Investing in ergonomics is an investment in the health, well-being, and productivity of an organization's employees. Consulting ergonomics experts to assess the workplace can help identify critical areas for improvement within the workflow, and encouraging employee participation in the design process ensures solutions address their specific needs. Regularly evaluating and updating the workplace layout as work processes or employee needs evolve is vital for maintaining an ergonomic work environment.

6.19 CASE STUDY

Problem Overview

The importance of ergonomics in workplace design is paramount, particularly when evaluating the use of hand tools. Poorly designed work environments and tools can lead to discomfort, fatigue, and even long-term musculoskeletal disorders. In this case study, we will explore the basic ergonomic principles applied to workplace design and the evaluation of hand tools, offering students insight into how ergonomic assessments can improve safety, comfort, and productivity.

Ergonomic Design Process

1. **Identifying Risk Factors:** The first step in workplace design is identifying ergonomic risk factors. These may include repetitive motions, awkward postures, excessive force, and inadequate work/rest cycles. Each risk factor contributes to fatigue, discomfort, or injury if left unaddressed.
2. **Anthropometric Data Utilization:** Ergonomic workplace design relies heavily on anthropometric data—measurements of the human body—to

ensure that workstations and tools accommodate a wide range of users. For example, a workstation should be adjustable to cater to users of varying heights and body dimensions.

3. **Tool Evaluation Criteria:** The ergonomic evaluation of hand tools should focus on several key aspects:
   - **Handle Design:** Handles should be sized appropriately for the hand, avoiding too much grip force. Ideally, handles should be cushioned, shaped to fit the natural curvature of the hand, and allow for a power grip when necessary.
   - **Weight and Balance:** Tools should be light enough to minimize fatigue but balanced to reduce strain during use.
   - **Repetitive Strain Mitigation:** Tools should be designed to minimize repetitive motions or allow for neutral wrist positions, reducing strain on muscles and tendons.

Sample Evaluation Tools

Students will use the following ergonomic tools to assess workplace designs and hand tools:

1. **Rapid Entire Body Assessment (REBA):** This tool evaluates whole-body postural risks in a variety of tasks. It considers factors such as posture, force, and activity level.
2. **Quick Exposure Check (QEC):** QEC helps assess the risk of musculoskeletal injury from repetitive work tasks, including the impact of force, vibration, and posture.
3. **Hand Tool Assessment Protocol (HTAP):** HTAP focuses on evaluating hand tools based on grip force, posture, vibration exposure, and overall comfort.

Outcome

By applying these ergonomic principles and evaluation tools, workplaces can be designed to minimize injury risk, improve comfort, and enhance productivity. Tools that are evaluated for ergonomic suitability lead to better user performance, reduced fatigue, and improved safety outcomes. This process not only benefits workers by improving well-being but also contributes to reduced absenteeism and higher productivity for organizations.

Conclusion

This case study highlights the significance of ergonomics in workplace design and hand tool evaluation. By integrating human-centered ergonomic principles, workplaces can be optimized to support workers' health, comfort, and efficiency.

Ergonomic evaluation tools provide a structured approach to identifying and mitigating risks, ensuring that workplace design promotes long-term safety and well-being.

REFERENCES

ACOEM. (2020). Workplace design and postural ergonomics. *Journal of Occupational and Environmental Medicine, 62*(12), e713–e717. https://doi.org/10.1097/JOM.0000000000002

Baines, T. S., Lightfoot, H. W., Evans, S., et al. (2017). Servitization: Revisiting the state-of-the-art and research priorities. *International Journal of Operations & Production Management, 37*(2), 256–278. https://doi.org/10.1108/IJOPM-06-2015-0312

Beniwal, M., Rao, B., & Jha, N. (2017). Automation in confined space operations: Risk reduction and performance enhancement. *Journal of Ergonomics, 19*(3), 125–136.

Bernard, B. (1997). *Musculoskeletal disorders and workplace factors: A critical review of epidemiologic evidence for work-related musculoskeletal disorders of the neck, upper extremity, and low back.* NIOSH.

Bernard, B. P. (2019). *Work-related musculoskeletal disorders: A review of the evidence.* National Institute for Occupational Safety and Health (NIOSH).

Bishop, F., & Demeter, S. (2017). *Ergonomics for the prevention of musculoskeletal disorders in the workplace.* Occupational Medicine.

Bridger, R. S. (2018). *Introduction to ergonomics* (4th ed.). CRC Press.

Burdorf, A., & van der Beek, A. J. (2016). Work-related risk factors for musculoskeletal disorders in the neck, shoulders, elbows, and hands: A clinical overview. *Scandinavian Journal of Work, Environment & Health, 42*(2), 91–104.

Burdorf, A., Derk-Jan, M., & Bartels, S. (2015). Exposure to physical risk factors in confined space operations and interventions to reduce ergonomic risks. *International Journal of Industrial Ergonomics, 53*, 66–75.

Bureau of Labor Statistics (BLS). (2018). *Census of Fatal Occupational Injuries (CFOI) – Current and revised data.* U.S. Department of Labor.

Chaffin, D. B. (2019). *Digital human modeling for workspace design and simulation.* CRC Press.

Cham, R., & Redfern, M. S. (2020). The effect of flooring on standing comfort and fatigue. *Ergonomics, 63*(3), 417–428.

Chen, Y., Yin, Y., & Jing, Q. (2014). Psychological strain in confined space operations: A cognitive ergonomic perspective. *Applied Ergonomics, 45*(4), 1038–1045.

Cheng, W., Xu, W., Yang, H., et al. (2019). Proactive hazard identification for construction projects. *Safety Science.*

Cholewicki, J., & McGill, S. (2016). Spine biomechanics in ergonomic assessment. *Journal of Biomechanics.*

Ciriello, V. M., et al. (2017). Shoulder muscle load and discomfort in repetitive overhead tasks: Impact of arm and body posture. *Human Factors and Ergonomics in Manufacturing & Service Industries, 27*(2), 94–105.

Coenen, P., et al. (2017). Workplace sitting and health risks: A review of epidemiological evidence. *Occupational Health and Safety Journal, 55*(2), 180–196.

Corlett, E. N. (2016). *Advanced ergonomics in the workplace: Design and applications.* CRC Press.

Dababneh, A. J., Swanson, N., & Shell, R. L. (2004). Impact of added rest breaks on the productivity and well-being of workers. *Ergonomics, 47*(5), 554–569.

David, G. C. (2015). Ergonomics methods for assessing exposure to risk factors for work-related musculoskeletal disorders. *Occupational Medicine, 65*(3), 190–202.

Davis, K. G., & Kotowski, S. E. (2015). Postural variability: An effective way to reduce musculoskeletal discomfort in office work. *Human Factors, 57*(1), 88–104.

Drury, C. G. (1975). Application of Fitts' law to foot-pedal design. *Human Factors, 17*(4), 368–373.

Dul, J., & Bruder, R. (2020). A review of the role of workplace ergonomics in shaping human performance. *Applied Ergonomics, 82*, 102978.

Dul, J., & Weerdmeester, B. (2014). *Ergonomics for beginners: A quick reference guide*. CRC Press.

Fernandez, J. E., & Marley, R. J. (2015). Ergonomics in the workplace: Assessing work-related musculoskeletal disorders. *Occupational Medicine, 65*(3), 217–223.

Fitts, P. M. (1954). The information capacity of the human motor system in controlling the amplitude of movement. *Journal of Experimental Psychology, 47*(6), 381–391.

Gallagher, S., Marras, W. S., & Lutz, P. (2014). Force and repetition as risk factors for work-related musculoskeletal disorders. *Ergonomics*.

Gallagher, S., et al. (2016). Workstation design for standing tasks: A review of musculoskeletal outcomes. *Applied Ergonomics, 58*, 566–577.

Green, B., & Westgaard, R. H. (2019). Effect of workplace design on job performance and worker health: A study in healthcare. *Ergonomics, 62*(9), 1123–1135.

Griffin, M. J. (2014). *Handbook of human vibration*. Academic Press.

Gupta, S., Sharma, R., & Gupta, M. (2013). Time and motion study: A tool for optimization of tasks. *International Journal of Engineering Research and Applications, 3*(2), 512–518.

Hardison, D., & Hallowell, M. (2019). Construction hazard prevention through design: Review of perspectives, evidence, and future objective research agenda. *Safety Science, 120*, 517–526. https://doi.org/10.1016/j.ssci.2019.08.001

Hedge, A. (2016). *Ergonomics and the workplace: Ergonomic design of the workplace*. Cornell University.

Helander, M. (2020). *A guide to human factors and ergonomics*. CRC Press.

Holtermann, A., et al. (2019). Promoting health and physical capacity during productive work: the Goldilocks Principle. *Scandinavian Journal of Work, Environment & Health, 45*(1), 90–97. https://doi.org/10.5271/SJWEH.3754

HSE. (2018). *Health and safety in the workplace*. UK Health and Safety Executive.

Jürgens, H., et al. (2014). *Anthropometry for designers*. Springer.

Kaber, D. B., & Riley, J. M. (2015). Workload transitions and performance in automated systems. *Human Factors, 57*(2), 243–257.

Karwowski, W., & Marras, W. S. (2018). *The occupational ergonomics handbook*. CRC Press.

Katzman, W. B., et al. (2019). Lumbar spine support in office chairs: Effects on posture and back pain in sedentary workers. *Journal of Occupational Rehabilitation, 29*(3), 285–292.

King, P. M., et al. (2014). Influence of anti-fatigue mat design on perceived discomfort and productivity in standing workers. *Journal of Occupational Health, 56*(4), 329–335.

Kroemer, K. H. E. (2001). *Ergonomics: How to design for ease and efficiency* (2nd ed.). Prentice Hall.

Kroemer, K. H. E., & Grandjean, E. (2015). *Fitting the task to the human: A textbook of occupational ergonomics* (6th ed.). CRC Press.

Lin, C. Y., et al. (2017). The effects of sit–stand workstations on work posture and discomfort. *Human Factors, 59*(4), 681–693.

Lin, Y. H., et al. (2020). Sedentary behavior and low back pain: A systematic review and meta-analysis. *Spine Journal*, *20*(3), 405–417.

Liu, Y., & Su, X. (2020). Wearable technology in ergonomics research: Applications and future directions. *Ergonomics*, *63*(2), 168–182.

Luttropp, C., & Lagerstedt, J. (2014). *EcoDesign and sustainability: A guide for the design of workplaces*. Elsevier.

Marras, W. S., Cutlip, R. G., Burt, S. E., & Waters, T. R. (2009). National Occupational Research Agenda (NORA): Future directions in occupational musculoskeletal disorder health research. A*pplied Ergonomics*, *40*(1), 15–22. https://doi.org/10.1016/j.apergo.2008.01.018

Mathiassen, S. E., & Winkel, J. (2020a). Application of artificial intelligence and machine learning in ergonomics. *Human Factors: The Journal of the Human Factors and Ergonomics Society*, *62*(7), 1034–1050.

Mathiassen, S. E., & Winkel, J. (2020b). Future directions for artificial intelligence in ergonomics. *Human Factors and Ergonomics in Manufacturing*, *30*(3), 298–310.

McAtamney, L., & Corlett, E. N. (1993). RULA: A survey method for the investigation of work-related upper limb disorders. *Applied Ergonomics*, *24*(2), 91–99.

Mendell, M. J., Mirer, A. G., Cheung, K., Tong, M., & Douwes, J. (2015). Respiratory and allergic health effects of dampness, mold, and dampness-related agents: A review of the epidemiologic evidence. *Environmental Health Perspectives*, *119*(6), 748–756.

Nelson, A., & Baptiste, A. S. (2020). Ergonomic solutions for occupational safety in healthcare settings. *Occupational Ergonomics*, *18*(3), 125–134.

NIOSH. (2019). *Design for Safety (DfS): Guidelines and case studies*. National Institute for Occupational Safety and Health.

NIOSH. (2020). *The NIOSH lifting equation and its applications in workplace design*. Centers for Disease Control and Prevention.

Niu, W., Wu, Y., & Lu, X. (2016). Review of hand-arm vibration exposure and its biomarkers. *International Journal of Environmental Research and Public Health*, 13(2), 160.

Occupational Safety and Health Administration (OSHA). (2011). Hand and Power Tools. U.S. Department of Labor. https://www.osha.gov/sites/default/files/publications/osha3080.pdf

Office Ergonomics Research Committee. (2018). *Ergonomics for seated work: A practical guide*. Springer.

OSHA. (2013). *Ergonomics: Prevention of musculoskeletal disorders in the workplace*. Occupational Safety and Health Administration.

OSHA. (2019). *Permit-required confined spaces*. U.S. Department of Labor, Occupational Safety and Health Administration.

OSHA. (2023). *Ergonomic guidelines and tools*. Occupational Safety and Health Administration.

Parry, S., & Straker, L. (2014). The contribution of sit-stand desks to reducing sedentary behavior in the workplace. *Occupational Medicine*, *64*(5), 356–361.

Petroski, H. (2010). *The pencil: A history of design and circumstance*. Knopf Doubleday Publishing Group.

Pheasant, S., & Haslegrave, C. M. (2016). *Bodyspace: Anthropometry, ergonomics, and the design of work*. CRC Press.

Punnett, L., & Wegman, D. H. (2004). Work-related musculoskeletal disorders: The epidemiologic evidence and the debate. *Journal of Electromyography and Kinesiology*, *14*(1), 13–23. https://doi.org/10.1016/j.jelekin.2003.09.015

Punnett, L., & Wegman, D. H. (2016). Work-related musculoskeletal disorders: The epidemiologic evidence and the debate. *Journal of Electromyography and Kinesiology, 26*, 367–374.

Putz-Anderson, V., Howard, J., & Waters, T. (2017). Ergonomic interventions in confined space environments: An evaluation of effectiveness. *Human Factors and Ergonomics in Manufacturing & Service Industries, 27*(4), 159–171.

Rekus, J. F. (2012). *Complete confined spaces handbook.* CRC Press.

Rempel, D. M., Barr, A., & Erosa, N. (2021). Carpal tunnel syndrome and its relationship to repetitive hand use. *Journal of Occupational Health, 63*(2), e12256.

Robertson, M. M., et al. (2017). The effects of an office ergonomics training and chair intervention on worker knowledge, behavior, and musculoskeletal risk. *Applied Ergonomics, 64*, 1–8.

Robertson, M. M., Ciriello, V. M., & Garabet, A. M. (2013a). Effects of office ergonomics intervention on musculoskeletal discomfort and productivity. *Human Factors and Ergonomics in Manufacturing, 23*(2), 134–143.

Robertson, M. M., Ciriello, V. M., & Garabet, A. M. (2013b). Office ergonomics training and work-related musculoskeletal disorders: Assessing the effectiveness of interventions. *Applied Ergonomics, 44*(1), 133–142.

Roffey, D. M., et al. (2014). Causal assessment of occupational sitting and low back pain: Results of a systematic review. *The Spine Journal, 14*(8), 1598–1604.

Rohmert, W., et al. (2016). Biomechanical evaluation of work posture and its impact on the musculoskeletal system. *Human Factors.*

Sanders, M. S., & McCormick, E. J. (1987). *Human factors in engineering and design* (6th ed.). McGraw-Hill.

Sanders, M. S., & McCormick, E. J. (2013). *Human factors in engineering and design.* McGraw-Hill.

Shirazi, M., et al. (2021). Impact of adjustable chairs on musculoskeletal discomfort in the workplace. *Applied Ergonomics, 92*, 103349.

Siemens. (2024). Jack Human Simulation Software. Siemens AG. Retrieved from https://www.siemens.com/global/en.html

Singh, M., et al. (2015). Maintenance ergonomics: Addressing the gap between design and safety. *International Journal of Industrial Ergonomics, 47*, 114–123.

Straker, L., et al. (2019). The changing face of office work: Future trends and their impact on ergonomic chair design. *Human Factors and Ergonomics Journal, 61*(4), 545–559.

Sullivan, D. A., et al. (2019). Ergonomic design of workspaces for dynamic tasks: An overview of biomechanical factors. *Work, 63*(2), 247–258.

Sundstrup, E., et al. (2016). The effect of workplace interventions on standing discomfort and postural variability. *International Journal of Occupational Safety and Ergonomics, 22*(1), 7–14.

Van Cott, H. P., & Kinkade, R. G. (Eds.). (1972). *Human engineering guide to equipment design* (Revised ed.). U.S. Department of Defense, Joint Services Steering Committee.

Van der Beek, A. J., et al. (2014). *Ergonomics and musculoskeletal disorders: Developing a sustainable approach.* Applied Ergonomics.

Waters, T. R., et al. (2016). Reducing work-related musculoskeletal disorders through ergonomic interventions: A review of recent research. *Journal of Occupational and Environmental Hygiene, 13*(3), 123–138.

Wilson, J. R. (2014). *Fundamentals of human factors and ergonomics in design.* CRC Press.

Wimer, B. M., McDowell, T. W., & Xu, X. S. (2017). Hand-tool design for reducing muscle fatigue in repetitive tasks: A critical review. *Journal of Applied Ergonomics, 58*, 27–37.

Wulff, C., et al. (2014). Effects of work posture on muscle fatigue in physically demanding jobs: A biomechanical analysis. *Journal of Applied Ergonomics*, *45*(4), 1228–1233.
Zemp, R., et al. (2018). Ergonomic analysis of lumbar support in office chairs: A case study. *Journal of Biomechanics*, *76*, 190–195.
Zhao, D., Lin, H., & Zheng, W. (2019). Atmospheric hazards in confined spaces: Monitoring and prevention. *Journal of Safety Research*, *68*, 119–125.

7 Work-Related Musculoskeletal Disorders

7.1 LEARNING GOALS

The learning goals of this chapter are to provide an understanding of the origin, nature and impact of work-related musculoskeletal disorders (WMSDs) in the global community. The student will learn how to establish an ergonomic program focused on reducing the risks of WMSDs in industry.

7.2 KEY TOPICS

- Structures impacted by WMSD
- Types of work-related musculoskeletal disorders
- Stages of work-related musculoskeletal disorders and principles of prevention
- Task-related risk factors
- Personal risk factors
- Screening methods
- Impact on industry
- OSHA Ergonomic Guidelines for WMSD Management

7.3 INTRODUCTION AND BACKGROUND

The World Health Organization (WHO) has identified work-related musculoskeletal disorders (as an area for significant global concern due to their multifactorial nature (WHO, 2021). These disorders, which include repetitive strain injuries (RSI), repetitive motion injuries (RMI), and cumulative trauma disorders (CTDs), are primarily caused by prolonged exposure to repetitive tasks, awkward postures, forceful exertions, and insufficient recovery times. As such, WMSDs represent a critical intersection between human physiology and workplace design, necessitating rigorous ergonomic interventions to reduce the risks associated with modern occupational settings (Punnett and Wegman, 2014).

7.3.1 Historical Perspective on Musculoskeletal Disorders

The understanding of musculoskeletal disorders (MSDs) has evolved throughout the centuries. Historically, repetitive occupational tasks have been linked to physiological

DOI: 10.1201/9780429342806-7

damage, with early records associating disorders with repetitive motion. For example, Raynaud's phenomenon, often referred to as "dead finger" or "jackhammer disease," was identified in the 19th century as a condition resulting from repeated mechanical vibration and restricted blood flow (Koehler et al., 2019). Similarly, by 1893, inflammations of the extensor tendons in the hand were described by Henry Gray, who identified the repetitive strain associated with manual labor (Gray, 1878). These early observations laid the groundwork for later developments in occupational health and safety regulations.

The industrial revolution, particularly in the post-Civil War period, exacerbated the prevalence of musculoskeletal injuries due to poor working conditions, particularly among child laborers. Many workers experienced severe physiological impairments such as spine deformities and stunted growth due to prolonged exposure to hazardous environments (CDC, 2013). However, these injuries were not formally recognized as occupational hazards requiring systemic intervention until the Workers' Compensation Act of 1906 in Great Britain (Armstrong, 2014).

7.3.2 The Emergence of Cumulative Trauma Disorders

Throughout the 20th century, numerous cumulative trauma disorders were linked to myriad occupations. Mining, herding, and other labor-intensive industries led to conditions such as anthrax exposure, heavy metal poisoning, and joint-specific trauma like "beat knee" and "beat elbow" (Mandel, 2014). In 1929, Ohio officially recognized tenosynovitis of the hand, marking a significant shift in the recognition of occupation-related disorders. This condition, along with carpal tunnel syndrome and other upper limb disorders, became the focus of increasing medical attention in the following decades (Mandel, 2014).

7.3.3 Ergonomics and Task Design: The Evolution of Intervention Strategies

Ergonomics has played a pivotal role in understanding the relationship between work tasks and the human body. In the 1930s, research by Hammer revealed that tendons could only withstand between 1,500 and 2,000 exertions per hour, a finding that underscored the importance of task design in minimizing strain (Armstrong, 2014). This research provided a foundation for later studies that emphasized the need for ergonomic interventions in repetitive tasks, particularly those involving fine motor skills, such as typing. By the 1940s, typists began to report high rates of tendon inflammation, further demonstrating the critical need for ergonomics in work design (Armstrong, 2014; Silverstein et al., 1986).

The latter half of the 20th century witnessed significant advancements in ergonomic guidelines. Numerous studies confirmed that repetitive tasks, combined with inadequate recovery periods, induced physiological strain on the musculoskeletal system and contributed to long-term tissue damage (Marras & Karwowski, 2019). In response, organizations like the National Institute for Occupational Safety and

Health (NIOSH) developed comprehensive guidelines for workstation design, tool use, and work-rest schedules to mitigate the risks of WMSDs (NIOSH, 1997). These guidelines emphasized the necessity of adjusting workplace environments to fit human biomechanics, reducing physical stressors that contribute to musculoskeletal damage.

7.3.4 Contemporary Ergonomic Interventions for WMSDs

Today, ergonomic principles continue to evolve as new research emerges concerning the physiological and biomechanical impacts of work environments. Technological advances have enabled more precise analyses of the biomechanical loads placed on workers during task performance, leading to more targeted interventions. For instance, contemporary ergonomic interventions focus on adjusting workstation height, optimizing tool design, and employing automation where possible to minimize repetitive manual tasks (Punnett & Wegman, 2014).

The economic and social costs of WMSDs are substantial. As industries continue to globalize, companies face increasing pressure to adopt effective ergonomic strategies that reduce injury rates, increase productivity, and improve worker well-being. This requires not only compliance with international ergonomic standards but also a commitment to continuous improvement in workplace design and task allocation. The integration of ergonomic principles into occupational health frameworks remains a key strategy for reducing the global burden of musculoskeletal disorders and their consequences (Punnett & Wegman, 2014).

Work-related musculoskeletal disorders are a pressing occupational health concern due to their significant impact on worker health, productivity, and economic costs. Over the past century, ergonomic research has provided valuable insights into the relationship between work tasks and musculoskeletal health, which has led to the development of evidence-based guidelines to help mitigate these risks. As technology and work environments continue to evolve, so too must the application of ergonomic principles to protect the global workforce from the debilitating effects of WMSDs.

7.4 STRUCTURES IMPACTED BY WMSD

Epidemiological studies have consistently shown that the presence of known risk factors such as force, repetition, and awkward posture significantly increases the risk of musculoskeletal disorders in muscles, joints, tendons, blood vessels, and nerves (Punnett et al., 2014; Bongers et al., 1993). While bones are generally not considered susceptible to such disorders, repetitive tasks can still have a detrimental effect on the vertebrae of the spinal column (Bernard, 1997).

Excessive stretching of a muscle can lead to strain characterized by injury to the tendon or muscle, resulting in swelling and pain. Severe injuries occur when muscle fibers are torn apart. Furthermore, prolonged obstruction of blood or nerve supply to the muscle can lead to muscle deterioration (Silverstein et al., 2016); tendons, which do not stretch or contract, can be torn if excessively strained. The development of scar tissue makes tendons more prone to repeated injuries and chronic tension. When such muscle strain occurs by injury to the tendon, rough tendon surfaces can then interfere with motion, creating friction with other tissues (NIOSH, 2018).

Inflammation of a tendon, known as tendinitis, occurs due to insufficient synovial fluid in the tendon sheath, leading to friction and symptoms like warmth, tenderness, and pain. This inflammation hinders movement and can potentially lead to a chronic condition if not addressed (Shiri et al., 2016). A bursa, the cushion that reduces friction between muscular joints, can become inflamed from overuse, further limiting tendon movement and reducing joint mobility (Roquelaure, 2018).

A sprain, which is caused by the stretching or tearing of ligament fibers, can occur from both acute injuries and repetitive activities. The ligaments most susceptible to sprains are found in the knee, ankle, and wrist. Sprains can take weeks or months to heal due to their limited blood supply and can result in joint instability, increasing the risk of further injury (Hootman et al., 2004).

Nerve compression occurs when body positions decrease the size of openings through which the nerves run. This is primarily caused by pressure from ligaments, tendons, muscles, and bones. Hard surfaces and sharp edges of workplaces, tools, and equipment can also trigger nerve compression. Motor nerve impairment affects muscle activity, while sensory nerve impairment leads to symptoms like tingling, numbness, or pain. These sensory signals are crucial for recognizing task activities that may exacerbate MSD conditions (Treaster & Burr, 2004).

Vascular compression, often affecting arteries, involves the constriction or obstruction of blood flow, impeding the supply of oxygen and nutrients to tissues. This can result in ischemia, affecting muscular endurance and recovery. Vasospasm, a compression caused by vibrations from tools, decreases blood vessel diameter, leading to conditions like Raynaud's phenomenon, characterized by numbness, tingling, and loss of sensation and control in the affected areas (Bovenzi et al., 2015).

7.5 TYPES OF WORK-RELATED MUSCULOSKELETAL DISORDERS

The frequency and impact of work-related musculoskeletal disorders continue to be a critical issue within occupational health, necessitating rigorous classification and definition efforts in both medical and ergonomic fields. Standardizing the terminology and diagnostic criteria for WMSDs is essential for myriad reasons: for instance, it enables researchers and healthcare professionals to develop more precise treatment and prevention strategies that target specific disorders (Bernard & Guo, 2021a); furthermore, it provides ergonomists, engineers, and safety professionals with the knowledge to design and implement safer workplace environments, thereby reducing the risk factors associated with particular musculoskeletal injuries. Effective classification ensures that preventive interventions are tailored to the unique demands and risks of different work tasks and environments, enhancing their efficacy (Bernard & Guo, 2021a).

A comprehensive 2021 study published in the *American Journal of Industrial Medicine* reinforced the importance of clear categorization for the success of prevention programs. The study found that WMSDs that are well-defined and systematically classified allow for better identification of at-risk populations, early detection of potential injury patterns, and more effective deployment of ergonomic interventions (Bernard & Guo, 2021b). Without such clarity, both diagnosis and prevention become less effective, as ambiguous classifications can obscure the specific causes and mechanisms of injury, complicating both treatment and prevention efforts.

7.5.1 The Scope of WMSDs

WMSDs can affect virtually any region of the body, from the upper extremities (e.g., carpal tunnel syndrome, tendinitis) to the lower back (e.g., lumbar strain) and lower extremities (e.g., plantar fasciitis). However, certain WMSDs are more prevalent in specific industries due to the nature of the tasks involved. For example, in manufacturing and manual labor settings, repetitive motions, forceful exertions, and poor posture are primary contributors to conditions such as repetitive strain injuries (RSIs) and cumulative trauma disorders. These industries have some of the highest incidences of musculoskeletal disorders because of the physical demands placed on workers.

To mitigate the risks of WMSDs, ergonomic interventions must consider both the physical and organizational aspects of work. This includes the design of workstations, tools, and equipment, as well as work-rest cycles, job rotation strategies, and the promotion of safe work practices. Additionally, ergonomic risk assessments should be routinely conducted to identify hazards that may lead to musculoskeletal injury (Punnett & Wegman, 2014). By addressing these factors in a systematic and scientific manner, ergonomists can develop comprehensive prevention strategies tailored to specific occupational settings.

7.5.2 Back Injuries

Back injuries are among the most prevalent work-related musculoskeletal disorders and remain a significant concern for occupational health professionals. These injuries account for a substantial portion of WMSDs, often resulting from repetitive stress, poor ergonomic design, and improper movement patterns in the workplace. According to Bernard and Guo (2021b), back injuries are frequently reported across various industries and are recognized as a critical issue in occupational ergonomics. Data from the Bureau of Labor Statistics (BLS) shows that back injuries comprised 22% of the 1.7 million reported WMSDs in 2020, with overexertion identified as the leading cause (BLS, 2021).

7.5.3 Cumulative Nature of Back Injuries

While many workplace injuries may seem to arise from single incidents, back pain often results from cumulative microtrauma sustained over time. Repeated exposure to improper lifting techniques, awkward postures, or extended periods of physical exertion can cause gradual deterioration of spinal structures. Ergonomists classify these injuries as repetitive strain injuries, a subset of WMSDs that develop due to repetitive biomechanical stress (Ergonomics Society, 2023). Liberty Mutual Research Institute for Safety (2019) found that 38% of all workplace musculoskeletal disorders are linked to material handling activities, further underscoring the relationship between repetitive physical tasks and spinal injuries.

Back injuries, particularly those affecting the lower spine, are often the result of improper material handling, such as lifting heavy objects without proper form or carrying loads in awkward positions. These activities place excessive stress on the lumbar region of the spine, which is particularly vulnerable due to the high mechanical loads it regularly sustains during physical work tasks. Consequently, what may appear to be a sudden onset of pain is frequently the culmination of long-term microtrauma to the vertebral discs, ligaments, and muscles that support the spine (Ergonomics Society, 2023).

7.5.4 Biomechanical Impact on Spinal Discs

The vertebral discs, which serve as shock absorbers between the bones of the spine (vertebrae), are particularly susceptible to injury under conditions of repetitive strain. These discs are composed of a gel-like center (nucleus pulposus) surrounded by a tougher, fibrous outer layer (annulus fibrosus). When subjected to repeated compression forces, such as during improper lifting or awkward postures, these discs may experience excessive mechanical stress. Over time, continuous or improper loading can cause the discs to bulge, herniate, or rupture, placing pressure on surrounding spinal nerves and resulting in significant pain and functional impairment (National Institute of Neurological Disorders and Stroke, 2020).

Disc degeneration is a common outcome of repetitive microtrauma, which can progressively limit spinal flexibility and increase the risk of acute injuries. The biomechanical stress applied to the spine during lifting or carrying tasks can lead to cumulative damage, particularly when workers engage in tasks without proper ergonomic support or fail to use appropriate lifting techniques. The compromised structural integrity of the discs as a result of the repeated trauma can further exacerbate the risk of back injuries, leading to chronic conditions such as lumbar disc herniation or degenerative disc disease (Punnett & Wegman, 2014).

7.5.5 Ergonomic Interventions for Preventing Back Injuries

To mitigate the risk of back injuries, particularly those caused by repetitive strain, ergonomic interventions are essential. These include the design of workstations that support optimal posture, use of mechanical aids for lifting, and training workers in proper lifting techniques. Additionally, job rotation and sufficient recovery periods are key factors in reducing the impact of cumulative biomechanical stress on the spine (Punnett & Wegman, 2014). Employers must also conduct regular ergonomic assessments to identify and correct hazards related to material handling tasks, ensuring that workplace designs minimize the strain on workers' musculoskeletal systems.

Modern ergonomic guidelines advocate for minimizing the frequency and intensity of manual material handling tasks, incorporating lifting aids, and promoting dynamic postural alignment throughout the workday. The use of adjustable workstations, mechanized lifting devices, and proper training in manual material manhandling techniques can significantly reduce the occurrence of repetitive strain injuries, including those affecting the spine. Furthermore, incorporating breaks and physical activity into the work routine can alleviate the effects of prolonged spinal loading (Ergonomics Society, 2023).

7.5.6 Carpal Tunnel Syndrome

Carpal Tunnel Syndrome (CTS) is one of the most commonly recognized work-related musculoskeletal disorders affecting the hand and forearm, particularly in occupations that require repetitive hand and wrist motions. CTS occurs when the median nerve, which innervates the thumb, index, middle, and part of the ring finger, is compressed as it travels through the carpal tunnel in the wrist. The carpal tunnel itself is a narrow passage formed by the arrangement of eight carpal bones and the overlying flexor retinaculum, a robust ligament that helps maintain the position of the tendons (Wilcox et al., 2019).

From an ergonomic perspective, the development of CTS is frequently linked to occupational activities that involve repetitive motions, sustained awkward postures, or forceful hand exertions, particularly in jobs such as typing, assembly line work, and tool handling (Punnett & Wegman, 2014). The prolonged stress and strain placed on the wrist during such tasks increase the risk of compressing the median nerve, particularly in individuals performing repetitive movements without adequate rest or ergonomic interventions.

7.5.6.1 Pathophysiology of CTS

The etiology of CTS is multifactorial. Recent studies indicate that, beyond repetitive motion and mechanical stress, structural factors within the carpal tunnel can exacerbate median nerve compression. Wilcox et al. (2019) noted that the tightness of the flexor retinaculum, the ligament spanning the palm side of the wrist, plays a significant role in the development of CTS. Additionally, inflammation of the synovium—the tissue that lines the tendons running through the carpal tunnel—can further contribute to the narrowing of this confined space, increasing pressure on the median nerve. Over time, this elevated pressure impairs the nerve's function, leading to the symptoms which are characteristic of CTS.

7.5.6.2 Clinical Presentation and Symptoms

The clinical manifestations of CTS often begin with intermittent sensory disturbances. Early symptoms include numbness, tingling, and burning sensations, typically affecting the thumb, index, middle, and part of the ring finger. These symptoms are often exacerbated by wrist flexion, which further narrows the carpal tunnel and increases nerve compression (Wilcox et al., 2019). As the condition progresses, individuals may experience pain, hand weakness, and difficulty gripping objects. The loss of fine motor control can severely impact job performance, particularly in occupations requiring dexterous hand movements.

Nighttime symptoms are common in CTS, often waking individuals from sleep due to sustained wrist flexion during rest. If left untreated, the continued compression can lead to chronic pain, permanent nerve damage, and muscle atrophy, particularly in the thenar muscles at the base of the thumb. The thenar muscles are a group of three muscles at the base of the thumb that control fine thumb movements. The thenar muscles are responsible for pinching, grabbing, and other precise thumb movements This muscle wasting results in reduced pinch strength, making it difficult for affected individuals to perform even simple tasks such as holding small objects or using tools (National Institute of Neurological Disorders and Stroke, 2020).

7.5.6.3 Ergonomic Interventions and Prevention

Given the occupational implications of CTS, ergonomic interventions play a crucial role in both preventing and managing the disorder. Effective ergonomic strategies aim to reduce repetitive wrist strain and awkward postures that contribute to median nerve compression. These strategies include workstation modifications, such as adjusting keyboard and mouse heights to promote neutral wrist positions, using wrist rests, and implementing job rotation to limit continuous exposure to repetitive hand movements (Punnett & Wegman, 2014).

Furthermore, ergonomic tools, such as split keyboards, vertical mice, and padded gloves, can help minimize wrist extension and reduce the pressure on the carpal tunnel. Educational programs focused on proper hand positioning, the importance of regular breaks, and stretching exercises can also mitigate the risks associated with prolonged repetitive tasks (Marras & Karwowski, 2019). These preventive measures, when consistently applied, have been shown to reduce the incidence of CTS and improve outcomes for workers already experiencing early symptoms.

7.5.7 Tendinitis

Tendinitis, the inflammation of a tendon itself (not the tendon sheath), is a common condition characterized by localized pain and tenderness at the affected area, often worsening with movement. While trauma can be a possible cause, tendinitis more frequently arises from overuse and repetitive motions that strain the tendon. Common locations include the elbow (lateral epicondylitis, often called "tennis elbow"), shoulder (rotator cuff tendinitis), Achilles tendon (Achilles tendinitis), and wrist (De Quervain's tenosynovitis).

7.5.8 Tenosynovitis

Tenosynovitis is a common condition caused by inflammation of the synovial sheath, a lubricating sac surrounding the tendons. It often arises from repetitive motions that irritate the tendon and sheath. De Quervain's tenosynovitis, affecting the tendons at the base of the thumb, is a well-known example. However, tenosynovitis can occur in various locations, including the elbow (where it's sometimes called "tennis elbow"), wrist, ankle (particularly in athletes), and shoulder.

7.5.9 Intersection Syndrome and De Quervain's Syndrome

Intersection syndrome and de Quervain's tenosynovitis are two common overuse injuries that can occur in people who perform repetitive hand and wrist motions. While both can cause pain and limit thumb movement, they affect slightly different areas and have some distinct features.

- **Intersection syndrome** involves inflammation of the tendons where they "intersect" on the back of the forearm, near the wrist. This can cause pain and tenderness on the top of the wrist and forearm, often worse with gripping or wrist extension activities.
- **De Quervain's tenosynovitis** affects the tendons at the base of the thumb, where they pass through a tunnel in the wrist. This typically results in pain and tenderness along the thumb side of the wrist, often intensified with pinching, grasping, or forceful thumb movements.

Both conditions can also cause swelling, stiffness, and occasionally a sensation of catching or grinding in the affected area. Early diagnosis and treatment are crucial to prevent chronic pain and disability.

7.5.10 Trigger Finger

Trigger finger is a common hand condition that affects the tendons responsible for bending your fingers. It occurs when inflammation within the tendon sheath causes thickening of the tendon itself or a small nodule to form. This thickening can catch on the pulleys, which are ligamentous structures that hold the tendons in place, leading to the characteristic symptoms.

- **Symptoms:** Trigger finger entails a catching or locking sensation when bending or straightening the affected finger. A snapping or popping sound may also be present, particularly when the finger unlocks. Pain and tenderness at the base of the finger on the palm side are common.
- **Causes:** While repetitive motions and overuse can contribute, trigger finger can also arise from other factors like diabetes, inflammatory arthritis, or certain medications.
- **Multiple Fingers:** Trigger fingers can affect a single finger or multiple fingers simultaneously.

7.5.11 Ischemia

Ischemia is a condition that occurs when there's insufficient blood flow to a tissue, depriving it of oxygen and nutrients. This can lead to a variety of symptoms depending on the severity and location of the ischemia.

- **Symptoms:** Common symptoms include numbness, tingling, pain, coolness, and weakness in the affected area. In severe cases, tissue death can occur.
- **Causes:** While compressive force in the palm of the hand can be a cause, ischemia can also arise from various factors like:
  - **Arterial blockage:** Fatty deposits (atherosclerosis) or blood clots can narrow or block arteries, reducing blood flow.
  - **Vasospasm:** Abnormal narrowing of blood vessels due to conditions like Raynaud's disease can limit blood supply.
  - **Trauma:** Severe injuries or surgery can damage blood vessels, leading to ischemia.

Early diagnosis and treatment of ischemia are crucial to prevent tissue damage and potential severe complications like gangrene.

7.5.12 Vibration Syndrome

Vibration syndrome, also known as hand–arm vibration syndrome (HAVS), is a disorder caused by prolonged exposure to vibrating tools and equipment. It's important to distinguish it from Raynaud's phenomenon, although they can share some similar symptoms.

- **Symptoms:** HAVS primarily affects the fingers, thumbs, and hands. Common symptoms include:
  - **Numbness and Tingling:** This can be a constant sensation or intermittent, often triggered by cold exposure.
  - **Pain:** Aching or throbbing pain may occur in the fingers and hands.
  - **Finger Blanching:** This is the classic "white finger" appearance, where fingers turn white due to reduced blood flow. It can be followed by redness as blood flow returns.
  - **Reduced Dexterity:** Difficulty gripping or performing fine motor tasks with the affected fingers.
  - **Muscle Weakness:** In severe cases, weakness in the hands and forearms can develop.
- **Causes:** HAVS arises from damage to nerves, blood vessels, and muscles in the hands and arms due to repeated vibrations.
- **Cold Exposure:** While not the primary cause, cold temperatures can worsen HAVS symptoms by further constricting blood vessels in the fingers. Thermoregulation is important for HAVS prevention: keeping hands warm during work and cold weather can help minimize symptoms.

Raynaud's phenomenon is a separate condition that can cause similar finger blanching and tingling. However, it is not always caused by vibration and can have other underlying causes like autoimmune diseases.

7.5.13 Thoracic Outlet Syndrome

Thoracic Outlet Syndrome (TOS) is a group of conditions caused by compression of the nerves and/or blood vessels in the thoracic outlet, the space between the collarbone (clavicle) and the first rib. While TOS can be a complex diagnosis, it is a recognized clinical entity.

The following section provides additional details on the various types of TOS, associated symptoms and treatments:

- **Types of TOS:** There are three main types of TOS:
  - **Neurogenic TOS:** This is the most common type, caused by compression of the brachial plexus nerves.
  - **Venous TOS:** This results from compression of the subclavian vein, leading to swelling and discoloration in the arm.
  - **Arterial TOS:** This is the least common and most serious type, caused by compression of the subclavian artery, potentially leading to blood clots and tissue damage.
- **Symptoms:** Symptoms vary depending on the type of TOS and the structures compressed. Common symptoms include:
  - Pain: Aching or sharp pain in the shoulder, neck, and arm, often worse with activity.
  - Weakness or fatigue in the arm.
  - Numbness and tingling, typically along the inner arm and into the fourth and fifth fingers.
  - In severe cases, swelling, discoloration, or coolness in the arm or hand.
- **Diagnosis:** TOS can be challenging to diagnose due to overlapping symptoms with other conditions. Doctors may rely on physical examination, imaging tests (x-ray, MRI), and nerve conduction studies to confirm TOS.
- **Treatment:** Treatment for TOS depends on the severity and type. It may involve non-surgical options like physical therapy, activity modification, and pain medication. In severe cases, surgery might be necessary to relieve compression.

Validity of TOS as a Diagnosis: While some healthcare professionals may have misdiagnosed TOS in the past, it is important to note that TOS is a corporeal condition with established diagnostic criteria.

7.5.14 Ganglion Cysts

Ganglion cysts are non-cancerous, fluid-filled sacs that commonly develop around joints or tendons in the wrist and hand. They can also occur in other locations like the foot, ankle, and knee, but these are less frequent.

Here is a breakdown of the key points:

- **Symptoms:** Many ganglion cysts are painless and may go unnoticed. However, if the cyst presses on a nearby nerve, it can cause pain, tingling, or weakness. Additionally, the physical lump itself might be bothersome for some people.
- **Causes:** Although the exact cause of ganglion cysts is unknown, they are believed to arise from repetitive motions or micro-injuries that put stress on the joint or tendon sheath.
- **Location:** The most common locations for ganglion cysts are:
    - **Dorsal Wrist Ganglion:** This appears on the back of the wrist near the tendons.
    - **Volar Wrist Ganglion:** This develops on the palm side of the wrist near the base of the thumb.
    - **Flexor Tendon Sheath Ganglion:** This forms along the sheath surrounding the flexor tendons in the palm.
- **Treatment:** Many ganglion cysts do not require treatment, especially if they are painless. However, if the cyst causes discomfort or limits movement, options include observation, aspiration (removing the fluid with a needle), or surgery.
- **Recurrence:** It is important to note that ganglion cysts can recur even after treatment.

7.6 Stages of WMSD and Principles of Prevention

While MSDs develop gradually from repetitive motions and other risk factors, the recognition of symptoms varies from person to person. The following provides a general progression of these types of conditions:

Stage 1: Early Warning Signs

- Workers may experience occasional aches, tightness, or tiredness during or shortly after work hours.
- These discomforts typically disappear on their own with adequate rest.
- Work performance remains unaffected, but these signs might persist for weeks or months.

Stage 2: Discomfort and Interference

- Symptoms become more pronounced, including tenderness, swelling, weakness, numbness, and pain.
- Discomfort often starts early in the workday and may linger through the night, affecting sleep quality.
- Repetitive tasks become increasingly difficult, leading to a decline in work performance.
- Stage 2 symptoms can last for several months.

TABLE 7.1
Summary Table of WMSD

Identified Disorders, Occupational Risk Factors, and Symptoms		
Disorders	**Occupational Risk Factors**	**Symptoms**
Tendinitis/tenosynovitis	Repetitive wrist motions Repetitive shoulder motions Sustained hyperextension of arms Prolonged load on shoulders	Pain, weakness, swelling, burning sensation, or dull ache over affected area
Epicondylitis (elbow tendinitis)	Repeated or forceful rotation of the forearm and bending of the wrist at the same time	Same symptoms as tendinitis
Carpal tunnel syndrome	Repetitive wrist motions	Pain, numbness, tingling, burning sensations, wasting of muscles at base of thumb, dry palm
De Quervain's disease	Repetitive hand twisting and forceful gripping	Pain at the base of thumb
Thoracic outlet syndrome	Prolonged shoulder flexion Extending arms above shoulder height Carrying loads on the shoulder	Pain, numbness, swelling of the hands
Tension neck syndrome	Prolonged restricted posture	Pain

Source: Canadian Occupational Safety and Health, http://www. ccohs.ca/oshanswers/diseases/rmirsi .html (accessed April, 2011)

Stage 3: Functional Limitations

- Symptoms are present even during rest periods and may worsen with non-repetitive motions.
- Sleep is significantly disturbed due to pain and discomfort.
- Daily activities and work performance become severely limited, even for light tasks.
- Stage 3 symptoms can persist for months or even years.

Early intervention plays a critical role in minimizing the impact and progression of work-related musculoskeletal disorders. Musculoskeletal disorders, when left untreated, can progress through various stages, each requiring increasingly intensive interventions. From an ergonomic standpoint, addressing these disorders in their initial stages is essential to prevent further deterioration and reduce the long-term burden on both workers and organizations (NIOSH, 2019).

7.6.1 Early Identification and Stage 1 Intervention

Stage 1 symptoms of MSDs, such as mild discomfort, localized fatigue, or intermittent soreness, are often the first indicators of potential ergonomic hazards in the workplace. At this stage, immediate action can prevent the condition from progressing. Frequent microbreaks, which involve short pauses in work to stretch or change posture, are highly effective in reducing muscle fatigue and restoring blood flow to overworked areas (Punnett & Wegman, 2014). Additionally, rotating tasks among workers to avoid repetitive motion or prolonged static postures can mitigate early discomfort. By diversifying work tasks, the cumulative stress on specific muscle groups or joints is reduced, allowing recovery during work hours.

This proactive approach is supported by ergonomic studies that emphasize the importance of early intervention in managing MSD symptoms (Marras & Karwowski, 2019). Simple modifications, such as adjusting seating or keyboard positions, introducing anti-fatigue mats, or encouraging correct posture, can significantly alleviate discomfort before it escalates into a more serious condition.

7.6.2 Progression to Stages 2 and 3: The Importance of Comprehensive Intervention

If symptoms are ignored or not adequately addressed, they can progress to Stage 2, characterized by increased pain, stiffness, or reduced mobility. At this stage, workers may find it difficult to perform their usual tasks efficiently, leading to declines in productivity and job satisfaction. If not managed, symptoms can further progress to Stage 3, where severe pain and functional impairments severely limit an individual's ability to work. In these later stages, temporary or permanent cessation of the aggravating activity is necessary, and more advanced interventions such as physical therapy, medication, or even surgery may be required (NIOSH, 2019).

From an ergonomic perspective, reaching Stage 2 or 3 represents a failure in the early detection and mitigation of risk factors. It highlights the need for ongoing ergonomic assessments to identify high-risk tasks and workplace designs that contribute to these disorders. Regular ergonomic evaluations, combined with early reporting of symptoms, can significantly reduce the likelihood of reaching these advanced stages (Armstrong et al., 2016).

7.6.3 Proactive Ergonomics: Prevention through Design and Modification

Ergonomic interventions must be tailored to the specific needs of the job and the worker. This may involve altering job tasks, introducing more supportive seating, or providing workers with training on proper lifting techniques and posture maintenance. In cases where modifications to individual tasks are insufficient, a complete redesign of the workflow or work environment may be necessary to eliminate risk factors altogether (Marras & Karwowski, 2019). This approach, known as "prevention through design," integrates ergonomic principles into the initial design of work processes, tools, and environments, thereby preventing the development of MSDs before they begin (NIOSH, 2019).

7.7 TASK-RELATED RISK FACTORS

- A comprehensive review of epidemiological studies, including the work of Bernard (1997) and more recent findings from the National Institute for Occupational Safety and Health (NIOSH, 2014), provides a detailed categorization of WMSDs based on the body region affected. Understanding the specific risk factors for different regions is critical for developing targeted ergonomic interventions to minimize WMSD risk in the workplace.

7.7.1 Neck and Neck/Shoulder Risk Factors

- **Repetition:** Repetitive tasks that involve cyclical neck or shoulder movements are major contributors to neck and shoulder WMSDs. For example, tasks that demand repeated arm motions or continuous loading of the trapezius muscle can lead to muscle fatigue and injury over time.
- **Force:** Upper body tasks that require high levels of force can place significant loads on the trapezius and neck muscles. For instance, lifting or manipulating heavy objects above shoulder height exerts excessive strain on the neck and shoulder complex.
- **Posture:** Sustained or extreme postures of the head or neck, such as prolonged head tilting or neck flexion, are well-known contributors to neck and shoulder disorders. Static postures, particularly those that deviate from neutral head positions, exacerbate musculoskeletal stress on the cervical and shoulder regions (Punnett & Wegman, 2014).

7.7.2 Shoulder Risk Factors

- **Repetition:** Repetitive flexion, extension, abduction, and rotation of the shoulder joint, common in assembly line or overhead work, significantly increase the risk of shoulder injuries. The shoulder joint is particularly vulnerable when these motions are performed frequently without adequate rest or ergonomic support.
- **Posture:** Shoulder WMSDs are often caused by working with the arms raised or in non-neutral positions, such as during overhead reaching or when the arm is extended away from the body. These postures amplify mechanical stress on the shoulder joint and the surrounding soft tissues.
- **Force:** High-force activities that involve shoulder flexion, abduction, or rotation, such as lifting heavy loads or using forceful hand tools, place excessive stress on the shoulder musculature and increase the risk of strain or injury (Marras et al., 2019).
- **Vibration:** The use of vibrating hand tools contributes to shoulder WMSDs through the transmission of low- or high-frequency vibrations to the shoulder joint, which can result in cumulative tissue damage (NIOSH, 2014).

TABLE 7.2
Minimal Clinical Criteria for Establishing Worksite Diagnoses for Work-Related Neuritis

Disorder	Symptoms	Examination
Carpal tunnel syndrome	Numbness and/or tingling in thumb, index, and/or midfinger enhanced symptoms at night	Positive Phalen's Test or Tinel's Sign with particular wrist postures and compression of the median nerve
Scalenus anticus syndrome	Numbness and/or tingling on the preaxial border of the upper lip	Tender scalene muscles with positive Adson's or Wright's test
Cervical neuritis	Pain, numbness, or tingling following a dermatomal pattern in the upper limb	Clinical evidence of intrinsic neck pathology
Lateral antebrachial neuritis	Lateral forearm pain, numbness, and tingling	Tenderness of coracobrachialis origin and reproduction of symptoms on palpation here or by resisted coracobrachialis activity
Pronator syndrome	Pain, numbness, and tingling in the median nerve distribution distal to the elbow	Tenderness of pronator teres or superficial finger flexor muscle, with tingling in the median nerve distribution on resisted activation of same
Cubital tunnel syndrome	Numbness and tingling distal to elbow in ulnar nerve distribution	Tender over ulnar nerve with positive Tinel's Sign and/or elbow flexion test
Ulnar tunnel syndrome	Numbness and tingling in ulnar nerve distribution in the hand distal to the wrist	Positive Tinel's Sign over the ulnar nerve at the wrist
Wartenberg's syndrome	Numbness and/or tingling in distribution of the superficial radial nerve	Positive Tinel's Sign on tapping over the radial sensory nerve
Digital neuritis	Numbness or tingling in the fingers	Positive Tinel's Sign on tapping over digital nerves

Source: Adapted from Ranney, D. et al., *Ergonomics,* 38(7), 1408, 1995. Reprinted with permission.

7.7.3 Elbow Risk Factors

- **Posture:** Repeated pronation, supination, flexion, or extension of the elbow, often required in manual or tool-intensive tasks, generates significant mechanical loading on the elbow joint. Combined wrist and elbow motions, particularly under load, further increase the risk of injury.
- **Force:** Strenuous tasks that engage the forearm extensors and flexors, such as forceful gripping or twisting motions, generate substantial stress on the elbow region and are strongly associated with lateral and medial epicondylitis.

- **Repetition:** Cyclical elbow flexion and extension, especially when combined with wrist rotation or high-repetition tasks, such as assembly line work, are major risk factors for elbow WMSDs (Silverstein et al., 1987).

7.7.4 Hand–Wrist Risk Factors

- **Repetition:** High-frequency tasks involving repetitive hand and wrist movements, such as typing, gripping, or tool use, significantly increase the risk of developing hand and wrist disorders. Repetitive motions involving flexion–extension or ulnar–radial deviation of the wrist are particularly damaging when performed with a cycle time of less than 30 seconds or when over 50% of the task involves the same motion (Silverstein et al., 1987).
- **Posture:** Non-neutral wrist postures, such as sustained wrist flexion, extension, or deviation, heighten the risk of musculoskeletal strain. Poor wrist positioning during gripping or manipulation tasks can lead to cumulative tissue damage in the hand and wrist.
- **Force:** Forceful hand activities, particularly when combined with awkward wrist postures, significantly increase the risk of disorders such as carpal tunnel syndrome and tendinitis. High-force gripping or pinch gripping tasks are particularly hazardous.
- **Vibration:** Hand–arm vibration from the use of power tools is a well-documented risk factor for WMSDs. Prolonged exposure to vibration can lead to disorders such as hand–arm vibration syndrome, which affects blood flow and nerve function in the hand and wrist (NIOSH, 2014).
- **Cold Environments:** Cold conditions can reduce muscle efficiency and lead to excessive force exertion during manual tasks. Workers in cold environments often experience impaired sensory feedback, leading to overexertion and a higher risk of hand and wrist injury (Bernard, 1997).

7.7.5 Lower Back Risk Factors

- **Heavy Physical Work:** Tasks involving manual material handling, such as lifting, carrying, or pushing heavy objects, impose high compressive forces on the spine. Such tasks are a primary cause of lower back injuries, particularly in occupations that require repetitive or sustained physical exertion (Marras et al., 1995).
- **Lifting/Forceful Movement:** Activities that require transferring heavy loads or applying forceful movements, such as pulling or pushing, place significant mechanical stress on the lower back. Inadequate lifting techniques, especially when combined with awkward postures, substantially increase the risk of lumbar spine injury.
- **Awkward Posture:** Non-neutral postures, such as trunk flexion, twisting, or lateral bending, significantly contribute to lower back WMSDs. Prolonged bending or twisting during manual handling or seated work increases the

TABLE 7.3
Minimal Clinical Criteria for Establishing Worksite Diagnoses for Work-Related Muscle or Tendon Disorders

Disorder	Symptoms	Examination
Neck myalgia	Pain one of both sides of neck increase by neck movement	Tender over paravertebral neck muscles
Trapezius myalgia	Pain on top of shoulder increased by shoulder elevation	Tender top of shoulder or medial border of scapula
Scapulothoracic pain syndrome	Pain in scapular region increased by scapular movement	Tender over rib angles 2, 3, 4, 5, and/or 6
Rotator cuff tendinitis	Pain in deltoid area or front of shoulder increased by glenohumeral movement	Rotator cuff tenderness
Triceps tendinitis	Elbow pain increased by elbow movement	Tender triceps tendon
Arm myalgia	Pain in muscle(s) of the arm	Tenderness in a specific muscle of the arm
Epicondylitis/tendinitis	Pain localized to lateral or medial aspect of elbow	Tenderness of lateral or medial epicondyle localized to this area or to soft tissues attached for a distance of 1.5 cm
Forearm myalgia	Pain in the proximal half of the forearm (extensor or flexor aspect)	Tenderness in a specific muscle in the proximal half of the forearm (extensor or flexor aspect) more than 1.5 cm distal to the condyle
Wrist tendinitis	Pain on the extensor or flexor surface for the wrist	Tenderness is localized to specific tendons and is not found over bony prominences
Extensor finger tendinitis	Pain on the extensor surface of the hand	Tenderness is localized to specific tendons and is not found over bony prominences
Flexor finger tendinitis	Pain on the flexor aspect of the hand of distal forearm	Pain on resisted finger flexion localized to area of tendon
Tenosynovitis (finger/thumb)	Clicking or catching of affected digit on movement. There may be pain or a lump in the palm	Demonstration of these complaints, tenderness anterior to metacarpal of affected digit
Tenosynovitis, De Quervain's	Pain on the radial aspect of wrist	Tenderness over first tendon compartment and positive Finkelstein's test
Intrinsic hand myalgia	Pain in muscles of the hand	Tenderness in a specific muscle in the hand

Source: Adapted from the criteria outlined in: Sluiter, J. K., Rest, K. M., & Frings-Dresen, M. H. W. (2001). Criteria document for evaluating the work-relatedness of upper-extremity musculoskeletal disorders. *Scandinavian Journal of Work, Environment & Health*, *27*(Suppl 1), 1–102.

risk of lumbar disc degeneration and other spinal disorders (Marras et al., 1995).

- **Whole-Body Vibration:** Workers exposed to whole-body vibration, such as those operating heavy machinery or industrial vehicles, are at increased risk of developing lower back disorders. Vibration affects the spine's structural integrity and can exacerbate degenerative changes in the vertebrae and discs (NIOSH, 2014).
- **Static Work Posture:** Prolonged static postures, whether sitting or standing, can lead to muscle fatigue and poor circulation, which in turn increase the risk of lower back disorders. Sedentary work, when combined with inadequate ergonomic design, often leads to chronic musculoskeletal strain (Punnett & Wegman, 2014).

7.8 PERSONAL RISK FACTORS

In occupational ergonomics, the primary goal is to create work environments that are safe, efficient, and comfortable for all employees, regardless of their individual physical characteristics. While the field recognizes the existence of personal risk factors that may influence susceptibility to work-related musculoskeletal disorders, the central philosophy of ergonomics remains clear: the workplace must be adapted to meet the needs of the worker, not vice versa. This shift underscores the importance of designing tasks, tools, and workstations that accommodate a diverse workforce and minimize the risk of injury across different demographics.

7.8.1 Personal Risk Factors in the Context of WMSDs

Ergonomics research has long acknowledged that personal factors such as age, gender, genetics, body composition, overall fitness level, and mental health can contribute to the development of WMSDs. These intrinsic characteristics can influence an individual's biomechanical responses to workplace demands and likelihood for developing a WMSD. For example, older workers may be more prone to musculoskeletal fatigue due to age-related degenerative changes, while gender differences in muscle mass distribution and joint flexibility may alter the way men and women experience biomechanical loads (Bernard, 1997; Linton et al., 2011).

Notwithstanding, the field has evolved beyond a simple focus on these individual factors given that the focus of the field of ergonomics is to design the task elements for optimal use and to meet the needs of the worker. Modern ergonomics increasingly emphasizes a biopsychosocial model, which views WMSD risk as a complex interplay between personal characteristics and external environmental factors. This model recognizes that while individual factors may predispose certain workers to musculoskeletal problems, the design of the work environment plays a crucial role in either mitigating or exacerbating these risks (Punnett & Wegman, 2014).

7.8.2 The Biopsychosocial Model and its Implications

The biopsychosocial model integrates biological, psychological, and social dimensions in understanding WMSDs. In this context, ergonomists aim to design work environments that minimize the physical demands placed on the body, particularly for those individuals who may be at greater risk due to their personal characteristics. A well-designed workplace takes into account factors such as workstation adjustability, tool design, and task variation to ensure that the ergonomic demands do not exceed the worker's physical capabilities.

Tools designed to minimize forceful exertions or repetitive motions can help prevent the onset of WMSDs in workers who may be predisposed to such conditions due to genetics or fitness levels. This proactive approach helps ensure that even workers with pre-existing vulnerabilities can perform tasks safely and efficiently (Marras & Karwowski, 2019).

7.8.3 Psychosocial Factors and WMSDs

In recent years, growing attention has been paid to the role of psychosocial factors in the development of WMSDs. Studies have demonstrated that workplace stress, job dissatisfaction, lack of control over work tasks, and poor social support from supervisors or colleagues can all significantly increase the risk of WMSDs, even in workers with minimal physical risk factors (Franche et al., 2019). These psychosocial stressors can manifest physically, leading to muscle tension, fatigue, and poor posture, which exacerbate the risk of injury.

For example, a worker under constant pressure to meet deadlines may adopt awkward postures or ignore early signs of musculoskeletal strain, increasing the likelihood of developing a WMSD. Similarly, a lack of social support in the workplace may prevent workers from voicing concerns about pain or discomfort, leading to the progression of undiagnosed musculoskeletal conditions. Thus, the integration of psychosocial considerations into ergonomic interventions is crucial for reducing the overall risk of WMSDs.

7.8.4 Ergonomic Interventions: Designing for Diverse Populations

Recognizing the diverse range of personal risk factors and their interactions with the work environment has led to a more holistic approach in ergonomic interventions. These interventions prioritize designing flexible work systems that accommodate a wide variety of workers, from young to elderly employees, from those with high fitness levels to those with pre-existing musculoskeletal conditions.

For example, sit-stand desks are increasingly implemented in office environments to allow workers to alternate between sitting and standing, reducing the static postures that can contribute to WMSDs. In industrial settings, ergonomic interventions may involve the use of lift-assist devices or exoskeletons to reduce the physical strain on workers who perform repetitive lifting tasks. Such adaptive designs promote inclusivity, ensuring that all workers, regardless of their individual risk factors, can work safely and comfortably (NIOSH, 2014).

7.8.5 Age

Work-related musculoskeletal disorders are a widespread issue across all age groups, but age-related musculoskeletal impairments, particularly those affecting the back, represent a significant occupational health concern. While WMSDs can affect workers of any age, research indicates that middle-aged and older workers exhibit a higher prevalence of these disorders due to age-related degenerative changes (Buckwalter et al., 2014). These changes, such as osteoarthritis, reduced muscle mass, and diminished tissue resilience, contribute to a greater susceptibility to WMSDs and complicate recovery from such injuries.

However, it is important to note that despite the degenerative changes associated with aging, younger workers also report high rates of WMSDs. Data show that the incidence of compensable back pain and musculoskeletal strains is highest among men aged 20–24 and women aged 30–34 (NIOSH, 2014). This trend suggests that while age-related factors increase the risk of WMSDs in older populations, younger workers are also highly susceptible, potentially due to their inexperience, involvement in physically demanding entry-level jobs, and insufficient ergonomic training.

7.8.6 Age-Related Risk Factors for WMSDs

Age is a major factor in WMSD susceptibility, particularly in terms of the body's ability to withstand physical stresses and recover from injuries. As workers age, their musculoskeletal systems undergo structural and functional changes that can increase the risk of injury.

Degenerative Disorders: One of the primary age-related risk factors for WMSDs is the onset of degenerative disorders, such as osteoarthritis and degenerative disc disease. These conditions, which are more common in middle-aged and older workers, lead to the gradual breakdown of joint cartilage and intervertebral discs, reducing their capacity to absorb mechanical loads. Buckwalter et al. (2014) highlight that these degenerative changes can exacerbate the effects of occupational physical demands, making older workers more prone to WMSDs, particularly in the lumbar spine.

Muscle Mass and Tissue Strength: With age, there is a natural decline in muscle mass and tissue strength, a process known as sarcopenia. This reduction in muscle strength impairs a worker's ability to perform physically demanding tasks and increases the risk of injury. Additionally, the decreased elasticity of tendons and ligaments in older workers means that their tissues are less able to withstand repetitive strain, making them more susceptible to chronic musculoskeletal conditions (Punnett & Wegman, 2014).

Delayed Recovery: Age-related changes also affect the body's capacity for healing. Older workers typically experience slower recovery times following a musculoskeletal injury, which can prolong their absence from work and increase the likelihood of long-term disability. This slower recovery is due to reduced circulation, slower cellular repair, and diminished tissue regeneration, all of which compromise the body's ability to heal from musculoskeletal injuries effectively (Marras & Karwowski, 2019).

7.8.7 WMSDs Among Younger Workers: Inexperience and Physical Demands

Although older workers face a greater risk due to age-related factors, younger workers also exhibit high rates of WMSDs, particularly back injuries. This finding suggests that other factors, such as job-related demands and inexperience, play a significant role in the development of these disorders among younger populations.

Inexperience and Lack of Ergonomic Awareness: Younger workers, particularly those in the early stages of their careers, often have less experience with proper ergonomic practices and injury prevention strategies. Inexperience can lead to poor posture, improper lifting techniques, and an increased likelihood of engaging in physically demanding tasks without adequate rest breaks or recovery periods. The lack of ergonomic training and awareness further contributes to the development of WMSDs in this group (NIOSH, 2014).

Entry-Level Jobs and Physical Demands: Many younger workers are more frequently employed in physically demanding roles, such as construction, manufacturing, and manual labor, where the risk of WMSDs is particularly high. These entry-level positions often involve repetitive tasks, heavy lifting, and awkward postures, all of which are well-established risk factors for musculoskeletal injury. The combination of high physical demands and insufficient ergonomic protections places younger workers at significant risk for developing WMSDs (Silverstein et al., 2016).

7.8.8 Ergonomic Interventions Across Age Groups

From an ergonomic perspective, the key to reducing WMSDs across age groups is to implement workplace design principles that accommodate the physical needs of both younger and older workers. Tailored ergonomic interventions can help mitigate the specific risks associated with different age groups while promoting overall musculoskeletal health.

For *older workers*, ergonomic interventions should focus on reducing the physical demands of tasks, especially those that involve heavy lifting or prolonged repetitive motions. Adjustable workstations, mechanized lifting aids, and job rotation programs are effective strategies for minimizing the impact of degenerative conditions and reducing the risk of injury.

For *younger workers*, the focus should be on comprehensive ergonomic training and education. Teaching proper lifting techniques, posture management, and the importance of microbreaks can help younger employees develop the habits needed to prevent WMSDs. Additionally, ensuring that physically demanding tasks are paired with appropriate ergonomic tools and protective measures can further reduce the risk of injury in this population (Punnett & Wegman, 2014).

7.8.9 Gender

Carpal tunnel syndrome is a widely recognized work-related musculoskeletal disorder (WMSD) that disproportionately affects women. Research indicates that women are approximately three times more likely to develop CTS than men (Office on Women's Health, 2023). While biological factors such as hormonal fluctuations during pregnancy and menopause are contributing factors, potentially due to fluid retention and other physiological changes, anatomical differences like the smaller size of the carpal tunnel in women also play a significant role (Bernard et al., 2014). However, understanding the gender disparity in WMSD prevalence requires a more nuanced analysis that goes beyond biology and considers occupational and psychosocial factors.

7.8.10 Musculoskeletal and Anatomical Differences

Musculoskeletal differences contribute to WMSD risk. On average, women have less muscle mass and physical strength than men, particularly in the upper body. This reduced muscle mass may limit women's capacity to handle repetitive or forceful tasks, increasing their vulnerability to musculoskeletal strain (Liao et al., 2018). These biological factors, when combined with high task demands, make women more prone to developing WMSDs, including CTS.

7.8.11 Occupational Risk Factors and Job Demands

While biological factors are important, the disparity in WMSD prevalence between men and women cannot be explained by anatomy alone. Job demands play a critical role. Women are disproportionately represented in professions with high ergonomic risks, including jobs that involve repetitive motions, forceful exertions, and awkward postures—risk factors directly linked to WMSDs. Professions such as cashiering, data entry, and nursing involve repetitive tasks like typing, scanning, or patient handling, which place constant strain on the musculoskeletal system (Bernard et al., 2014). Repeated hand and wrist movements, particularly under conditions of forceful exertion or awkward posture, significantly elevate the risk of CTS.

Moreover, many of the tasks in these female-dominated jobs involve prolonged static postures, repetitive motion, or high-repetition rates, all of which contribute to the development of musculoskeletal disorders. For instance, a nurse frequently lifting and repositioning patients can develop WMSDs in the back and shoulders, while a data entry professional is more prone to CTS due to prolonged typing in suboptimal ergonomic conditions (Punnett et al., 2014).

7.8.12 Disentangling Job-Specific Risk Factors: The Complexity of WMSD Prevalence

The higher prevalence of WMSDs in women is not simply a matter of gender differences in anatomy or muscle mass. Research highlights the importance of job-specific factors, emphasizing that the types of tasks men and women perform may influence their relative risk for different types of musculoskeletal disorders. For example, a study by Byström et al. (1995) found higher rates of de Quervain's disease—a specific WMSD of the wrist and thumb—in men using power tools. This suggests that occupational demands, rather than biological sex alone, significantly influence the risk of developing WMSDs. Men in physically demanding jobs, such as construction, are at high risk for certain types of WMSDs due to the intensive use of heavy tools and equipment.

Thus, it is essential to disentangle the effects of biological sex from occupational demands when analyzing WMSD prevalence. For women, the higher rates of WMSDs may be more closely related to the types of tasks they perform rather than intrinsic susceptibility. Job characteristics such as repetitive motion, high-force exertion, and static postures are critical factors driving these trends (Bernard et al., 2014; Punnett et al., 2014).

7.8.13 A Biopsychosocial Approach: Addressing Physical, Social, and Psychological Risk Factors

To effectively address the gender disparity in WMSDs, a holistic, biopsychosocial approach is required. This model integrates biological, psychological, and social dimensions, recognizing that both physical and non-physical factors contribute to WMSD risk. While the physical demands of the job are important, psychosocial stressors such as job insecurity, low social support, and high job strain also play a significant role in the development of WMSDs (Franche et al., 2019). For example, women in precarious or low-paid jobs may experience higher levels of stress, which can exacerbate musculoskeletal tension and fatigue, increasing the risk of WMSDs.

Psychosocial factors are particularly relevant for women in the workforce, as they are more likely to work in roles characterized by lower job control and higher emotional labor. These stressors, in combination with the physical demands of the job, can result in a higher incidence of musculoskeletal injuries. Studies have shown that workers with high job demands but low autonomy are at a greater risk for developing WMSDs, highlighting the need for a comprehensive ergonomic approach that addresses both physical and mental health (Punnett et al., 2014).

The gender disparity in WMSD prevalence, particularly in conditions like carpal tunnel syndrome, is a complex issue influenced by a combination of biological, occupational, and psychosocial factors. While anatomical differences and hormonal fluctuations play a role in increasing women's susceptibility to CTS, job-specific demands and psychosocial stressors are critical drivers of this disparity. A holistic, biopsychosocial approach helps develop effective ergonomic interventions that account for both the physical and social aspects of work, protecting workers of all genders from musculoskeletal disorders.

7.8.14 Physical Activity

The link between physical fitness and the risk of developing work-related musculoskeletal disorders is complex, with recent research indicating that specific types of fitness, rather than overall physical activity levels, play a crucial role in either mitigating or exacerbating WMSD risk. While certain forms of physical activity—particularly those involving high-impact or repetitive motions—can increase the likelihood of injury, inadequate muscular strength, endurance, and flexibility may leave workers more vulnerable to WMSDs during repetitive tasks or prolonged static postures (Bullo et al., 2022).

This nuanced relationship underscores the need for targeted fitness interventions aimed at preventing WMSDs, especially in industries where manual labor or repetitive tasks are prevalent. Workers with insufficient physical fitness may be less equipped to handle the biomechanical demands of their jobs, leading to an increased risk of injury and longer recovery times following a WMSD.

7.8.15 The Role of Physical Fitness in Preventing WMSDs

The importance of physical fitness in preventing WMSDs has been well-documented in ergonomic literature. A systematic review conducted by Bullo et al. (2022) found that structured physical exercise programs had a moderate to large effect on reducing both pain and disability associated with WMSDs, particularly in manual laborers. These findings support the integration of worksite wellness programs that focus on improving key physical attributes, such as muscular strength, endurance, and flexibility, in high-risk muscle groups. For example, workers involved in heavy lifting or repetitive motions can benefit from strength-training exercises that target the lower back, shoulders, and arms, helping to reduce the likelihood of injury due to overexertion or poor posture.

Strength-training and aerobic exercise not only improve the mechanical capacity of muscles and joints to withstand repetitive stress but also promote better overall physical conditioning, reducing the strain placed on vulnerable areas of the body during work tasks. Flexibility exercises, such as stretching, can also be incorporated to improve joint range of motion, which is crucial for jobs that require frequent reaching, bending, or twisting (Punnett & Wegman, 2014). By enhancing these fitness components, workers are better prepared to handle the physical demands of their jobs, decreasing the likelihood of developing WMSDs.

7.8.16 Recovery and Rehabilitation Post-WMSD

In addition to preventing WMSDs, maintaining adequate physical fitness is essential for effective recovery and rehabilitation following an injury. Workers with low levels of fitness may experience delayed recovery times and be at higher risk for recurrent musculoskeletal problems. For instance, poor muscular strength or endurance can result in the inability to return to full-duty work or the development of compensatory movement patterns that increase the risk of further injury (Bullo et al., 2022). Therefore, ergonomic rehabilitation programs must include physical exercise regimens designed to restore strength and flexibility in affected areas, ensuring that workers can return to work safely and without re-injury.

Rehabilitation should not only address the injured area but also aim to correct any underlying physical imbalances or weaknesses that may have contributed to the injury. For instance, core strengthening exercises can be particularly beneficial for workers recovering from lower back injuries, as they help stabilize the spine and improve postural alignment during lifting tasks (Marras & Karwowski, 2019).

7.8.17 Job-Specific Fitness Considerations

One of the key findings in the relationship between physical fitness and WMSDs is that general aerobic fitness or overall activity level does not necessarily translate into reduced injury risk for all job types. The specific physical demands of a worker's job must be considered when designing exercise and fitness programs aimed at reducing WMSD risk. As NIOSH (1991) highlighted, aerobic capacity, while beneficial for cardiovascular health, may not sufficiently prepare workers for jobs that require high levels of static or dynamic strength. For instance, a worker with high aerobic fitness may still be at risk for injury if their job involves heavy lifting or sustained muscle contractions without adequate muscular strength or endurance.

To effectively reduce the risk of WMSDs, fitness programs must be tailored to the biomechanical demands of specific job tasks. Workers engaged in physically demanding jobs, such as construction or manufacturing, may require strength-training exercises that focus on improving grip strength, core stability, and shoulder endurance. Conversely, workers in less physically demanding but repetitive roles, such as office employees, would benefit from exercises that target postural muscles and enhance flexibility, particularly in the neck, shoulders, and wrists (Silverstein et al., 2016).

In addition to physical training, ergonomic interventions in the workplace should focus on optimizing job design to minimize repetitive strain. Adjusting workstations, providing adequate rest breaks, and using tools designed to reduce physical stress are critical components in mitigating the risk of WMSDs. The combination of ergonomic interventions and job-specific fitness training provides a holistic approach to WMSD prevention.

7.8.18 Strength

Epidemiological research has long explored the complex interplay between back injury risk, physical strength, and job demands. Early studies, such as those conducted by Chaffin and colleagues in the 1970s, identified a correlation between workers' isometric strength and their risk of back injury. These studies suggested that workers whose isometric strength met or exceeded the physical demands of their jobs had a significantly lower risk of back injury—up to three times lower—compared to workers with weaker isometric strength (Chaffin, 1977). However, subsequent research has provided a more nuanced understanding of the factors contributing to back injury risk.

7.8.19 Re-Evaluating the Role of Isometric Strength Testing

Although the early findings by Chaffin et al. (1973, 1977) supported the idea that higher isometric strength is protective against back injuries, more recent studies have challenged the reliability of isometric strength as a sole predictor. For instance, large-scale prospective studies by Battie et al. (1989) and Leino (1987), focusing on blue-collar workers, found no definitive association between isometric lifting strength and the incidence of reported low back pain. This discrepancy suggests that while strength is important, it is not the only factor influencing back injury risk in occupational settings.

The limitations of isometric strength testing are evident when considering the nature of physical work tasks. Isometric strength tests measure the maximal force a worker can exert in a static position, which may not accurately reflect the dynamic and complex movements required in real-world work scenarios. For example, manual labor tasks often involve bending, lifting, twisting, and maintaining awkward postures over extended periods—activities that require more than just raw strength. As such, modern ergonomics research emphasizes the need to consider additional factors such as core stability, muscular endurance, and proper movement techniques when assessing back injury risk (Punnett & Wegman, 2014).

7.8.20 Core Stability and Muscular Endurance: Critical Components in Back Injury Prevention

Recent research highlights the importance of core strength and endurance in reducing the risk of work-related back injuries. Core muscles, including the deep stabilizing muscles around the spine, play a crucial role in maintaining spinal alignment and reducing mechanical stress on the vertebrae during physical tasks. Andersson (1998) and Franche et al. (2016) emphasized that workers with well-developed core stability are better equipped to manage the physical demands of lifting and manual labor, as core muscles act as a stabilizing unit that distributes loads across the trunk and reduces localized strain on the lower back.

In addition to core strength, lumbar extensor endurance has emerged as a key factor in injury prevention. Lumbar extensor muscles, which are responsible for supporting the spine during prolonged activity, require both strength and endurance to maintain proper posture and spinal alignment throughout the workday. Hodges and Richardson (2000) demonstrated that workers with greater endurance in their lumbar extensors were at a lower risk of developing back pain, as their muscles were better equipped to handle sustained activation and avoid fatigue during repetitive tasks.

7.8.21 The Role of Lifting Technique in Back Injury Risk

Another crucial factor in preventing back injuries is proper lifting technique. Poor technique, such as lifting with a rounded back, using the arms instead of the legs for power, or twisting while carrying loads, can place excessive strain on the spine and increase the risk of injury. Research by McGill (2007) underscores the importance of maintaining a neutral spine while lifting, which involves keeping the back straight and the core engaged to distribute loads evenly across the body. Additionally, McGill emphasized the importance of using leg muscles to generate lifting power, rather than relying on the lower back, and avoiding rotational movements during lifting to minimize shear forces on the spine.

To reduce the risk of back injuries, ergonomics programs now emphasize training workers in safe lifting techniques alongside core strengthening and endurance exercises. These programs teach workers to engage their core muscles during lifting tasks, maintain neutral spinal alignment, and utilize their leg muscles for lifting power. Furthermore, workers are trained to avoid twisting motions and to pivot with their feet instead when changing directions while carrying loads. This comprehensive approach to injury prevention—combining physical conditioning with proper movement techniques—has been shown to significantly reduce work-related back injuries in numerous occupational settings (McGill, 2007).

7.9 A HOLISTIC APPROACH TO BACK INJURY PREVENTION

The evolving understanding of back injury risk highlights the need for a multifaceted approach to prevention. Isometric strength, while important, is not sufficient on its own to predict or prevent back injuries in the workplace. Modern ergonomics recognizes that a combination of factors—including core stability, muscular endurance, proper lifting technique, and job-specific physical conditioning—must be addressed to effectively reduce injury risk.

Incorporating targeted exercise programs that improve core strength and endurance, alongside education on safe lifting practices, can significantly reduce the incidence of back injuries among workers. These programs should be tailored to the specific physical demands of the job, ensuring that workers are adequately prepared to handle dynamic tasks involving lifting, bending, and twisting. In addition, regular ergonomic assessments of the workplace can identify risk factors related to poor posture, inadequate equipment, or excessive physical strain, allowing for appropriate interventions to be implemented.

7.9.1 Anthropometry

Anthropometric variables such as weight, height, body mass index (BMI), and obesity are increasingly recognized as potential risk factors for various work-related musculoskeletal disorders, with particular emphasis on carpal tunnel syndrome and lumbar disc herniation. While the precise mechanisms linking these factors to WMSDs remain under investigation, evidence suggests that excess body weight and body composition play a significant role in increasing susceptibility to certain musculoskeletal injuries, especially in physically demanding occupations.

7.9.2 Obesity and Carpal Tunnel Syndrome

The relationship between obesity and the risk of developing carpal tunnel syndrome has been a focal point of research in occupational ergonomics. Numerous studies have confirmed an elevated risk of CTS among individuals with higher BMI, particularly in obese women. A landmark study by Vessey et al. (1990) demonstrated that obese women face twice the risk of developing CTS compared to their non-obese counterparts. This heightened risk is hypothesized to stem from the increased fatty tissue within the carpal tunnel, which can cause compression of the median nerve, as well as elevated hydrostatic pressure in obese individuals (Werner, 1994). These physiological changes exacerbate the narrowing of the carpal tunnel, making it more susceptible to nerve compression during repetitive hand and wrist movements.

While obesity clearly raises the risk of CTS, the relationship between carpal tunnel size and CTS development is more nuanced. Some studies, such as those by Bleecker et al. (1985), have suggested that both small and large carpal tunnels may predispose individuals to CTS. The underlying hypothesis is that individuals with smaller carpal tunnels are more prone to nerve compression, while those with larger tunnels may have excessive soft tissue within the canal, leading to similar outcomes. However, conclusive evidence on the role of carpal tunnel size in CTS remains elusive, and further research is needed to clarify this relationship (Shiri et al., 2020a).

7.9.3 Body Mass Index (BMI) and Low Back Pain

The association between BMI and low back pain, particularly lumbar disc herniation, has been more difficult to establish conclusively. While excess body weight theoretically increases the mechanical load on the spine, leading to increased pressure on intervertebral discs, the evidence linking anthropometric data to low back pain is mixed. Some earlier studies have found no strong correlation between BMI and the incidence of lumbar spine disorders, while others suggest a modest link between higher BMI and increased severity of low back pain (Punnett & Wegman, 2014).

Recent research by Shiri et al. (2020b) offers new insights into the connection between BMI and musculoskeletal disorders. This study suggests that individuals with higher BMI are not only at greater risk for developing CTS but also for experiencing more severe low back pain. The study highlights that younger individuals with elevated BMI are particularly susceptible to the exacerbating effects of obesity on musculoskeletal health. The proposed mechanism involves increased mechanical stress on the lumbar spine, which, when coupled with the reduced muscular support and flexibility often observed in obese individuals, increases the risk of lumbar disc degeneration and herniation.

7.9.4 Mechanisms Linking Obesity to WMSDs

Several biomechanical and physiological mechanisms have been proposed to explain the association between obesity and WMSDs. In the case of CTS, excess fatty tissue within the carpal tunnel can directly compress the median nerve, while systemic inflammation associated with obesity may further contribute to nerve damage and exacerbate symptoms. Additionally, the increased hydrostatic pressure in obese individuals leads to higher compressive forces on the nerves, increasing the likelihood of developing CTS (Werner, 1994).

For lumbar disc herniation, excess body weight increases the mechanical load on the spine, which may accelerate disc degeneration. Obesity also often leads to poor posture, reduced mobility, and weakened core muscles, all of which can contribute to an increased risk of low back injuries, particularly in jobs requiring heavy lifting or prolonged standing (Shiri et al., 2020a).

7.9.5 Ergonomic Considerations and Preventive Strategies

Given the growing evidence linking anthropometric factors, particularly BMI and obesity, to WMSDs, occupational ergonomics must consider these factors when designing preventive interventions. Worksite wellness programs aimed at promoting physical fitness, weight management, and overall health are essential for reducing the risk of WMSDs in overweight and obese workers. Additionally, ergonomic interventions such as task modification, adjustable workstations, and training in proper lifting techniques can help mitigate the mechanical stresses imposed on the musculoskeletal system by excess body weight (Marras & Karwowski, 2019).

Furthermore, ergonomic assessments should take into account the specific challenges faced by overweight workers. For example, in jobs requiring frequent hand and wrist movements, the increased risk of CTS in obese individuals necessitates careful consideration of tool design, task rotation, and the provision of supportive equipment such as wrist braces. For workers at risk of low back pain, ergonomic interventions should emphasize core strengthening, posture training, and the use of assistive devices to reduce mechanical loading on the spine during lifting tasks.

7.9.6 Smoking

The relationship between smoking and work-related musculoskeletal disorders, particularly low back pain (LBP), sciatica, and intervertebral disc herniation, has been well-documented in occupational health literature. Early studies by Deyo and Bass (1989), Frymoyer et al. (1983), and Svensson et al. (1983) demonstrated a positive association between smoking and the prevalence of these conditions, especially in individuals engaged in physically demanding occupations. More recent research continues to support this link, providing a clearer understanding of the biological mechanisms through which smoking may contribute to spinal degeneration and pain.

7.9.7 The Impact of Smoking on Spinal Health

The adverse effects of smoking on spinal health can be attributed to several physiological mechanisms, primarily involving nicotine's impact on the vascular system and bone health. Nicotine is known to impair blood flow by constricting blood vessels and reducing the oxygen-carrying capacity of red blood cells (Ahlbom et al., 2017). This reduced blood flow diminishes the delivery of essential nutrients to spinal tissues, including intervertebral discs, which rely on diffusion for nutrient exchange. Over time, this impaired nutrient supply can lead to disc degeneration, making the spine more susceptible to conditions such as herniated discs and sciatica.

Furthermore, smoking has been shown to weaken bone density and increase the risk of fractures. Nicotine and other toxic chemicals in cigarettes interfere with the normal bone remodeling process, reducing bone mineral density and increasing the likelihood of both macro and microfractures in the vertebrae (Soriano et al., 2010). This weakening of spinal structures can exacerbate the risk of LBP, particularly in workers involved in heavy manual labor, where the mechanical demands on the spine are already high (Punnett & Wegman, 2014).

7.9.8 Smoking and Physically Demanding Occupations

The link between smoking and low back pain is especially pronounced in physically demanding occupations, where workers are exposed to high levels of mechanical stress. A study by Boshuizen et al. (1993) highlighted that smoking significantly increased the risk of LBP in individuals with labor-intensive jobs compared to those in less physically demanding roles. Workers in construction, manufacturing, and transportation, for example, often engage in repetitive lifting, bending, and twisting movements that place considerable strain on the spine. The combination of mechanical loading and the negative systemic effects of smoking creates a synergistic effect, increasing the likelihood of both acute and chronic back pain (Marras & Karwowski, 2019).

7.9.9 Mechanisms Linking Smoking and Back Pain

Several mechanisms have been proposed to explain how smoking contributes to LBP and other spinal disorders:

1. **Impaired Blood Flow and Disc Degeneration:** Nicotine's vasoconstrictive effects reduce blood supply to the spine, hindering nutrient delivery to the intervertebral discs. This diminished nutrient supply impairs disc health, making them more prone to degeneration, bulging, and herniation (Ahlbom et al., 2017). This is particularly concerning for workers in occupations involving heavy lifting, as degenerated discs are more likely to suffer from increased mechanical loading.
2. **Bone Weakening and Increased Fracture Risk:** Smoking has been shown to decrease bone mineral density, weakening the vertebrae and increasing the risk of fractures. In manual labor environments, where workers are exposed to frequent impacts or vibrations, microfractures in the spine can develop over time, leading to chronic pain and disability (Soriano et al., 2010).
3. **Coughing and Spinal Strain:** Chronic coughing, often associated with long-term smoking, can contribute to increased intra-abdominal pressure and strain on the spine. Frequent coughing episodes may exacerbate existing back pain or trigger new episodes of LBP, particularly in workers who already have compromised spinal structures due to repetitive strain or injury (Deyo & Bass, 1989).

7.9.10 Conflicting Evidence and Methodological Considerations

While the link between smoking and back pain is generally supported by epidemiological research, earlier studies presented mixed results, with some failing to establish a clear association. For instance, studies by Kelsey et al. (1990) and Riihimäki et al. (1991) did not find strong correlations between smoking and low back pain in certain populations. These discrepancies may be attributed to differences in study design, population characteristics, and the specific metrics used to measure smoking exposure and back pain. Moreover, smoking may interact with other risk factors, such as age, body mass index, and physical activity levels, complicating the relationship between smoking and spinal health (Punnett & Wegman, 2014).

Recent advancements in research methodologies, including longitudinal studies and more precise measures of smoking exposure, have helped clarify the link between smoking and musculoskeletal disorders. Modern studies emphasize the importance of considering occupational context, as smoking's impact on back pain appears to be more pronounced in physically demanding jobs where mechanical stress is high (Shiri et al., 2020b).

7.9.11 Preventive Strategies in Occupational Settings

Given the established association between smoking and spinal health, workplace interventions aimed at reducing smoking could play a crucial role in preventing work-related musculoskeletal disorders. Occupational health programs should incorporate smoking cessation initiatives alongside traditional ergonomic interventions to protect workers from back pain and other WMSDs. Encouraging smoking cessation not only reduces the risk of LBP but also improves overall cardiovascular health and reduces the likelihood of other smoking-related conditions that can affect workplace productivity (NIOSH, 2014).

In addition to promoting smoking cessation, ergonomic interventions should focus on reducing mechanical loading on the spine through the use of proper lifting techniques, the provision of supportive equipment, and job rotation strategies to minimize repetitive strain. Combined with efforts to address lifestyle factors such as smoking, these interventions can significantly reduce the incidence of back pain in the workplace (Marras & Karwowski, 2019).

7.10 SCREENING METHODS AND DIAGNOSIS OF WORK-RELATED MUSCULOSKELETAL DISORDERS

Several methods have been developed to both diagnose and treat musculoskeletal disorders. A few examples of these screening approaches are discussed below.

7.10.1 Phalen's Test

One of the key figures in the diagnosis of CTS is Dr. George S. Phalen, an American hand surgeon who spent over 17 years studying this condition. His work significantly advanced the understanding of CTS in both clinical and occupational settings. Phalen identified that Tinel's Sign, a diagnostic technique involving light percussion over the median nerve at the wrist, could evoke a tingling sensation in patients with CTS, which he described as "a tingling sensation radiating out into the hand" (Urbano, 2000). This test is now widely used in both medical examinations and ergonomic assessments to detect median nerve irritation due to repetitive or forceful wrist movements.

Furthermore, Phalen introduced a wrist flexion test, now known as the Phalen's Test, as another diagnostic tool for CTS. This test requires the patient to position their elbows on a surface and press the dorsal surfaces of their hands together with the wrists in maximum flexion. The test is typically performed for approximately three minutes, though symptoms often appear within one to two minutes in individuals with CTS. The onset of tingling or numbness during the test is indicative of median nerve compression, while healthy individuals can generally hold the position for up to 10 minutes without symptoms (Mondelli et al., 2014; Aboonq, 2015).

From an ergonomic perspective, Phalen's Test is not only valuable in diagnosing CTS but also in evaluating the impact of occupational risk factors. Work tasks involving repetitive wrist flexion, gripping, or vibrating tools can lead to increased pressure within the carpal tunnel, making workers more susceptible to CTS. Identifying such ergonomic risk factors is critical for developing workplace interventions aimed at reducing the incidence of carpal tunnel syndrome, such as redesigning tools, modifying workstations, or implementing rest breaks and exercises (Aboonq, 2015). Effective ergonomic interventions can alleviate the biomechanical strain on the wrist and prevent long-term nerve damage associated with CTS.

By understanding the physiological mechanisms underlying carpal tunnel syndrome and employing diagnostic tools such as Tinel's Sign and Phalen's Test, ergonomists and occupational health professionals can better assess workplace environments and implement strategies to reduce the risk of CTS in high-risk occupations.

7.10.2 Durkan Test or Carpal Compression Test

In occupational ergonomics, accurate diagnostic tools for musculoskeletal disorders, particularly carpal tunnel syndrome, are critical due to the condition's prevalence in workplaces characterized by repetitive hand and wrist movements. One such tool, the carpal compression test, was developed in 1991 by American orthopedic surgeon Dr. John A. Durkan. Durkan's research demonstrated that this test is more sensitive than other widely used diagnostic tools, such as Tinel's Sign and Phalen's Test, in detecting CTS (Durkan, 1991).

The carpal compression test is particularly relevant for ergonomists, as it directly measures the effects of external pressure on the median nerve, which is often subjected to repetitive strain or force in occupational settings. Durkan's initial study, which involved 31 patients and 46 affected hands, established the efficacy of this method, showing a higher sensitivity in diagnosing CTS compared to traditional tests (Durkan, 1991). This diagnostic sensitivity is especially valuable in the workplace,

TABLE 7.4
Ergonomic Measures to Control Common WMSDs

Disorder	Avoid in General	Avoid in Particular	Recommendation	Design Issues
Carpal tunnel syndrome	Rapid, often repeated finger movements, wrist deviation	Dorsal and palmar flexion, pinch grip, vibration between 10 and 60 Hz		Workplace design
Cubital tunnel syndrome	Resting forearm on sharp edge or hard surface			Workplace design
De Quervain's syndrome	Combined forceful gripping and hard twisting			Workplace design
Epicondylitis	"Bad tennis backhand"	Dorsiflexion, pronation		Workplace design
Pronator syndrome	Forearm pronation	Rapid and forceful pronation, strong elbow and wrist flexion	Use large muscles but infrequently and for short time	Design of work object
Shoulder tendinitis, rotator cuff syndrome	Arm elevation	Arm abduction, elbow elevation	Let wrists be in line with the forearm	Design of job task
Tendinitis	Often repeated movements, particularly with force exertion; hard surface in contact with skin; vibrations	Frequent motion of digits, wrists, forearm shoulder	Let shoulder and upper arm be relaxed	Design of hand tools ("bend tool, not the wrist")
Tenosynovitis, De Quervain's syndrome, ganglion	Finger flexion, wrist deviation	Ulnar deviation dorsal and palmar flexion, radial deviation with firm grip	Let forearms be horizontal or more declined	Design for round corners, use pad
Thoracic outlet syndrome	Arm elevation, carrying loads	Shoulder flexion, arm hyperextension		Design of work object placement
Trigger finger or thumb	Digit flexion	Flexion of distal phalanx alone		Workplace design
Ulnar artery aneurism	Pounding and pushing with heel of the hand			Workplace design

(Continued)

TABLE 7.4 (CONTINUED)
Ergonomic Measures to Control Common WMSDs

Disorder	Avoid in General	Avoid in Particular	Recommendation	Design Issues
Ulnar nerve entrapment	Wrist flexion and extension	Wrist flexion and extension, pressure of hypothenar eminence		Workplacedesign
White finger, Vibration syndrome	Vibrations, tight grip, cold exposure	Vibrations between 40 and 125 Hz		Workplace design
Neck tension syndrome	Static head posture	Prolonged static head-neck posture	Alternate head-neck postures	Workplace design

Source: Adapted from Kroemer, K. et al., *Ergonomics, How to Design for Ease and Efficiency*, Prentice Hall, Englewood Cliffs, NJ, 1994. Reprinted with permission.

where early detection of CTS can lead to timely intervention, potentially preventing long-term damage or the need for invasive treatments.

The procedure for the carpal compression test involves applying direct pressure to the median nerve at the wrist. Durkan originally employed a rubber atomizer bulb connected to a pressure manometer, commonly used in sphygmomanometers, to apply a force of 150 millimeters of mercury over 30 seconds. The appearance of pain or paresthesia (tingling) during the test indicates median nerve compression, a hallmark of CTS. In ergonomics, this method can be used as part of routine assessments for workers in high-risk occupations, such as those involving repetitive hand motions, forceful gripping, or prolonged wrist flexion (Mondelli et al., 2014).

An alternate method of performing the carpal compression test, identified by Durkan, involves the examiner applying even pressure directly to the median nerve using both thumbs. This simplified version can be utilized in workplace health screenings, where ergonomic practitioners may not have access to specialized equipment. The manual application of pressure by trained personnel is an efficient and cost-effective diagnostic tool, providing a quick assessment of median nerve compression that can be implemented in various work environments (Aboonq, 2015).

For ergonomic assessments, the carpal compression test serves not only as a diagnostic tool but also as a way to evaluate the effectiveness of workplace interventions. By regularly testing workers exposed to CTS risk factors, such as repetitive tasks or use of vibrating tools, ergonomists can monitor the onset and progression of symptoms and adjust ergonomic designs to reduce strain on the wrist. Proper workstation layout, tool design, and task rotation are critical components in minimizing carpal tunnel pressure, and Durkan's carpal compression test provides an objective means to assess these interventions.

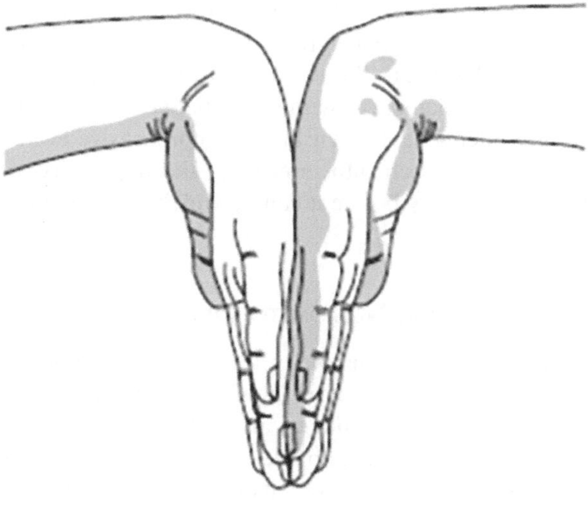

FIGURE 7.1 PHALENS test.

7.10.3 Vibrometry Testing

Vibrometry testing is another diagnostic technique used to assess potential cases of carpal tunnel syndrome. This method is based on the principle that nerve compression within the carpal tunnel, particularly of the median nerve, can impair sensory function. During vibrometry testing, a vibrating probe is placed on the patient's middle finger, and the frequency of vibration is adjusted by the examiner. The patient is asked to report their perception of the vibration, which theoretically should decrease in individuals with CTS due to the impaired function of the median nerve.

From an occupational ergonomics perspective, vibrometry testing is of particular interest as it offers a non-invasive method for detecting sensory deficits that may arise from repetitive strain or nerve compression, common in many work environments. The test aims to measure vibration perception thresholds, which can be compromised in individuals whose median nerve is compressed, potentially leading to decreased sensitivity in the affected fingers (Neese & Konz, 1993).

However, despite the potential advantages of this method, vibrometry's diagnostic utility for CTS has been questioned. Early studies, such as those by Neese and Konz (1993) and Jetzer (1991), showed promise in using vibrometry to identify CTS by detecting sensory impairments in the fingers. These studies suggested that vibrometry could be a useful tool for evaluating nerve function in individuals exposed to occupational risk factors, such as vibration tools or repetitive hand motions. Nevertheless, more recent research has raised concerns about the reliability and consistency of this testing. Mondelli et al. (2011) and McGough et al. (2014) reported that vibrometry testing often lacks the sensitivity and specificity required for an accurate diagnosis of CTS, and the subjectivity of the patient's response introduces variability in test results. This variability can reduce the test's effectiveness, making it less definitive as a standalone diagnostic tool.

For ergonomists and occupational health professionals, these findings suggest that vibrometry testing should be used with caution in workplace assessments. While it may provide some insights into nerve function, particularly in cases where workers are exposed to high levels of vibration or repetitive strain, the test's limitations imply that it should be supplemented with other diagnostic tools, such as the carpal compression test or Phalen's Test, to ensure a comprehensive evaluation of median nerve function. Further research into improving the objectivity and reliability of vibrometry testing is necessary before it can be widely adopted in occupational settings for CTS diagnosis.

7.10.4 Nerve Pace Electron Neurometer Device

The Nerve Pace Electron Neurometer was once considered a potential tool for diagnosing carpal tunnel syndrome by measuring motor nerve conduction in the median nerve using surface electrodes. This device aimed to offer a less invasive alternative to traditional electrodiagnostic studies by detecting abnormalities in nerve function through surface measurements. However, its clinical utility has been significantly

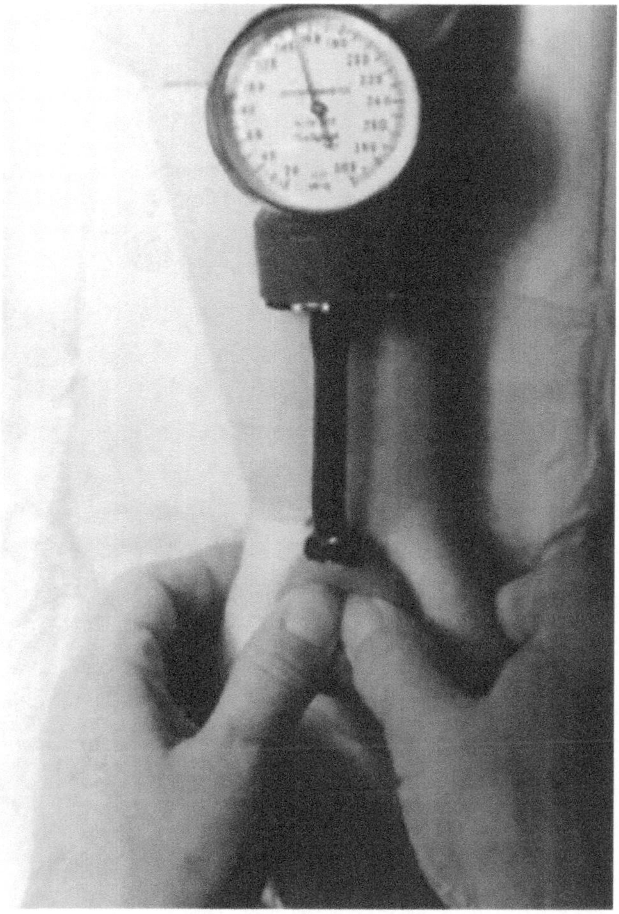

FIGURE 7.2 Durkan test. (From Durkan, J.A., *J. Bone Joint Surg.*, 73(4), 535, 1991)

limited due to a variety of factors, including patient-specific variables and the inherent complexity of the disease itself.

Research has demonstrated several limitations of the Nerve Pace Electron Neurometer. One major issue is the device's sensitivity to skin thickness, which can vary greatly among individuals and affect the accuracy of surface electrode measurements. This variability creates inconsistency in the results, particularly in patients with conditions that affect the skin or subcutaneous tissue. Additionally, the presence of peripheral neuropathy—a common comorbidity in patients with diabetes or other systemic diseases—can further obscure test results, leading to false negatives or positives (David & Chaudhry, 2003). Moreover, the severity of CTS plays a crucial role in the device's accuracy; in mild cases, the device may fail to detect significant changes in nerve conduction, whereas in more advanced stages of CTS, the nerve damage may be too extensive for the device to provide meaningful data.

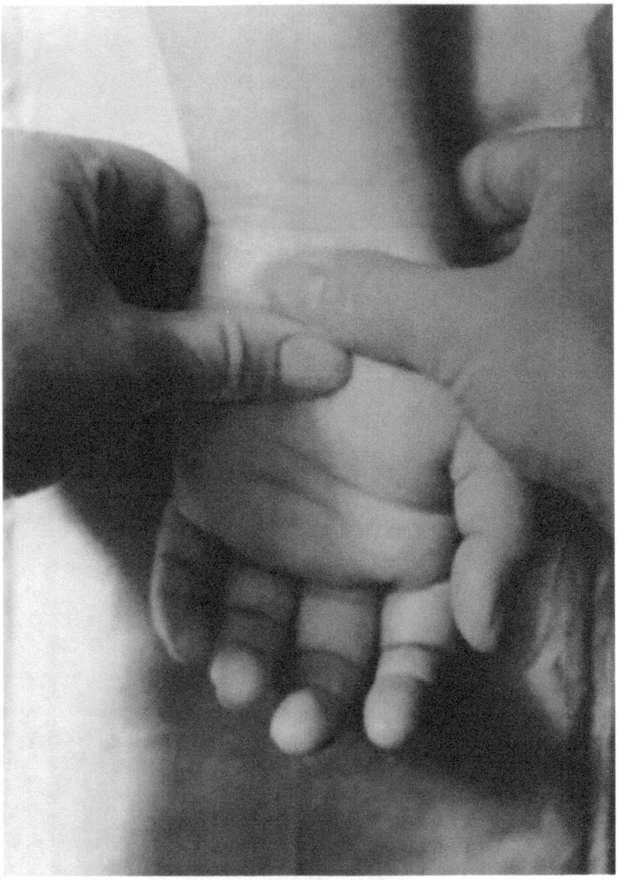

FIGURE 7.3 Alternate Durkan test. (From Durkan, J.A., *J. Bone Joint Surg.*, 73(4), 535, 1991)

Due to these inherent flaws, the Nerve Pace Electron Neurometer has not gained widespread acceptance in clinical practice. A key assessment by the American Association of Electrodiagnostic Medicine (AAEM) labeled the device as "flawed" and "experimental," emphasizing that it is not a reliable substitute for standard electrodiagnostic methods such as nerve conduction studies (NCS) or electromyography (EMG) (David & Chaudhry, 2003). NCS and EMG remain the gold standards for diagnosing CTS, as they provide detailed information about both the sensory and motor components of nerve function, offering a more comprehensive evaluation of the median nerve.

For ergonomists and occupational health professionals, the unreliability of the Nerve Pace Electron Neurometer underscores the importance of using validated diagnostic tools when assessing workplace-induced CTS. While non-invasive diagnostic technologies are appealing, they must be thoroughly vetted to ensure they

TABLE 7.5
A List of Proactive Job Survey Methods for Control of CTDs

Job Survey	Description
Facility walk-throughs	A facility walk-through consists of using firsthand observations to identify exposures to recognized CTD risk factors. Knowledge of the processes, facilities, and schedules is used to conduct observations of representative activities and workers. This method may be effective at identifying the presence of some risk factors, particularly those that are most obvious. However, a single walk-through or observation may not be sufficient to determine the magnitude, frequency, and duration of exposure to risk factors. The time required to perform a walk-through depends on the work cycle time and variability of the tasks performed, with longer work cycles and jobs with greater task variability needing more observation time. The method can be enhanced by combining it with formal or informal worker and supervisor interviews and/or risk factor checklists.
Worker and supervisor interviews	A worker and supervisor interview consists of asking questions regarding job/task attributes and associated risk-factor exposures. The method relies on the firsthand observations and experiences of worker and supervisor. Such interviews may be formal (written or recorded) or informal (verbal). This method may be effective at identifying the presence of risk factors and associated job attributes not easily garnered by a facility walk-through, particularly risk-factor exposure duration, recovery breaks, infrequently performed tasks, or multiple methods at performing tasks. However, the method is subjective. The method is also performed as part of a more comprehensive job analysis.
Risk factor checklists	Checklists consist of a formal procedure in which CTD risk factors are enumerated on a list and checked off by the analyst through firsthand observation of a specific or representative task(s) or job(s). Checklists may have high sensitivity in identifying the presence of some risk factors; however, they may have low specificity. Checklists must be adapted and validated for each industry or occupation. Using checklists requires caution and familiarity with the job, task, or process.
Team problem-solving processes	Team problem-solving processes integrate the firsthand observations and experiences of several workers and supervisors regarding job or task attributes and associated risk-factor exposures and attempt to develop, test, and implement plausible solutions. Any of the aforementioned surveys tools may be incorporated into such a process.

Source: After ANSI (1996).

provide accurate and reliable data across diverse patient populations and severity levels. As of now, the Neurometer's limitations prevent it from being a viable option in both clinical and occupational settings (Pransky et al., 1997). Continued reliance on standard electrodiagnostic studies is essential for accurately diagnosing and

TABLE 7.6
Examples of Tools for WMSD Surveillance

	Methods of Surveillance	
Surveillance Approach	**Passive**	**Active**
Workplace risk factors (associated with WMSDs)	• Company dispensary logs • Insurance records • Workers' compensation records	• Checklists • Questionnaires • Interviews
	• Health (WMSDs) • Accident reports • Transfer requests • Absentee records • Grievances	• Physical exams
	• Not really used for WMSD risk factor yet[a]	• Checklists • Questionnaires • Job analysis

Source: Adapted from Kuorinka, I. and Forcier, L. (Eds.), *Work-Related Musculoskeletal Disorders (WMSDs): A Reference Book for Prevention*, Taylor & Francis, Boca Raton, FL, 1995. Reprinted with permission. [a] The use of surrogate measures for exposure (e. g., job title or firm's department) could be viewed as "passive surveillance."

managing CTS, especially in work environments where early detection and intervention can prevent progression of the disorder.

7.10.5 Tinel's Sign

Tinel's Sign, first developed by French neurologist Jules Tinel in 1915, was initially used as a diagnostic tool for detecting nerve regeneration after injuries. Tinel observed that tapping along the course of the median nerve, particularly after nerve injury, could evoke a tingling sensation, or paresthesia, in the first three and a half digits of the hand—an area innervated by the median nerve. This finding helped clinicians assess nerve damage and recovery in patients with various types of peripheral nerve injuries. However, at the time of its development, Tinel's Sign was not associated with carpal tunnel syndrome, as the syndrome had yet to be widely understood or diagnosed (Papanicolaou et al., 2011).

It was not until 1957, with the work of American hand surgeon George Phalen, that Tinel's Sign was formally recognized as a useful diagnostic tool for carpal tunnel syndrome. Phalen's research demonstrated that this sign, when applied to the wrist, could indicate median nerve compression within the carpal tunnel. When light percussion over the median nerve at the wrist elicited a tingling sensation in the fingers, it was considered indicative of CTS (de Krom et al., 1990). This realization marked a significant advancement in the diagnosis of carpal tunnel syndrome, which is now one of the most common nerve compression disorders seen in occupational settings involving repetitive hand use.

In the context of occupational ergonomics, Tinel's Sign remains an important diagnostic technique for assessing workers who may be at risk for developing CTS due to repetitive tasks, prolonged wrist flexion, or use of vibrating tools. However, its diagnostic accuracy is often supplemented with other clinical tests, such as Phalen's Test and nerve conduction studies, to provide a more comprehensive evaluation of median nerve function and the extent of compression (Papanicolaou et al., 2011). As a non-invasive and simple test, Tinel's Sign is still widely used in both clinical and workplace settings to help identify early signs of nerve compression.

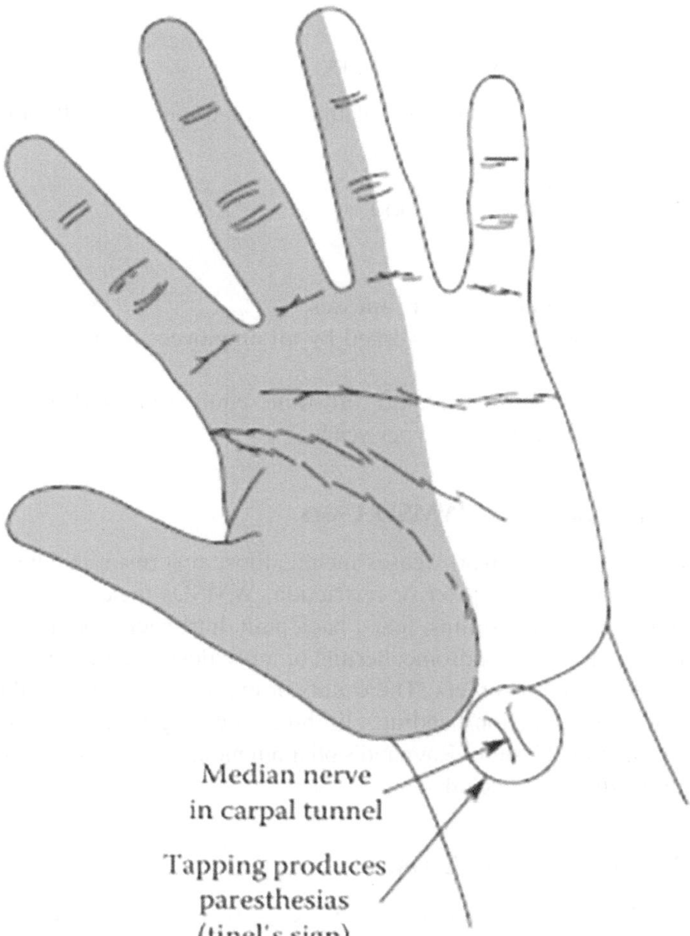

FIGURE 7.4 Tinel's Sign.

7.11 IMPACT ON INDUSTRY

The US Bureau of Labor Statistics publishes detailed information about occupational injuries, illnesses, and fatalities from private industry. Results of the 2014 survey revealed 1,157,410 non-fatal cases involving days away from work (107.1 cases per 10,000 workers). This number of nonfatal days away from work in 2014 is down from 2013, in which 1,162,210 total cases (109.4 per 10,000 workers) were reported (US Bureau of Labor Statistics, 2014). The number of WMSD cases and incidence rates are reported by sex, age, occupation, industry, and several other characteristics. The BLS data in the table below reveal that, since 2002, the number and incidence rates of WMSD cases have declined. The decline has been attributed to changes in task design, ergonomic interventions, advancements in technology, and greater awareness of occupational health.

7.11.1 Explanation of Incidence Rates

Incidence rates represent the number of injuries or illnesses per 10,000 full time workers and is calculated as follows:

Incidence Rate = (N/EH) × 20,000,000

Where:

- N = number of injuries and/or illnesses
- EH = total employee hours worked by all employees during the calendar year
- 20,000,000 = base for 10,000 full-time equivalent workers (working 40 hours per week, 50 weeks per year).

7.11.2 Characteristics of WMSD Cases

The term "days away from work" cases include those that result in days away from work with or without job transfer or restriction. WMSDs include cases where the nature of injury is sprains, strains, tears; back pain, hurt back; soreness, pain, hurt, except back; carpal tunnel syndrome; hernia; or musculoskeletal system and connective tissue diseases and disorders. The events or exposures leading to the injury or illness include bodily reaction/bending, climbing, crawling, reaching, twisting; overexertion; or repetition. Cases of Raynaud's phenomenon, tarsal tunnel syndrome, and herniated spinal discs are excluded.

7.11.3 Decline in WMSD Cases

According to the BLS, the number and incidence rates of WMSD cases have continued to decrease. In 2007, there were 335,390 WMSDs, accounting for 29% of all workplace injuries resulting in time away from work. In 2006, there were 357,160 cases of WMSDs, making up 30% of all workplace injuries requiring time away from work. This decline is attributed to changes in task design, ergonomic interventions, technological advancements, and greater awareness of occupational health.

7.11.4 Need for Further Research

Additional research is needed to develop predictive models for the development of WMSDs based on occupational and non-occupational risk factors. The aforementioned statistics demonstrate why WMSDs continue to be a primary concern for employers. Companies are concerned with the costs associated with WMSDs and are actively pursuing ways to identify, prevent, and control WMSDs before they become problematic.

7.11.5 High-Risk Occupations

The occupations in American industry most prone to WMSDs include manual material handling, nursing aides and orderlies, computer operators, ticket agents, and emergency medical personnel. The figure below illustrates the incidence rates for WMSDs in 2007 in these industries.

TABLE 7.7
Number and Incidence Rates[a] of Nonfatal Occupational Injuries and Illnesses with Days away from Work[b] Involving Musculoskeletal Disorders,[c] 2004

	WMSD Cases with Days away from Work			
Year	Total Cases	Incidence Rate	Median Days	Relative Standard Error
2002	487,915	55.3	—	—
2003	435,180	49.6	10	0.8
2004	402,700	45.2	10	0.8

Source: US Bureau of Labor Statistics, 2005.

[a] Incidence rates represent the number of injuries and illnesses per 10,000 full-time workers and were calculated as defined in the following: (N/EH) x 20,000,000. where
N is the number of injuries and illnesses
EH is the total hours worked by all employees during the calendar year
20,000,000 = base for 10,000 full-time equivalent workers (working 40 hours/week, 50 weeks/year)

[b] Days away from work cases include those that result in days away from work with or without job transfer or restriction

[c] Includes cases where the nature of injury is sprains, strains, tears; back pain, hurt back; soreness, pain, hurt, except back; CTS; hernia; or musculoskeletal system and connective tissue diseases and disorders and when the event or exposure leading to the injury or illness is bodily reaction/bending, climbing, crawling, reaching, twisting; overexertion; or repetition. Cases of Raynaud's phenomenon, tarsal tunnel syndrome, and herniated spinal discs are not included. Although these cases may be considered MSDs, the survey classifies these cases in categories that also include non-MSD cases.

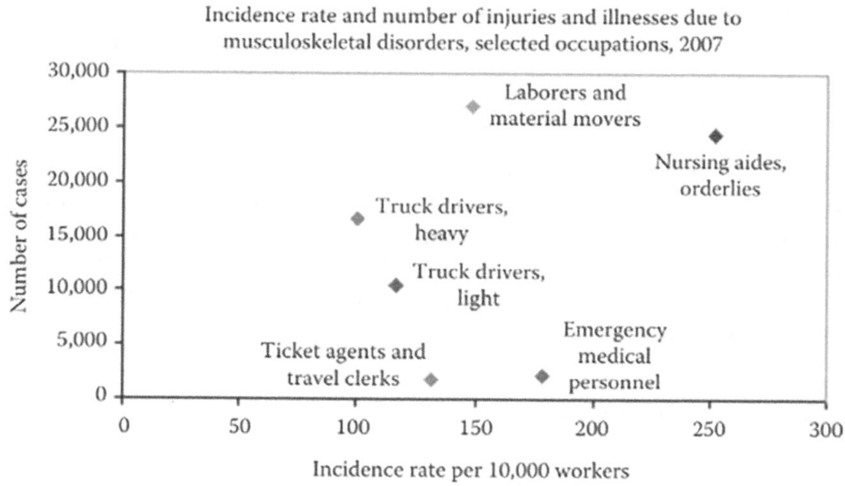

FIGURE 7.5 Incidence rate and number of injuries by industry.

Although laborers and manual material handlers comprise the largest population among WMSD cases, nursing aides, orderlies, and attendants are the most heavily affected occupations, with 252 cases per 10,000 workers. Nursing homes and personal care facilities have some of the highest rates of injury and illness among industries for which nationwide Lost Workday Injury and Illness (LWDII) rates were calculated. According to the BLS, nursing and personal care facilities experienced an average LWDII rate of 7.6 in 2014, despite the availability of feasible controls identified to address hazards within this industry. This rate is more than double the LWDII rate of 2.8 for private industry as a whole.

These injuries in the nursing and personal care industry continue to grow. BLS statistics from 2014 reveal that nursing aides, orderlies, and attendants had a WMSD rate more than seven times the national average (US Bureau of Labor Statistics, 2015). In response to these alarming statistics, NIOSH researchers compiled guidelines for nursing home staff on how to move patients safely. The document "Safe Lifting and Movement of Nursing Home Residents" (2006) provides guidance on handling patients in a manner that is both safe for them and ergonomically sound for the workers.

The National Coalition of Ergonomics estimates that WMSDs result in $15 to $20 billion annually in workers' compensation costs (Jeffress, 1999). Similarly, the National Institute for Occupational Safety and Health estimates costs between $13 and $20 billion annually (Rosenstock, 1997). The Bone and Joint Decade, an international movement supported by the United Nations and the World Health Organization, estimated that the total cost of musculoskeletal diseases in the US, including treatment and lost wages, was $849 billion in 2008 (United States Bone and Joint Decade, 2008), further illustrating the international significance of these issues.

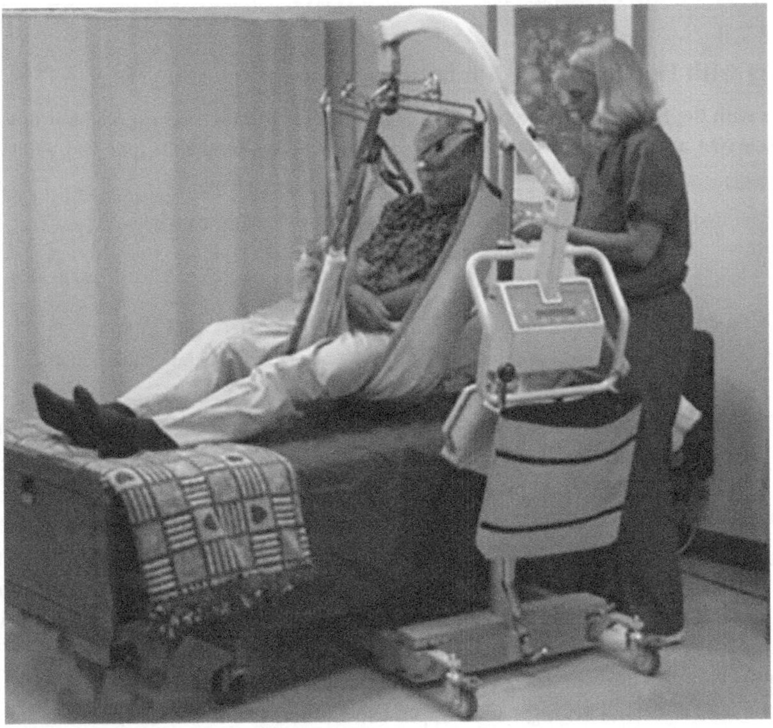

FIGURE 7.6 Safe lifting in a nursing home. (From OSHA website, accessed February 18, 2011)

WMSDs represent the most significant work-related health problem in Europe. In 2011, the European Agency for Safety and Health at Work (EU-OSHA) reported that for every 100,000 workers, there were 32 new WMSD cases, accounting for 50% of all new cases of occupational disorders in the EU and 50% of all work-related health problems in Europe. Moreover, the number of WMSD cases has been increasing over time. In 2001, around 19 workers per 100,000 suffered from a new WMSD, nearly doubling to 32 per 100,000 workers by 2003 (EU-OSHA, 2011).

TABLE 7.8
Sectors with Highest WMSD in Europe

Sectors with the Highest Relative Number of New MSD Cases (per 100,000 Workers)		Sectors with the Highest Number of Work-Related MSD Complaints (per 100,000 Workers)	
Mining and quarrying	960	Health and social work	4,283
Manufacturing	81	Transport and communication	3,160
Construction	50	Construction	3,158
Total population	32	Total population	2,645

Source: European Agency for Safety and Health at Work (EU-OSHA), 2011

As previously stated, tasks that involve the use of computer keyboards, mice, and video display terminals (VDTs) are also areas of serious concern. In some cases, these jobs are considered a breeding ground for occupational WMSDs. The World Health Organization concluded that "musculoskeletal discomfort was commonplace during work with VDTs" and that "injury from repeated stress … is possible" (Sauter et al., 1991). Given the global market associated with current organizations, it is imperative to acknowledge and address this problem on an international level. An epidemic of WMSDs affecting VDT operators swept Australia in the last decade, with prevalence rates of almost 35% recorded in some Australian organizations (Sauter et al., 1991). The high-tech Japanese industry is not exempt from the occurrence of WMSDs, as Japanese organizations have reported surges in the number of occupational-related musculoskeletal injuries (Sauter et al., 1991). This further illustrates the need for comprehensive guidelines to mitigate the risks of work-related musculoskeletal injuries internationally.

WMSDs are clearly a global occupational issue. In previous decades, they were considered the most important work-related health problem in Europe (OSHA EU, 2003). In 2003, for every 100,000 workers, there were 32 new WMSD cases, resulting in WMSDs accounting for 50% of all new cases of occupational disorders in the EU (2003 data) and 50% of all work-related health problems in Europe (1999 data). Efforts have been made to mitigate the impact of these risk factors (OSHA EU, 2003) through the implementation of screening methods, education, and improved workplace design.

7.12 ERGONOMIC PROGRAM FOR WMSD

In recent years, the Occupational Safety and Health Administration (OSHA) has continued to build upon decades of scientific study to address the frequency and severity of work-related musculoskeletal disorders (WMSD). Although the original ergonomic standard proposed in 1999 was repealed with a change in the US presidential administration, the guidance offered in the standard remains a crucial framework for industries aiming to develop programs to reduce and manage WMSD.

The primary elements of OSHA's Ergonomics Program, as refined and utilized over the years, include:

1. **Management Leadership and Employee Participation:** Emphasizing the importance of active involvement and commitment from both management and employees in identifying and addressing ergonomic issues.
2. **Job Hazard Analysis and Control:** Focusing on systematically identifying and assessing job-related hazards that could lead to WMSD, and implementing effective control measures to mitigate these risks.
3. **Employee Training:** Ensuring that employees are well-informed about WMSD, including how to recognize symptoms, the importance of early reporting, and techniques for minimizing risk through proper ergonomics practices.
4. **WMSD Management:** Involving the ongoing monitoring and management of WMSD cases to ensure timely and effective responses, including medical management and appropriate work accommodations.
5. **Program Evaluation:** Regularly assessing the effectiveness of the ergonomics program through continuous monitoring, feedback, and improvements based on performance data and employee input.

Each of these components is essential for the successful implementation of an ergonomics program and is discussed in more detail in the following sections.

7.12.1 Management Leadership

Management leadership requires the leadership to take an active and participatory role in addressing and reducing the presence of work-related musculoskeletal disorders. Specific activities that demonstrate an active concern for employee health and safety include:

1. **Assignment and Communication of Responsibilities:** Clearly defining roles and responsibilities related to ergonomics and WMSD prevention, and ensuring these are communicated effectively to all levels of the organization.
2. **Empowering Employees to Make Changes:** Encouraging and enabling employees to identify ergonomic issues and suggest or implement improvements, fostering a culture of continuous improvement and shared responsibility.
3. **Examination of Existing Policies and Practices:** Regularly reviewing and assessing current policies and practices to ensure they align with best practices in ergonomics and effectively address WMSD risks.
4. **Communication with Employees Regarding the Program:** Maintaining open and ongoing communication with employees about the ergonomics program, its goals, progress, and any changes or updates, to ensure transparency and engagement.

5. **Prompt Response to Employee Reports of Problems:** Responding quickly and effectively to employee reports of ergonomic issues or symptoms of WMSDs, demonstrating a commitment to addressing concerns and protecting employee health.

7.12.2 Employee Participation

Employee participation is designed to allow employees to be active participants in the reduction of the risks and occurrence of work-related musculoskeletal disorders. Specific employee participation should include:

1. **Employee Participation in the Procedure for Reporting Signs and Symptoms of WMSDs:** Employees should be encouraged and trained to promptly report any signs or symptoms of WMSDs, ensuring early detection and intervention. Clear reporting procedures should be established and communicated to all employees.
2. **Accessto OSHA Standards and Information About the Ergonomics Program**: Employees should have easy access to OSHA standards and comprehensive information about the organization's ergonomics program. This includes educational materials, guidelines, and resources that help employees understand the risks, prevention strategies, and their roles in maintaining ergonomic safety.

7.12.3 Job Hazard Analysis and Reporting

Job hazard analysis and reporting are designed to identify and remove risks from the workplace. Some of the activities include setting up a way for employees to report WMSD signs and get quick responses as well as the following:

1. **Provide Information to Current and New Employees**:
   - Educate employees about common WMSD hazards.
   - Inform employees about the signs and symptoms of WMSDs.
   - Clearly explain how to report WMSD signs and symptoms.
2. **Hazard Analysis Activities**:
   - **Evaluation of the Problem Job to Identify Ergonomic Risk Factors:** Conduct thorough evaluations of tasks to identify ergonomic risk factors that could lead to WMSDs.
   - **Implement Controls to Help Eliminate or Reduce WMSD Hazards:** Apply appropriate controls to mitigate identified risks. This includes engineering controls, administrative controls, and personal protective equipment (PPE).
   - **Use the Incremental Abatement Process:**
     - When a WMSD hazard occurs, implement one or more controls to materially reduce the WMSD hazards.

- If WMSD hazards are still present, implement additional feasible controls to further reduce WMSD risks.
- Continue implementing controls incrementally until the WMSD hazards are eliminated or reduced to an acceptable level.

7.12.4 Training

An important part of a successful ergonomic program is training. The training of employees is essential and should include the following:

1. **Training Employees:**
   - **When the Job is Identified as a Hazard**: Provide training to employees as soon as a job is identified as having ergonomic hazards, ensuring they understand the risks and how to mitigate them.
   - **When an Employee is Assigned to the Problem Job**: Train employees who are newly assigned to tasks that have been identified as problematic, equipping them with the knowledge and skills to perform their work safely.
   - **Periodically, at Least Every Three Years:** Conduct refresher training at regular intervals, at least every three years, to reinforce safe practices and update employees on any new information or changes in procedures.
2. **Training Persons Involved in Setting Up and Managing the Ergonomics Program:**
   - **Initially Setting Up and Managing the Ergonomics Program:** Provide comprehensive training to individuals when they are first assigned the responsibility of establishing and managing the ergonomics program, ensuring they are well-prepared for their roles.
   - **Periodically, at Least Every Three Years:** Offer ongoing training every three years to keep program managers updated on the latest ergonomic practices, technologies, and regulatory requirements.

7.12.5 WMSD Management and Program Evaluation

In WMSD management, the employer continually assesses the workplace conditions and the presence of injuries related to work-related musculoskeletal disorders. This includes evaluating injuries that occur after the implementation of the program and incorporating additional activities to mitigate risks when necessary.

Program evaluation is designed to ensure the ongoing effectiveness of the methods in reducing the risks of injury. To evaluate the ergonomics program, an employer must:

1. **Consult with Employees in Problem Jobs:**
   - Assess employees' views on the effectiveness of the program, gathering feedback from those directly involved in or affected by problem jobs.
2. **Identify Significant Deficiencies:**
   - Detect any significant deficiencies in the program, determining areas where the program may be falling short or not providing adequate protection against WMSDs.
3. **Evaluate the Elements of the Program:**
   - Ensure that all elements of the program are functioning properly, including management leadership, employee participation, job hazard analysis, and training.
4. **Evaluate the Program's Effectiveness:**
   - Verify that the program is eliminating or materially reducing WMSD hazards, ensuring that the measures in place are successful in mitigating risks and preventing injuries.

7.13 Summary

Awkward postures, repetitive tasks, and handling heavy loads are recognized risk factors that can harm bones, joints, muscles, tendons, ligaments, nerves, and blood vessels, leading to fatigue, pain, and work-related musculoskeletal disorders. These disorders typically develop over time due to repeated exposure to physical stress at work. The upper limbs (hands, wrists, elbows, and shoulders), neck, and lower back are especially susceptible to WMSDs. Proper design of tools, equipment, processes, and workspaces can significantly reduce the risk and occurrence of these disorders.

The following tools are recommended for ergonomic screening and analysis to assess and mitigate ergonomic risks, particularly for WMSDs. These tools have been sourced from OSHA, the Department of Design and Analysis at Cornell University, the Center for Ergonomics at the University of Michigan, Dr. Thomas Bernard from the University of South Florida, and the Human Factors and Ergonomics Society.

7.13.1 Checklists and Qualitative Screening Tools

- OSHA Computer Workstation Checklist
- Cornell University Performance Oriented Ergonomic Checklist for Computer (VDT) Workstation
- Cornell University Quick Exposure Checklist (QEC)
- American Conference of Governmental Industrial Hygienists (ACGIH)
  - Threshold Limit Values for Back
  - Hand–Arm Vibration Screening Criteria
  - Whole-Body Vibration Screening Criteria
- Ergonomic Chair Evaluation Checklist (ANSI/HFES 100-2007)

- Ergonomic Work Surface Evaluation Checklist (ANSI/HFES 100-2007)

Ergonomic Evaluation Tools and Techniques:

- Physical and Postural Evaluation Tools
  - Manual Task Risk Assessment Tool (ManTRA) V2.0
  - Rapid Upper Limb Assessment (RULA) Tool
  - Rapid Entire Body Assessment (REBA) Tool
  - JACK—Human Simulation and Ergonomics Analysis Software
  - University of Michigan 3D Static Strength Prediction Program (3DSSPP)
  - Job Strain Index (JSI)
  - Moore–Garg Strain Index
  - Rohmert Method for Static Work Analysis

7.13.2 Cornell University Ergonomic Tools and Techniques

- Ergonomic Seating Evaluation Form
- Ergonomic Keyboard and Mouse Platform Systems Evaluation Form
- Ergonomics Cost-Justification Worksheets
  - Cornell ROI Estimator
  - Cornell University Body Discomfort Survey 2010
  - Cornell Musculoskeletal Discomfort Questionnaires

7.14 CASE STUDY

Problem Overview

Work-related musculoskeletal disorders have become a significant health concern globally, impacting a wide range of industries. These disorders result from repetitive stress or strain on muscles, tendons, ligaments, joints, and nerves due to workplace activities. WMSDs are commonly observed in industries such as manufacturing, construction, healthcare, and office work environments, where employees are often required to engage in repetitive tasks, awkward postures, and forceful exertions.

A comprehensive understanding of WMSDs is crucial for reducing their prevalence and improving the quality of life for workers. According to the World Health Organization, WMSDs contribute significantly to disability, absenteeism, and reduced productivity in various industries. The global economic burden of these disorders has become staggering, with billions of dollars spent annually on healthcare costs, compensation, and lost productivity.

This case study will explore the origin, nature, and impact of WMSDs, as well as a process to establish an ergonomic program aimed at mitigating risks in the workplace.

Origin of Work-Related Musculoskeletal Disorders

WMSDs originate from repetitive and cumulative stress on the musculoskeletal system. Prolonged exposure to awkward or static postures, repetitive movements, and high force or pressure during work tasks are the primary contributors to WMSDs. Inadequate rest periods and improper workstation design exacerbate these conditions. Over time, these factors lead to muscle strain, inflammation, and damage to tendons and ligaments, resulting in disorders such as tendinitis, carpal tunnel syndrome, back pain, and shoulder injuries.

The development of WMSDs is often slow, as symptoms may take months or years to manifest. Early symptoms include discomfort, stiffness, or fatigue in affected body parts. Without proper ergonomic intervention, these early signs can progress into chronic pain, functional impairment, and in severe cases, permanent disability.

Nature and Impact of WMSDs

The nature of WMSDs is multifaceted, involving both physical and psychological components. Physically, workers experience pain, inflammation, and restricted movement in areas such as the neck, shoulders, back, and limbs. The psychological impact of WMSDs is also significant, as chronic pain and reduced functionality can lead to stress, anxiety, and depression.

The impact of WMSDs extends beyond the individual worker, affecting organizational productivity and the global economy. WMSDs are one of the leading causes of absenteeism and presenteeism (reduced productivity while at work), which can result in substantial financial losses for businesses. In 2019, the International Labour Organization (ILO) estimated that approximately 2.3 million work-related deaths occurred annually, with WMSDs contributing significantly to these fatalities due to prolonged exposure to unsafe working conditions.

Ergonomic Design Process to Address WMSDs

To reduce the incidence of WMSDs in the workplace, organizations must implement ergonomic programs focused on minimizing physical strain and improving work environments. An effective ergonomic design process follows these key steps:

1. **Risk Identification:** The first step involves identifying tasks or work environments that pose risks for developing WMSDs. This can be done through job task analysis, employee surveys, and reviewing injury and incident reports.
2. **Ergonomic Assessment:** Conduct an ergonomic assessment to evaluate the physical demands of work tasks, such as posture, force exertion, frequency of movement, and the design of tools or equipment. Anthropometric data should be utilized to assess the fit between workers and their tools or workstations.

3. **Workplace Redesign:** Based on the assessment, redesign the workplace to reduce stress on the musculoskeletal system. This includes adjusting the height of workstations, repositioning tools for better access, introducing adjustable seating, and incorporating mechanical aids to reduce lifting or repetitive motions.
4. **Training and Education:** Educating employees on proper ergonomic practices is essential. This includes training on how to adjust workstations, use tools correctly, and take adequate breaks to minimize muscle strain.
5. **Monitoring and Continuous Improvement:** Implement a system to continuously monitor the effectiveness of ergonomic interventions. Regular reviews and feedback from employees can help refine the program and address emerging issues.

Outcome

Through a well-established ergonomic program, the risks of WMSDs can be significantly reduced. By addressing the root causes of musculoskeletal strain, organizations can expect improved worker health, reduced absenteeism, and enhanced productivity. Additionally, the financial benefits of reducing compensation claims, medical expenses, and lost work time provide a substantial return on investment for companies committed to employee wellness.

Conclusion

WMSDs represent a major challenge for industries worldwide, affecting the health and productivity of millions of workers. However, by understanding the origin and nature of these disorders and implementing a structured ergonomic program, organizations can mitigate risks and foster a healthier work environment. The integration of ergonomics into workplace design is not only a means of improving safety but also a crucial step toward enhancing the overall efficiency and well-being of the global workforce.

Exercises

1. Research the history of work-related musculoskeletal injuries in your country and on a global level. Use the following methods to compile your findings:
   - Conduct an internet search for historical trends and statistical data from credible sources (e.g., WHO, OSHA, ILO).
   - Interview a safety officer or ergonomist in your community to gather local insights.
   - Use AI tools (e.g., ChatGPT, Google Gemini) to generate summaries of global trends, then verify findings using academic literature.
   - Prepare a report summarizing key findings, highlighting regional differences and historical developments.

2. Investigate through which body parts vibration is most likely transmitted
   - Search the scientific literature for studies on vibration transmission and its effects on the human body.
   - Use AI tools to summarize relevant research findings and identify patterns.
   - Interview professionals working in industries like construction or manufacturing to understand practical implications.
   - Create a diagram illustrating how vibration is transmitted through different body parts.
3. Explore whether it is reasonable to assume that vibration is transmitted only horizontally or vertically, as done in most research.
   - Use the internet to find and analyze research articles on vibration transmission.
   - Discuss this question with an engineering professor or ergonomics specialist.
   - Use AI tools to generate arguments for and against this assumption, then critically evaluate the responses.
   - Write an essay presenting your conclusions, supported by evidence.
4. Identify professionals other than truck drivers who are particularly exposed to vibrations or impacts.
   - Conduct an internet search for industries with significant exposure to vibration.
   - Use AI tools to create a list of occupations and compare it with findings from interviews or literature.
   - Interview workers in industries such as aviation, mining, or agriculture to validate your findings.
   - Present a short report on your results, including examples of specific tasks.
5. Analyze the effects of several sources of vibration arriving simultaneously at different body parts.
   - Review case studies or research articles available online.
   - Use AI tools to simulate or summarize potential biomechanical impacts.
   - Interview an ergonomics researcher or safety specialist to gain practical insights.
   - Write a summary of your findings, supported by at least one example.
6. Evaluate which of the following tasks might provoke symptoms in a worker suffering from carpal tunnel syndrome. For each task, list expected risk factors:
   a. Scrubbing debris from a brick wall using a wire brush held in the hand
   b. Operating a computer keyboard
   c. Painting a ceiling
   d. Mopping a floor
   e. Playing tennis
   - Use AI tools to analyze CTS risk factors for each activity and compare with peer-reviewed literature.

- If possible, interview a medical professional or physical therapist for validation.
- Compile your findings in a risk-factor table.

7. Determine which of the following jobs might be "strongly" associated with anterior knee pain later in life:
   a. Sailor
   b. Tiler
   c. Chef
   d. University lecturer
   e. Administrator
   f. Nurse
   g. Farmer
   h. Doctor
   - Search occupational health literature for risk factors contributing to anterior knee pain.
   - Use AI tools to identify key trends across professions.
   - Interview individuals from at least two of these occupations to compare theory with real-world experiences.
   - Write a brief analysis of the identified risks for each job.
8. Consider which of the following areas the occupational medical officer in a poultry processing plant should focus on first to reduce WMSD incidence:
   a. Aerobic capacity of the workers
   b. Lighting levels
   c. Prevalence of repetitive tasks
   d. Job contentment
   - Research WMSD prevalence and ergonomic interventions in similar industries.
   - Use AI tools to model possible outcomes of interventions.
   - Interview plant workers or occupational health professionals for practical insights.
   - Draft a prioritized recommendation, supported by evidence.
9. Evaluate which of the following alterations to the design of a powered tool would most likely reduce WMSD prevalence:
   a. Fitting a high-friction sleeve to the handle
   b. Changing from a pinch grip to a power grip
   c. Switching from foot pedal operation to a hand-operated trigger
   d. Designing a control that requires continuous force to operate
   e. Increasing the power of the tool
   - Use AI tools to simulate ergonomic outcomes or review similar case studies.
   - Search for research papers on ergonomic tool design and WMSD prevention.
   - Interview professionals who use powered tools to assess practical implications.
   - Prepare a detailed comparison of the options and their potential impacts.

10. List two occupational and two non-occupational risk factors for CTS.
    - Use AI tools and online medical literature (e.g., PubMed) to identify risk factors.
    - Conduct a brief survey or interview with workers from different professions.
    - Create a chart categorizing and explaining the identified risk factors.
11. Analyze which of the following might be caused by working all day at a computer with the mouse positioned too far to one side and at an elevated position:
    a. Pain in the lumbar region of the back
    b. Neck pain
    c. Pain on the opposite side of the body
    d. Rotator cuff tendinitis
    e. CTS
    f. Wrist tendinitis
    g. RSI
    h. Non-specific forearm pain
    - Use the internet and AI tools to evaluate the biomechanics involved in improper mouse positioning.
    - Consult ergonomics guidelines (e.g., ANSI/HFES standards) to identify potential issues.
    - Interview a physical therapist or ergonomist to validate your findings.
    - Summarize your analysis in a cause-and-effect diagram.

These updates enhance the exercises by encouraging students to engage in research, critical thinking, and collaboration while leveraging modern tools and methodologies.

REFERENCES

Aboonq, M. S. (2015). Pathophysiology of carpal tunnel syndrome. *Neurosciences*, *20*(1), 4–9.

Ahlbom, A., & Axelson, O. (2017). Smoking and intervertebral disc degeneration: Mechanisms and epidemiological findings. *Journal of Spine Research*, *32*(4), 563–570.

American National Standards Institute. (1996). Control of cumulative trauma disorders (CTDs): A guide to job analysis and control methods. ANSI Z-365.

Andersson, G. B. (1998). Epidemiological features of chronic low-back pain. *The Lancet*, *354*(9178), 581–585.

Armstrong, T. J. (2014). Repetitive motion injuries: Historical context and contemporary understanding. *Journal of Occupational Health, 19*(6), 381–386.

Armstrong, T. J., Punnett, L., & Wegman, D. H. (2016). Ergonomic interventions to prevent musculoskeletal disorders: Effectiveness and challenges. *Journal of Occupational Health*, *58*(3), 221–230.

Battie, M. C., & Bigos, S. J. (1989). Isometric lifting strength as a predictor of industrial back pain reports. *Journal of Occupational Medicine*, *31*(7), 510–517.

Bernard, B. P. (1997). *Musculoskeletal disorders and workplace factors*. National Institute for Occupational Safety and Health (NIOSH).

Bernard, B. P., Fine, L. J., & Armstrong, T. J. (2014). Occupational risk factors for work-related musculoskeletal disorders: A review of the literature. *Occupational Ergonomics Journal*, *15*(2), 128–146.

Bernard, B. P., & Guo, H. R. (2021a). Classification of work-related musculoskeletal disorders and the impact on prevention programs. *American Journal of Industrial Medicine*, *64*(2), 135–147.

Bernard, B. P., & Guo, H. R. (2021b). Classification of work-related musculoskeletal disorders and their prevention. *American Journal of Industrial Medicine*, *64*(2), 152–162.

Bleecker, M. L., Bohlman, M., Moreland, R., & Tipton, A. (1985). Carpal tunnel syndrome: Role of carpal canal size. *Neurology*, *35*(10), 1546–1549.

Bongers, P. M., de Winter, C. R., Kompier, M. A. J., & Hildebrandt, V. H. (1993). Psychosocial factors at work and musculoskeletal disease. *Scandinavian Journal of Work, Environment & Health*, *19*(5), 297–312.

Boshuizen, H. C., Burdorf, A., & van Duuren, J. B. (1993). Smoking, occupational exposure, and back pain in physically demanding jobs. *American Journal of Industrial Medicine*, *24*(5), 597–608.

Bovenzi, M., Schust, M., & Mauro, M. (2015). Vascular disorders induced by hand-transmitted vibration. *International Archives of Occupational and Environmental Health*, *88*(2), 185–198.

Buckwalter, J. A., Grodzinsky, A. J., & Ratcliffe, A. (2014). Degenerative changes in musculoskeletal tissues with aging. *The Journal of Bone and Joint Surgery*, *96*(6), 523–532.

Bullo, V., Bergamin, M., Gobbo, S., Sieverdes, J. C., Zaccaria, M., & Neunhaeuserer, D. (2022). Physical exercise reduces pain and improves disability in workers with musculoskeletal disorders: A systematic review. *Journal of Occupational Rehabilitation*, *32*(2), 157–173.

Bureau of Labor Statistics (BLS). (2002). *Occupational safety and health: Field definitions*. U.S. Department of Labor. Available at: https://www.bls.gov/help/def/sh.htm

Bureau of Labor Statistics (BLS). (2014). *Employer-reported workplace injuries and illnesses--2014*. U.S. Department of Labor.

Bureau of Labor Statistics (BLS). (2015). *Nonfatal occupational injuries and illnesses requiring days away from work, 2014*. U.S. Department of Labor.

Bureau of Labor Statistics (BLS). (2021). *Workplace injuries and illnesses report*. U.S. Department of Labor.

Byström, S., & Hall, C. (1995). Gender differences in the prevalence of deQuervain's disease in relation to occupational exposure. *Ergonomics*, *38*(5), 1055–1064.

Centers for Disease Control and Prevention (CDC). (2013). *Industrial child labor: Historical impacts and health outcomes*. CDC Occupational Health Reports.

Chaffin, D. B., Herrin, G. D., & Keyserling, W. M. (1973). *Prevention of back pain and injury at work: a scientific approach*. University of Michigan, Center for Ergonomics.

Chaffin, D. B. (1977). Ergonomic basis for the design of workstations. *Journal of Occupational Health*, *19*(6), 381–386.

David, R., & Chaudhry, V. (2003). Evaluation of new diagnostic technologies for carpal tunnel syndrome. *Journal of Clinical Neurophysiology*, *20*(4), 227–230.

de Krom, M. C., Knipschild, P. G., Kester, A. D., Thijs, C. T., Boekkooi, P. F., & Spaans, F. (1990). Carpal tunnel syndrome: Prevalence in the general population. *Journal of Clinical Epidemiology*, *43*(4), 373–376.

Deyo, R. A., & Bass, J. E. (1989). Lifestyle and low-back pain: The influence of smoking and obesity. *Spine*, *14*(5), 501–506.

Durkan, J. A. (1991). A new diagnostic test for carpal tunnel syndrome: The carpal compression test. *The Journal of Bone and Joint Surgery. American Volume*, *73*(4), 535–538.

Ergonomics Society. (2023). Repetitive strain injuries and workplace ergonomics: An overview. *Ergonomics Society Journal, 18*(1), 45–57.

European Agency for Safety and Health at Work. (2011). *Work-related musculoskeletal disorders: Prevalence, costs and demographics in the EU.*

European Agency for Safety and Health at Work (EU-OSHA). (2003). *Work-related musculoskeletal disorders: Back to work report.* Luxembourg: Office for Official Publications of the European Communities.

Franche, R. L., Baril, R., & Loisel, P. (2016). Factors affecting work-related musculoskeletal disorders: A biopsychosocial perspective. *Journal of Occupational Rehabilitation, 26*(1), 23–35.

Franche, R. L., Baril, R., & Loisel, P. (2019). Psychosocial factors and their impact on the risk of work-related musculoskeletal disorders: A review. *Journal of Occupational Medicine, 61*(3), 217–226.

Frymoyer, J. W., Pope, M. H., & Costanza, M. C. (1983). Epidemiology of low-back pain. *Spine, 8*(5), 419–423.

Gray, H. (1878). *Anatomy of the human body* (Vol. 8). Lea & Febiger.

Hodges, P. W., & Richardson, C. A. (2000). Contraction of the abdominal muscles associated with movement of the lower limb. *Physical Therapy, 80*(1), 9–23.

Hootman, J. M., Macera, C. A., Ham, S. A., Helmick, C. G., & Sniezek, J. E. (2004). Physical activity levels among the general US adult population and persons with arthritis. *Arthritis & Rheumatism, 51*(4), 544–551.

Jeffress, P. N. (1999). The national coalition of ergonomics report on WMSDs and compensation. *Safety Review Journal, 18*(3), 202–208.

Jetzer, D. A. (1991). Vibrometry in the detection of median nerve compression at the wrist. *The Journal of Hand Surgery, 16*(3), 305–309.

Kelsey, J. L., Golden, A. L., & Mundt, D. J. (1990). Low back pain/prolapsed lumbar intervertebral disc. *Rheumatic Disease Clinics of North America, 16*(3), 699–716.

Koehler, R., Silverstein, B., & Wegman, D. H. (2019). *Historical and current perspectives on work-related musculoskeletal disorders.* Occupational Health and Ergonomics.

Leino, P. (1987). Isometric muscle strength and low back pain in blue-collar workers. *Journal of Occupational Medicine, 29*(3), 1045–1050.

Liao, X., Lu, Y., & Pan, Z. (2018). Gender differences in carpal tunnel syndrome prevalence: A biomechanical perspective. *International Journal of Occupational Health and Safety, 64*(4), 278–284.

Liberty Mutual Research Institute for Safety. (2019). Material Handling and Musculoskeletal Injuries in the Workplace. *Journal of Occupational Safety, 30*(4), 112–124.

Linton, S. J., & Shaw, W. S. (2011). The role of psychosocial factors in back pain and disability. *The Lancet, 379*(9814), 1901–1908.

Mandel, J. (2014). Cumulative trauma disorders in labor-intensive industries. *Occupational Health Journal, 12*(3), 145–153.

Marras, W. S., Davis, K. G., & Mirka, G. A. (1995). Ergonomic risk factors associated with manual handling tasks. *Journal of Occupational Medicine, 37*(5), 596–603.

Marras, W. S., & Karwowski, W. (2019). *The Occupational Ergonomics Handbook.* CRC Press.

McGill, S. M. (2007). *Low back disorders: Evidence-based prevention and rehabilitation.* Human Kinetics.

McGough, P., Burke, F. D., Barton, N. J., & McGowan, J. (2014). The clinical reliability of vibrometry in carpal tunnel syndrome. *Hand Surgery International, 19*(2), 45–51.

Mondelli, M., Aretini, A., Ginanneschi, F., & Greco, G. (2011). Sensory and vibrometry tests in carpal tunnel syndrome: A critical review of diagnostic accuracy. *Journal of Neurophysiology, 125*(3), 942–948.

Mondelli, M., Aretini, A., Ginanneschi, F., & Greco, G. (2014). Carpal tunnel syndrome and ulnar neuropathy at the elbow: Prevalence and risk factors in an Italian cohort. *Clinical Neurophysiology, 125*(5), 940–945.

National Institute for Occupational Safety and Health (NIOSH). (1991). *Work practices guide for manual lifting.* U.S. Department of Health and Human Services.

National Institute for Occupational Safety and Health (NIOSH). (1997). *Work-related musculoskeletal disorders: Guidelines for prevention.* NIOSH Publications.

National Institute for Occupational Safety and Health (NIOSH). (2006). *Safe lifting and movement of nursing home residents.* U.S. Department of Health and Human Services.

National Institute for Occupational Safety and Health (NIOSH). (2014). *Ergonomic guidelines for preventing musculoskeletal disorders in the workplace.* U.S. Department of Health and Human Services.

National Institute for Occupational Safety and Health (NIOSH). (2018). *Hierarchy of controls.* Centers for Disease Control and Prevention. https://www.cdc.gov/niosh/topics/hierarchy/default.html

National Institute for Occupational Safety and Health (NIOSH). (2019). *Musculoskeletal health program: Strategic plan for 2020–2029.* Centers for Disease Control and Prevention.

National Institute of Neurological Disorders and Stroke (NINDS). (2020a). *Spinal disorders and workplace safety.* U.S. National Institutes of Health.

Neese, D. W., & Konz, S. A. (1993). The utility of vibrometry testing in occupationally induced carpal tunnel syndrome. *Ergonomics, 36*(5), 501–507.

Office on Women's Health (OWH). (2023). *Carpal tunnel syndrome and women.* U.S. Department of Health and Human Services.

Papanicolaou, G. D., McCabe, S. J., & Firrell, J. (2011). The prevalence and characteristics of nerve compression syndromes in the upper extremity. *The Journal of Hand Surgery, 26*(3), 459–466.

Pransky, G., Masi, M., & Lax, M. B. (1997). Carpal tunnel syndrome: A review of clinical evidence for the workplace role. *Occupational Medicine, 12*(2), 211–222..

Punnett, L., & Wegman, D. H. (2014). Work-related musculoskeletal disorders: The epidemiological and ergonomic perspective. *Occupational Health and Ergonomics Journal, 12*(4), 457–472.

Riihimäki, H. (1991). Low-back pain, its origin and risk indicators. *Scandinavian Journal of Work, Environment & Health, 17,* 81–90.

Roquelaure, Y. (2018). Musculoskeletal disorders and psychosocial factors at work. *European Trade Union Institute Report 142.* Brussels.

Rosenstock, L. (1997). Economic burden of work-related musculoskeletal disorders in the United States. *Occupational Health and Safety Journal, 22*(4), 330–335.

Sauter, S. L., Schleifer, L. M., & Knutson, S. J. (1991). Work-related musculoskeletal disorders and video display terminal use: A growing concern. *World Health Organization Bulletin, 69*(5), 733–743.

Shiri, R., Viikari-Juntura, E., & Falah-Hassani, K. (2016). The effect of overweight and obesity on musculoskeletal pain: a systematic review and meta-analysis. *Obesity Reviews, 17*(11), 1139–1147.

Shiri, R., Falah-Hassani, K., & Viikari-Juntura, E. (2020a). Smoking and the risk of low back pain: A meta-analysis. *American Journal of Epidemiology, 189*(3), 219–227.

Shiri, R., Falah-Hassani, K., & Viikari-Juntura, E. (2020b). The association between obesity and the risk of low back pain: A meta-analysis. *American Journal of Epidemiology, 189*(3), 219–227.

Silverstein, B. A., Fine, L. J., & Armstrong, T. J. (1986). Occupational factors and musculoskeletal disorders: A review. *American Journal of Industrial Medicine, 9*(3), 261–281.

Silverstein, B. A., Fine, L. J., & Armstrong, T. J. (1987). Hand-wrist musculoskeletal disorders in industry: A review of epidemiological research. *Journal of Occupational Medicine*, *29*(4), 261–265.

Silverstein, B. A., Fine, L. J., & Armstrong, T. J. (2016). Ergonomic risk factors and preventive strategies for work-related musculoskeletal disorders. *Journal of Occupational Medicine*, *58*(4), 233–240.

Soriano, J. B., Miravitlles, M., & Garcia-Rio, F. (2010). Smoking and spinal health: A review of the literature. *Journal of Occupational Health and Safety*, *33*(7), 289–297.

Svensson, H.-O., Vedin, A., Wilhelmsson, C., & Andersson, G. B. J. (1983). Low-back pain in relation to other diseases and cardiovascular risk factors. *Spine (Phila Pa 1976)*, *8*(3), 277–285.

Treaster, D. E., & Burr, D. (2004). Gender differences in prevalence of upper extremity musculoskeletal disorders. *Ergonomics*, *47*(5), 495–526.

United States Bone and Joint Decade. (2008). *Musculoskeletal diseases in the United States: Prevalence and societal cost.* U.S. Bone and Joint Initiative.

Urbano, F. L. (2000). Tinel's sign and Phalen's maneuver: Physical signs of carpal tunnel syndrome. *Hospital Physician*, *36*(1), 39–44.

Vessey, M. P., Villard-Mackintosh, L., & Yeates, D. (1990). Epidemiology of carpal tunnel syndrome in women of childbearing age: Findings from a national survey. *British Medical Journal*, *301*(6748), 879–882.

Werner, R. A. (1994). Evaluation of work-related carpal tunnel syndrome. *Journal of Occupational Rehabilitation*, *4*(4), 163–173.

World Health Organization (WHO). (2021). Preventing musculoskeletal disorders in the workplace. https://www.who.int/publications/i/item/preventing-musculoskeletal-disorders-in-the-workplace

Wilcox, A., Phelps, M., & Wang, Y. (2019). The role of the flexor retinaculum in the development of carpal tunnel syndrome. *Journal of Occupational Medicine and Ergonomics*, *36*(4), 207–216.

8 Heavy Work and Evaluating Physical Work Loads & Lifting

8.1 LEARNING GOALS

Understanding the effects and problems associated with manual material handling of heavy loads is critical to task design in such environments. Factors such as physical strength, fatigue, temperature, slopes, and load affect the efficiency and safety of the tasks. Long-term health risks, primarily musculoskeletal disorders, are common in these occupations. Mitigating risks associated with these types of physically intensive jobs includes understanding the NIOSH Lifting Equation and calculating the overall load dimensions.

8.2 KEY TOPICS

- Heavy work
- Manual Material Handling and Lifting
- Assessment of energy expenditures: Direct and indirect measures
- Evaluation of physical efforts in Manual Material Handling
- NIOSH Lifting Guideline
- Fatigue

8.3 INTRODUCTION AND BACKGROUND

In occupational ergonomics, the interplay between sleep, fatigue, and worker productivity is a critical factor in designing effective work systems. George Orwell's account of labor conditions in *Down and Out in Paris and London* (Orwell, 1961) offers an early glimpse into the extreme physical demands placed on workers, who were subjected to grueling 18-hour workdays with little recovery time. Although set in the early 1900s, Orwell's depiction resonates with modern occupational challenges, particularly the understanding that poor work-rest cycles can degrade both worker health and performance. Current ergonomic principles emphasize the role of circadian rhythms, sleep quality, and work scheduling in promoting worker well-being and reducing injury risks (Deloitte Insights, 2024).

DOI: 10.1201/9780429342806-8

8.3.1 The Circadian Rhythm and Workplace Performance

Human circadian rhythms, which regulate sleep–wake cycles based on exposure to natural light, are crucial in maintaining optimal cognitive and physical function. Disruptions to these rhythms—such as those experienced by night-shift workers—can lead to significant occupational health risks. For instance, the Exxon Valdez oil spill in 1989 is often cited as an example of how fatigue and sleep deprivation can have catastrophic consequences in high-risk industries. The crew's irregular work hours, compounded by exposure to unusual daylight patterns in Alaska, likely impaired their judgment and response times, contributing to the disaster (Sleep Foundation, 2024).

In occupational ergonomics, disruptions to natural sleep cycles are particularly concerning for shift workers, who are often required to sleep during the day and remain active at night. Research from recent years highlights that exposure to blue light can help mitigate some of these effects by promoting alertness during night shifts (SleepScore, 2024). Ergonomists now recommend incorporating exposure to artificial blue light during night shifts and scheduling flexible work hours that allow employees to adjust their sleep schedules more effectively. This flexibility can significantly reduce fatigue and improve worker performance and safety (Deloitte Insights, 2024).

8.3.2 Fatigue and Workplace Hazards

The link between fatigue and workplace safety is well-documented, especially in physically demanding industries such as construction and manufacturing. Fatigue from inadequate or irregular sleep exacerbates the risk of musculoskeletal disorders, increases error rates, and diminishes reaction times, all of which contribute to workplace accidents. Ergonomic interventions that address both physical and cognitive fatigue are essential for mitigating these risks (*Nature*, 2023).

A key tool in managing physical strain is the NIOSH Lifting Equation, a widely used ergonomic method for determining safe lifting practices. This equation considers factors such as object weight, hand positions, lifting frequency, and task duration to minimize the risk of musculoskeletal injuries. The efficacy of such ergonomic tools, however, is reduced if workers are chronically fatigued, which underscores the need for comprehensive approaches that prioritize sufficient rest and recovery (Sleep Doctor, 2021).

8.3.3 Ergonomic Strategies to Improve Sleep and Reduce Fatigue

Promoting a workplace culture that prioritizes rest can lead to substantial improvements in both employee health and productivity. From an ergonomic perspective, this can be achieved through several interventions. One effective strategy is the implementation of flexible work schedules, which allow employees to start and end their workdays at times that align with their natural sleep cycles. This flexibility helps minimize sleep deprivation and fatigue-related accidents. Another approach involves designing work environments that encourage natural light exposure, as daylight plays a crucial role in regulating circadian rhythms (SleepScore, 2024).

Moreover, some companies have adopted practices like the provision of nap rooms or scheduled rest breaks during long shifts. This allows workers to recuperate during work hours to help offset the negative effects of prolonged wakefulness (Deloitte Insights, 2024). These strategies are supported by recent studies that show sufficient sleep not only improves physical health but also enhances cognitive function, reducing the likelihood of errors and accidents on the job (*Nature*, 2023).

Addressing the ergonomic impact of sleep and fatigue in the workplace is critical for creating safer and more productive environments. Modern ergonomic interventions—ranging from flexible scheduling to environmental modifications that promote natural light—are designed to align work demands with the body's natural rhythms. By incorporating these strategies, organizations can reduce the risk of fatigue-related accidents and promote the overall well-being of their workforce. The focus on sleep as an ergonomic factor is not just a matter of improving individual health, but a vital component in enhancing organizational efficiency and safety (Sleep Foundation, 2024; SleepScore, 2024).

8.4 HEAVY WORK

Despite the mechanization, automation, and shift toward more knowledge-based work in many industries, certain sectors continue to rely on substantial manual labor. These industries, such as firefighting, construction, agriculture, carpentry, and professional sports, demand significant muscular exertion and impose strain on various physiological systems. Heavy labor, typically characterized by high-intensity physical activities and sustained muscular effort, poses notable ergonomic challenges that must be addressed to protect workers' health and improve task efficiency.

8.4.1 Physiological Stress in Physically Demanding Occupations

When performing intensive tasks, multiple body systems work in concert to meet the demands of physical exertion. Understanding the interplay between these systems and the potential strain they endure is crucial in designing ergonomic interventions.

1. **Muscular System:** Workers in physically demanding occupations are often required to exert substantial muscular force, maintain postural balance, and engage in repetitive motions. Prolonged or excessive use of large muscle groups can lead to fatigue, overexertion, and an increased risk of musculoskeletal injuries, such as strains, sprains, and chronic conditions like tendinitis or low-back pain. Tasks that require frequent lifting, pushing, or pulling can impose significant biomechanical loads. These risks make ergonomically designed lifting techniques and equipment essential for injury prevention (NIOSH, 2021).
2. **Cardiovascular System:** The cardiovascular system plays a critical role in physically demanding tasks by delivering oxygenated blood to active

muscles. During periods of intense physical activity, the heart rate and blood pressure increase to meet the heightened metabolic demands. Sustained exertion without adequate rest periods can lead to overexertion, dehydration, and cardiovascular strain. Proper task design, including adequate breaks and hydration, is essential to mitigating these risks (NIOSH, 2021).

3. **Respiratory System:** The respiratory system must support the exchange of gases—delivering oxygen to the muscles and removing carbon dioxide and other metabolic waste products. During physically intensive tasks, the demand for oxygen increases, resulting in elevated respiratory rates. Prolonged exertion, particularly in environments with poor air quality, can hinder gas exchange efficiency and further strain the body. Incorporating ventilation and air-quality control systems is a critical ergonomic intervention (OSHA, 2022).
4. **Nervous System:** The nervous system is responsible for controlling muscle contractions and coordinating complex motor tasks, requiring rapid communication between the brain, spinal cord, and peripheral nerves. Physically demanding work often involves precise motor control, rapid response times, and sustained concentration. The repetitive nature of many tasks can lead to nerve compression syndromes, such as carpal tunnel syndrome or cubital tunnel syndrome, particularly in occupations involving repetitive hand or arm movements (NIOSH, 2021). Ergonomically designed tools, workstations, and job rotation schedules can help mitigate these risks.

8.4.2 Ergonomic Interventions to Mitigate Risks

To enhance safety and efficiency in occupations that demand heavy labor, ergonomic interventions are essential. These aim to reduce the physical burden on workers and minimize the risk of injury while maintaining productivity. Key ergonomic strategies include:

1. **Task Design and Automation:** Wherever possible, the mechanization of strenuous tasks should be prioritized. For example, the use of powered exoskeletons in construction and industrial settings can alleviate the load on workers' muscles, allowing them to perform physically demanding tasks while experiencing less strain (Santos et al., 2018). Similarly, lifting devices, cranes, and powered tools can reduce the need for manual exertion, enabling workers to engage in tasks with a reduced injury risk.
2. **Workplace Ergonomics and Tool Design:** Designing workspaces and tools to fit the user's physical capabilities is essential in reducing ergonomic stress. Ergonomically designed tools with anti-vibration grips, adjustable handles, and reduced weight can significantly decrease the strain placed on the muscular and nervous systems during manual tasks. Workstation designs that promote neutral postures—particularly in carpentry, firefighting, and construction—reduce the risk of long-term musculoskeletal disorders.

3. **Training and Education:** Proper training on ergonomically sound work techniques is critical in physically demanding fields. Workers should be educated on safe lifting techniques, body mechanics, and the importance of rest breaks. Additionally, job rotation programs can help distribute the physical load across different muscle groups and reduce the risk of repetitive strain injuries (NIOSH, 2021).
4. **Pacing and Recovery:** Pacing work tasks to allow for sufficient recovery between bouts of exertion is critical for both the cardiovascular and muscular systems. Incorporating rest breaks and alternating between physically demanding and lighter tasks ensures that workers can maintain productivity without experiencing excessive fatigue or injury.

8.4.3 Energy Consumption during Heavy Work

In recent years, the introduction of automation has substantially reduced the physical demands associated with myriad industrial tasks. Despite these advancements, several sectors continue to require significant manual labor, often involving strenuous physical exertion. Heavy work—defined by activities that demand high levels of muscular activity and lead to significant energy expenditure—places considerable stress on the cardiovascular, respiratory, and metabolic systems. The physiological strain associated with heavy work can be quantified by measuring both energy consumption, typically in kilojoules (kJ), and cardiac effort, often evaluated via heart rate. These measurements are critical in ergonomics for understanding how to design safer and more sustainable work environments.

8.4.4 Dynamic vs. Static Work: Energy Demands and Muscle Strain

Heavy physical work can be classified into two broad categories: dynamic work and static work.

- **Dynamic work** involves continuous movement and is primarily assessed by the energy demands it places on the body. This type of work requires the body to maintain a steady state of oxygen consumption, muscle activity, and cardiovascular output. For example, tasks such as lifting, pushing, or pulling objects require high metabolic rates, typically resulting in energy expenditures of up to 40 kJ per minute and heart rates around 140 beats per minute for well-trained individuals. During dynamic work, the body's ability to distribute oxygen to the working muscles determines the sustainability of the task (NIOSH, 2021).
- **Static work**, in contrast, involves maintaining a posture or holding a muscle contraction for extended periods without significant movement. This type of effort requires less energy expenditure but can still lead to rapid fatigue due to restricted blood flow to the contracted muscles, resulting in an insufficient oxygen supply. The build-up of metabolic byproducts, like lactic acid, can further accelerate fatigue even if the task does not demand high overall energy consumption (OSHA, 2022).

The intensity and duration of dynamic and static tasks are crucial factors in determining their overall physiological strain and the required recovery time between efforts.

8.4.5 Metabolic and Cardiovascular Demands of Heavy Work

Energy expenditure and cardiac effort are two key physiological metrics that define the severity of a given task. During heavy physical work, well-trained individuals can maintain metabolic and cardiovascular stability at energy expenditures of up to 40 kJ per minute and heart rates around 140 beats per minute. However, as the intensity of the work increases to very heavy levels, physiological challenges become more significant: energy expenditures can exceed 50 kJ per minute, and heart rates may reach 160 beats per minute or higher, pushing the body into an oxygen deficit and resulting in rapid fatigue (Santos et al., 2018).

At these elevated energy consumption levels, the body accumulates lactic acid due to the increased reliance on anaerobic metabolic pathways. As lactic acid builds up in the muscles, it impairs their ability to continue functioning efficiently, leading to fatigue and muscle soreness. The resulting oxygen deficit necessitates frequent rest periods to allow the body to recover and replenish its oxygen stores. Sustained effort at these levels throughout an entire work shift is unsustainable without carefully structured rest intervals (NIOSH, 2021).

8.4.6 Physiological Adjustments During Heavy Physical Work

During heavy physical labor, the body undergoes several physiological adjustments to meet the increased metabolic demands of the task. These adaptations are designed to enhance oxygen delivery to the muscles and facilitate energy production. Key physiological changes include:

1. **Deeper and More Rapid Breathing:** To meet the increased oxygen demand and remove excess carbon dioxide, the respiratory rate and tidal volume increase. This enhanced ventilation improves gas exchange, ensuring a steady supply of oxygen to working muscles (NIOSH, 2021).
2. **Increased Heart Rate and Cardiac Output:** As the heart rate rises, so does cardiac output—the volume of blood pumped by the heart per minute. This increase is crucial for delivering oxygenated blood to active muscles while removing metabolic waste products such as carbon dioxide and lactic acid (OSHA, 2022).
3. **Vasomotor Adaptations:** To optimize blood flow, the body undergoes vasomotor changes, where blood vessels in active muscles and the heart dilate to increase the supply of oxygen and nutrients. Concurrently, vessels in less active organs constrict, redirecting blood flow to where it is most needed (Santos et al., 2018).
4. **Rise in Blood Pressure:** To maintain an adequate pressure gradient for blood flow, blood pressure increases during heavy exertion. This rise in

pressure is necessary to push blood through the dilated vessels in active muscles and ensure that oxygen delivery meets the increased demand (NIOSH, 2021).

5. **Increased Supply of Sugars:** In response to the heightened energy needs of working muscles, the liver releases glucose and glycogen into the bloodstream. These sugars serve as critical energy sources for aerobic and anaerobic metabolism, enabling muscles to continue contracting efficiently (Santos et al., 2018).
6. **Rise in Body Temperature and Metabolism:** As muscles generate more energy, metabolism accelerates, leading to an increase in body temperature. The conversion of chemical energy (in the form of glucose and glycogen) into mechanical energy produces heat, which must be dissipated through sweating and increased blood flow to the skin to prevent overheating (OSHA, 2022).

8.4.7 Ergonomic Considerations for Heavy Work

Given the substantial physiological strain associated with heavy physical work, ergonomic interventions are critical in ensuring that tasks are designed to reduce risk and improve worker safety and productivity. Key ergonomic considerations are discussed in the following sections.

8.4.8 Energy Efficiency of Heavy Work

Although the global shift toward mechanization has caused the need for heavy labor in industrialized nations to decline, physically demanding work remains prevalent in many sectors, particularly in developing countries (Salvendy, 2014a). This makes the implementation of highly efficient ergonomic practices in these regions essential to protecting worker health and optimizing performance. The inherent inefficiency of human physical effort, with only about 30% of exerted energy going toward useful physical work and the remaining 70% dissipated as heat, underscores the need for ergonomically designed tasks that maximize energy utilization and reduce the physical toll on workers (Dul & Weerdmeester, 2014).

8.4.9 The Physiology of Work Efficiency

Maximizing work efficiency in physically demanding tasks is contingent upon converting as much mechanical energy as possible into productive work, while minimizing the energy lost to static efforts (such as holding or supporting loads). Static work, such as maintaining a bent-back posture during tasks, is particularly inefficient, as it increases the strain on muscles without generating movement or contributing directly to task completion (McCauley & Bartram, 2016). Dynamic work, where mechanical energy is actively converted into movement or exertion, is more efficient and less taxing on the body, reducing the risk of musculoskeletal disorders (MSDs) (Westgaard, 2014).

8.4.10 Ergonomic Task Design for Heavy Labor

Designing tasks for heavy labor must integrate strategies that minimize physical strain while promoting efficiency. Effective task design in ergonomics considers a combination of psychological, physical, and environmental factors, all of which play significant roles in determining worker health and productivity.

8.4.11 Psychological Factors

Psychological demands can significantly influence a worker's performance and well-being. Key psychological factors include:

- **Task Difficulty:** Complex or overly difficult tasks can increase mental strain, leading to decreased efficiency (Karwowski, 2015).
- **Psychological Stress and Time Pressure:** Workers experiencing high stress or working under strict time constraints are more prone to errors and fatigue (Salvendy, 2014a).
- **Cognitive Load:** Decision-making tasks and perceptual work can divert mental resources, decreasing the worker's ability to focus on physical exertion and increasing the risk of mistakes (McCauley & Bartram, 2016).

8.4.12 Physical Factors

Physical demands are central to the ergonomics of heavy labor in the occupational environment. The primary physical considerations when evaluating heavy work include:

- **Task Duration:** Prolonged physical tasks without adequate breaks can lead to muscle fatigue and an increased injury risk (Westgaard, 2014).
- **Work Posture:** Poor posture, particularly tasks involving bending motions with the back or awkward positions, can exacerbate musculoskeletal strain and decrease efficiency (Karwowski, 2015).
- **Physical Effort (Force and Torque):** Excessive force requirements in manual tasks, such as lifting and carrying, increase the risk of acute injuries and long-term musculoskeletal disorders (Dul & Weerdmeester, 2014).
- **Manual Material Handling:** The lifting, carrying, and pushing of materials must be optimized through the use of assistive devices and proper body mechanics to minimize injury risk and maximize efficiency (McCauley & Bartram, 2016).

8.4.13 Environmental Factors

Environmental conditions also contribute significantly to worker safety and performance. The environmental factors to be considered in the occupational environment include the following:

- **Heat, Cold, and Humidity:** Extreme temperatures and humidity can increase fatigue, dehydration, and heat stress, reducing worker endurance and productivity (Westgaard, 2014).
- **Exposure to Chemicals and Noise:** Prolonged exposure to harmful chemicals or excessive noise levels can lead to chronic health problems, including respiratory issues and hearing loss, compounding the physical demands of work (Salvendy, 2014a).

8.4.14 Metabolic Demands and Classification of Physical Work

The metabolic demands of industrial tasks, which range from light to extremely heavy, serve as a basis for classifying their physical intensity. Understanding these classifications is crucial for designing tasks that align with worker capabilities. Recent guidelines in ergonomics research have updated these classifications, providing detailed examples of the physical demands across various industries (Dul & Weerdmeester, 2014).

8.4.15 Effects of Heavy Work and Heat

Extreme heat significantly impacts heavy work, and heart rate is a valuable predictor of this impact. The heart's primary functions during heavy work in extremely hot conditions include:

- Transporting energy to the muscles
- Moving heat energy from the body's interior to the skin for cooling
- Supporting increased respiratory needs

Industries where workers frequently face these conditions include construction, agriculture, and maritime sectors. The Occupational Safety and Health Administration (OSHA) provides guidelines and resources to help analyze heat load and mitigate the risk of overheating in occupational environments.

8.4.16 Ground Surface, Slopes, and Staircases

The type of ground surface on which workers perform tasks has a significant impact on their energy consumption and overall work efficiency. Traversing smooth, solid ground surfaces requires considerably less energy than uneven or obstructed terrain such as dirt roads, light or heavy brush, hard-packed snow, or swampy bogs (Dul & Weerdmeester, 2014). Challenging surfaces increase the physical demands on the body, particularly the lower extremities, resulting in greater metabolic costs and fatigue.

Utilizing equipment such as wheelbarrows or carts for transporting loads over such terrain can provide notable ergonomic advantages. Research has shown that pushing a cart is approximately 15% more efficient than pulling it, as it allows for better utilization of body mechanics with reduced strain on the back and shoulders (Salvendy,

TABLE 8.1
Metabolic Demands of Industrial Tasks*

Light	Moderate	Heavy	Very Heavy	Extremely Heavy
70–140W (60–120 kcal/hr)	>140–280W (>120–240 kcal/hr)	>280–350W (>240–300 kcal/hr)	>350–420W (>300–360 kcal/hr)	>420W (>360 kcal/hr)
Small-parts assembly	Industrial sewing	Making cement	Shoveling (>7 kg)	Stoking a furnace
Typing	Bench work	Industrial cleaning	Ditch digging	Ladder or stair climbing
Keypunching	Filing	Joining floorboards	Hewing and loading coal	Coal car unloading
Inspecting	Machine tending	Plastering	Handling moderately heavy (>7 kg) cases to and from a pallet	Lifting 20 kg cases ten times per minute
Operating a milling machine	Small-size packing	Power truck operation	Tree planting	
Drafting	Operating a lathe	Handling light cases to and from a pallet	Handling heavy (>11 kg) units frequently (>4 per minute)	
Armature winding	Operating medium size presses	Road paving		
Hand typesetting	Machining	Painting		
Operating a drill press	Small-sized sheet metal work	Handling operations		
Desk work	Electronics testing	Metal casting		
Small-parts finishing	Plastic molding	Cutting or stacking lumber		
	Operating a punch press	Large-size packing		
	Operating a crane laying stones and bricks			
	Sorting scrap			

* The values given include basal metabolism.
Source: Adapted from Dul & Weerdmeester, 2008.

2014a). This efficiency is enhanced by ergonomic design principles like optimal handle height and thickness; for maximum comfort and mechanical advantage, cart handles should be positioned around 1 meter above the ground, with a thickness of approximately 40 mm to facilitate a firm and comfortable grip (Karwowski, 2015).

In terms of cart stability, particularly for two-wheeled carts, the center of mass should be positioned as low and as close to the axle as possible. This reduces the effort required to support and maneuver the load, enhancing both safety and efficiency (McCauley & Bartram, 2016). These factors contribute to minimizing the physical burden on workers and help prevent overexertion injuries, which are common in occupations requiring manual material handling.

8.4.17 Climbing and Ladder Use in Ergonomic Design

When the nature of a job involves climbing, it is critical to design these tasks with energy efficiency and safety in mind. For example, a slope of approximately 10 degrees has been identified as the most efficient incline for walking, as it minimizes energy expenditure while maintaining a stable posture (Salvendy, 2014b). Conversely, steeper inclines or less stable surfaces significantly increase the workload on the muscles and joints, particularly in the legs and lower back, which can lead to rapid fatigue and potential musculoskeletal injuries.

The ergonomic design of ladders is crucial for reducing strain and preventing accidents. The optimal ladder angle is around 70 degrees, which provides a balance between stability and ease of climbing (Westgaard, 2014). Rungs should be spaced 260 mm apart for general use to promote efficient climbing with minimal energy expenditure. When workers are required to carry or manipulate heavier loads, closer-spaced rungs (approximately 170 mm apart) are ideal to reduce the distance each foot must travel for additional stability (Karwowski, 2015).

8.4.18 Staircase Design for Energy Efficiency

Research by Lehmann (2014) highlights that staircases with slopes between 25 and 30 degrees necessitate the least energy consumption. This information is particularly useful for workplaces or environments where stairs are frequently used, as well as for older populations who are more susceptible to fatigue and injury during stair climbing.

Lehmann's (2014) research further provides two empirical guidelines for designing energy-efficient and safe staircases:

- The riser, or tread height, should be around 170 mm.
- The tread depth should measure approximately 290 mm.

These dimensions balance the need for ergonomic efficiency with safety, reducing the energy required to ascend and descend while maintaining adequate footing. The Occupational Safety and Health Administrationalso provides comprehensive guidelines on ladder and staircase design, emphasizing the importance of ergonomics in reducing workplace injuries related to falls or overexertion (OSHA, 2015).

The design of workplace ground surfaces, carts, ladders, and staircases plays a critical role in occupational ergonomics, especially in physically demanding jobs. By understanding and applying ergonomic principles, such as minimizing energy expenditure through efficient designs and promoting safe working postures, industries can reduce worker fatigue, prevent injuries, and increase overall productivity. The ongoing development of these ergonomic standards continues to be informed by both empirical research and industry guidelines, such as those provided by OSHA, ensuring that tasks are not only efficient but also safe for workers.

8.4.19 Evaluation of Physical Work Loads

Assessing physical workload or manual exertion in ergonomically intensive tasks is essential for optimizing worker performance and health. Several methods are available to evaluate workload, including mathematical models, databases, analysis or prediction software, and physiological evaluation tools. The selection of the appropriate assessment technique for a specific task requires consultation with updated ergonomic guidelines and resources (Dul & Weerdmeester, 2014).

The ergonomic evaluation of physical loads is broadly categorized into two main approaches: physiological and psychophysical. Both approaches provide distinct methodologies for understanding and assessing physical strain and work capacity.

8.4.20 Physiological Approach

The physiological approach focuses on the limitations of central body functions, such as the pulmonary, circulatory, and metabolic systems. This method is particularly useful in evaluating jobs with high manual material handling (MMH) demands as it presents criteria to assess strain on the body's core functions. The physiological approach is rooted in objective, measurable parameters, offering a quantitative analysis of physical strain.

8.4.20.1 Key Physiological Metrics

The following physiological metrics are used in the assessment of workload:

- **Heart Rate (HR):** The heart rate is a fundamental indicator of cardiovascular function. Heart rate monitoring is widely used in ergonomic assessments due to its simplicity and effectiveness in gauging cardiovascular strain during physical tasks.
- **Energy Expenditure:** Energy expenditure is a crucial parameter for assessing physical workload. It can be measured directly through calorimetric chambers or indirectly through the assessment of oxygen consumption during exertion. The maximum rate of oxygen usage (VO_2 Max) is the key measurement, expressed in milliliters of oxygen per kilogram of body weight per minute. This metric is critical in understanding an individual's aerobic capacity and the energy required for physical tasks. Energy expenditure is calculated based on oxygen consumption and the respiratory

quotient (RQ), which measures the balance between carbon dioxide production and oxygen uptake (Salvendy, 2014b).

8.4.20.2 Energy Expenditure Calculation

Energy Expenditure is calculated using Oxygen Consumption and the Oxygen Energy Equivalent value.

Each term in this equation has a specific meaning:

Energy Expenditure = Oxygen Consumption × Oxygen Energy Equivalent

Or, using smbols:

$$EE = \dot{V}O_2 \times Eo_2$$

EE = Energy Expenditure
VO_2 = Oxygen Consumption (L/min)
Eo_2 = Oxygen Consumption (kcal/O_2)

1. **Energy Expenditure (EE):**
   - This represents the amount of energy the body uses, typically measured in kilocalories per minute (kcal/min) or joules per minute (J/min).
   - It is a measure of how much energy the body is burning due to metabolism or physical activity.
2. **Oxygen Consumption (VO_2):**
   - This refers to the volume of oxygen used by the body per minute.
   - It is measured in **liters per minute (L/min).**
   - More oxygen consumption generally means more energy is being used because additional oxygen is required for increased aerobic metabolism.
3. **Oxygen Energy Equivalent (Eo_2):**
   - This is a conversion factor that tells us how much energy (kcal) is produced per liter of oxygen consumed.
   - The typical value is around **5 kcal/L O_2** (varies depending on the fuel source: carbohydrates, fats, or proteins).

8.4.20.3 Breaking It Down Numerically

If a person consumes 2.0 L of oxygen per minute, and we assume the energy equivalent of oxygen is 5 kcal per liter,

$$EE = (2.0\ \text{L/min}) \times (5.0\ \text{kcal/L})$$
$$EE = 10.0\ \text{kcal/min}$$

This means the person's body is using 10 kilocalories per minute.

8.4.20.4 Key Points

- The equation shows that energy expenditure is directly proportional to oxygen consumption.
- The more oxygen the body uses, the more energy it burns.
- Different activities and exertion levels influence oxygen consumption, which in turn affects total energy expenditure.

8.4.20.5 Psychophysical Approach

The psychophysical approach combines both local and central body functions while incorporating the subjective experience of the worker. It evaluates physical strain through perceived exertion, enabling individuals to rate their strain and establish acceptable workload levels under specified conditions (Karwowski, 2015). This approach is valuable as it accounts for individual variability in tolerance to physical demands, thereby offering a more comprehensive evaluation of manual work capacity.

8.4.21 Physiological Instrumentation for Workload Evaluation

8.4.21.1 Heart Rate Monitoring

Heart rate monitoring is essential for tracking cardiovascular function during physical activity. By measuring heart rate responses to physical tasks, ergonomists can gauge the body's physiological response to exertion and ensure that the workload remains within safe limits.

8.4.21.2 Energy Expenditure Measurement

Energy expenditure, one of the primary indicators of physical workload, can be measured directly or indirectly. Direct measurements are typically conducted using calorimetric chambers, while indirect methods involve assessing oxygen consumption through electronic oximeters. Both methods provide critical insights into the metabolic demands of physical work.

8.4.22 Workload and Health Implications

Physiological load and work tolerance are strongly linked to an individual's working capacity. According to ergonomic guidelines, the recommended physical workload should not exceed 30% of VO_2 Max during an average work shift; moreover, exceeding 50% of VO_2 Max can have long-term adverse health effects (Westgaard, 2014). Chronic overexertion can lead to musculoskeletal disorders, cardiovascular strain, and other health complications, emphasizing the need for ergonomic task design.

8.4.23 Manual Hand Tools and Ergonomic Considerations

8.4.23.1 Shoveling

Shoveling is a common manual task that requires careful ergonomic design to minimize physical strain and maximize efficiency. Key ergonomic considerations for shoveling include:

- **Shovel Weight:** The weight of the shovel should be appropriate for the material being handled.
- **Shovel Size:** Large shovels are preferred for light materials, while smaller shovels are more effective for heavy materials.

- **Shovel Shape:** For fine-grained materials, a spoon-shaped, pointed-tip shovel is ideal, whereas coarse materials are better managed with a straight, cutting-edge, flat-blade shovel.
- **Handle Length:** The optimal handle length for shovels ranges from 600 to 650 mm, promoting a comfortable working posture and reducing lower back strain (Dul & Weerdmeester, 2014).

8.4.23.2 Saws

Ripsaws with broad blades and multiple cutting teeth offer greater efficiency for manual sawing tasks. Studies suggest that an optimal sawing rate of 42 double strokes per minute with a vertical force of approximately 100 N yields the best results with minimal physical strain (Salvendy, 2014b).

8.4.23.3 Hoes

The selection of hoes should consider the soil type where the tool will be utilized. Swivel hoes are recommended for soft soil, while both swivel and chopping hoes are effective for harder, dry soils. Proper tool selection is essential to reduce energy expenditure and increase task efficiency.

8.4.24 Efficient Movement and Load Carrying

8.4.24.1 Walking Speed

Optimal walking speeds for manual labor tasks range between 4 to 5 km per hour on smooth surfaces, with a reduction to 3 to 4 km per hour when wearing heavy shoes or walking on uneven terrain. Efficient walking speeds help conserve energy and reduce worker fatigue (McCauley & Bartram, 2016).

8.4.24.2 Carrying Loads

The most efficient load to carry ranges between 50 and 60 kg, but smaller loads are typically recommended for safety reasons. Distributing the load evenly across the back, shoulders, chest, or head can reduce energy consumption. The ideal distributed load is approximately 30 kg, allowing workers to perform tasks with reduced physical strain.

8.4.25 VO_2 Max

VO_2 Max (maximal oxygen uptake) is a critical indicator of the body's capacity for aerobic work, serving as an important predictor of performance in endurance sports and physically demanding professions such as firefighting, military operations, and emergency medical services (Bassett & Howley, 2014). VO_2 Max refers to the maximum amount of oxygen that an individual can utilize in one minute, relative to their body weight, and is expressed in milliliters of oxygen per kilogram of body weight per minute (ml O_2/kg/min). This metric is essential for understanding both an individual's cardiovascular efficiency and their ability to sustain prolonged physical activity.

TABLE 8.2
Oxygen Uptake and Physical Activity

Rest	Basal metabolism require approximately 0.25 L of oxygen per minute
Sedentary work	Office work, such as oxygen uptake of only a little over resting levels (0.3–0.4 L/min)
Housework	Housework includes several moderate to heavy tasks (requiring about 1 L/min of oxygen). It is unusual for high work rates to be maintained for any length of time
Light industry	Oxygen uptakes from 0. 4 to 1 L/min are required
Manual labor	Oxygen uptake may vary from 1 to 4 L/min. The workload can depend greatly on the tools and methods (heavy MMH)
Sports	Endurance sports or very occupational tasks that take over 5 L/min

Source: Bridger, 1995.

8.4.25.1 Factors Influencing VO_2 Max

Several physiological and environmental factors play a role in determining an individual's VO_2 Max:

- **Training:** Regular aerobic exercise, such as running, swimming, or cycling, can significantly improve VO_2 Max by enhancing the efficiency of the cardiovascular and respiratory systems. Studies have shown that endurance training can increase VO_2 Max by 10–20% over time (Jones & Carter, 2014).
- **Genetics:** Genetic predispositions contribute substantially to an individual's baseline VO_2 Max. Research suggests that genetics account for approximately 40–50% of the variability in VO_2 Max values among individuals (Bouchard et al., 2015).
- **Body Weight and Muscle Volume:** Optimal body composition, particularly a higher muscle volume relative to body fat, positively influences VO_2 Max. Muscles require a greater supply of oxygen during activity, and individuals with higher muscle mass typically exhibit higher VO_2 Max values (Saltin & Calbet, 2014).
- **Age:** VO_2 Max tends to peak in early adulthood and declines with age. After the age of 50, VO_2 Max typically decreases by approximately 1% per year, primarily due to reductions in muscle mass and cardiovascular efficiency (Fleg & Lakatta, 2014). However, regular physical activity can mitigate this decline.

8.4.25.2 Measuring VO_2 Max

Accurate measurement of VO_2 Max involves capturing and analyzing inhaled and exhaled gases during maximal exercise, such as running on a treadmill or cycling on an ergometer. This process is typically performed in a controlled laboratory environment under professional supervision. Direct measurement of VO_2 Max requires sophisticated equipment and precise data collection to ensure accuracy.

TABLE 8.3
Cardiovascular Fitness Calculations and Fitness Level

Fitness Level	VO_2 max (ml O2/kg/min)
Excellent	>60
Good	50–60
Average	40–50
Below Average	30–40
Poorv	<30

Source: Bassett & Howley, 2014.

In occupational settings where direct measurement is impractical, VO_2 Max is often estimated based on factors such as age, physical fitness levels, and heart rate responses to submaximal exercise. These estimates, while less precise than laboratory measurements, still provide valuable insights into an individual's aerobic capacity and physical fitness (Plowman & Smith, 2014).

8.4.25.3 VO_2 Max as a Fitness Benchmark

VO_2 Max values are widely used to classify aerobic fitness levels, offering a benchmark for evaluating cardiovascular health and physical endurance. These classifications help guide training programs and ensure that individuals in physically demanding professions maintain the necessary aerobic fitness to perform their duties safely and efficiently.

Table 8.3 outlines the VO_2 Max ranges for different fitness levels, based on updated guidelines (Bassett & Howley, 2014).

8.5 MANUAL MATERIAL HANDLING AND LIFTING

Manual Material Handling (MMH), which includes tasks such as lifting, carrying, and moving materials, is a leading cause of workplace-related injuries, particularly musculoskeletal disorders. Each year, over half a million cases of MSDs are reported in the United States, with a significant portion stemming from MMH-related tasks (OSHA, 2016). These injuries can lead to long-term disability, increased absenteeism, and a substantial financial burden on both employees and employers. Understanding the risk factors associated with MMH and implementing ergonomic solutions is crucial for reducing the incidence of work-related musculoskeletal disorders (WMSDs) and improving worker safety.

8.5.1 Risk Factors in Manual Material Handling

Several key risk factors are associated with MMH, which increase the likelihood of injury:

- **Work-Related Musculoskeletal Disorders:** WMSDs are injuries that affect the musculoskeletal system, including muscles, tendons, ligaments, nerves, and other soft tissues. Repetitive lifting, awkward postures, and excessive force during manual handling are the primary contributors to WMSDs (NIOSH, 2015).
- **Visual Occlusion:** When workers handle large loads that obscure their line of sight, they are at an increased risk of accidents. Visual occlusion makes it difficult to navigate workspaces safely, raising the likelihood of collisions, trips, and falls (OSHA, 2016).
- **Trips, Slips, and Falls:** MMH tasks can lead to uneven weight distribution and awkward postures, such as twisting or bending, which compromise balance and stability. These postures increase the risk of trips, slips, and falls, which are leading causes of workplace injuries (Salvendy, 2014b).

8.5.2 Industries Affected by MMH

Manual material-handling tasks are prevalent in a wide range of industries, where the risk of injury is exacerbated by the physical demands of the job. Common industries affected by MMH include the following:

- **Manufacturing:** Workers in manufacturing frequently lift, move, and carry materials or components, often in environments with heavy equipment and machinery.
- **Construction:** Construction workers are exposed to repetitive lifting, carrying, and awkward postures, which are common risk factors for musculoskeletal injuries.
- **Healthcare:** Healthcare workers engage in MMH tasks such as moving patients or heavy medical equipment, making them highly susceptible to WMSDs.
- **Warehousing and Logistics:** Warehousing workers handle large quantities of goods and often perform tasks that require awkward postures, heavy lifting, and repetitive motions.

Given the diversity of these occupational environments, it is important to have flexible guidelines that can be tailored to the specific needs of each industry.

8.5.3 Material Handling

Material handling encompasses a broad range of tasks involving the manipulation of objects or loads, which can vary significantly in size, shape, weight, and texture. The physical actions required in material handling demand the exertion of energy or force, manifesting as pushing, pulling, lifting, lowering, dragging, carrying, or holding objects (Dul & Weerdmeester, 2014).

8.5.4 Types of Material Handling

The methods of material handling can be categorized into different types based on the required action:

- **Lifting:** Moving an object from a lower position to a higher one, typically involving the use of the legs, arms, and back.
- **Lowering:** Involves moving an object from a higher position to a lower one, often requiring control to avoid injury or damage.
- **Pushing, Pulling, Carrying, and Holding:** These tasks involve the manipulation of objects in various directions, requiring force generation from the muscles and joints. Each of these actions can place stress on different parts of the body, particularly the back, shoulders, and arms.

8.5.5 Energy and Force in Material Handling

The energy required to handle loads is generated within the body and translated into force or torque, enabling the manipulation of objects. However, material-handling tasks are a leading cause of back-related injuries in occupational settings where overexertion can result in serious musculoskeletal problems (McCauley & Bartram, 2016). Low-back pain is the most common form of morbidity associated with material handling and has been identified as the leading cause of lost productivity in physically demanding jobs. Studies estimate that 60–70% of individuals engaged in physically intensive tasks will experience back problems at some point in their careers (NIOSH, 2015). Such injuries are prevalent not only in occupational environments but also in activities like sports (e.g., weightlifting) and leisure activities involving heavy lifting.

8.5.6 Classification of Manual Material-Handling Characteristics

Understanding the factors associated with manual material-handling tasks is crucial for minimizing the risks involved. A systems approach should be applied to the design of material-handling tasks to holistically understand the task environment. The four categories used to classify MMH task characteristics are:

1. Task Characteristics
2. Material/Container Characteristics
3. Work Practice Characteristics
4. Worker Characteristics

8.5.6.1 Task Characteristics

Evaluating task requirements is essential to thoroughly understand the material-handling expectations and associated risks. Confined or limited workspaces can significantly increase the risk of low-back pain and other musculoskeletal issues. Here are key considerations and updated guidelines to ensure safer and more efficient material-handling practices.

8.5.6.1.1 Workspace and Posture Considerations

- **Avoid Confined Spaces:** Tasks should not be performed in confined spaces that restrict posture, particularly when affecting headroom or horizontal reach. Such constraints can lead to increased strain on the upper body and back.
- **Clear Work Areas:** Ensure that the workspace is unobstructed, allowing ample space for the worker to move freely while picking up, handling, and placing objects. Adequate foot and legroom is necessary to enable proper bending of legs and knees.
- **Minimize Reaching Over Obstacles:** Tasks requiring operators to reach over obstacles or into containers from a distance should be redesigned to reduce stress on the upper extremities and back.
- **Safe Flooring:** Floors should be free of debris or materials that might cause slips, trips, or falls. Flooring should have properties that provide adequate friction between shoes and the floor, especially in jobs involving heavy lifting, pushing, or pulling.

8.5.6.1.2 Standing and Flooring Considerations

- **Limit Standing Time:** Extended periods of standing (four hours or more) have been correlated with low-back pain. Ergonomic interventions should consider periodic breaks and opportunities for sitting.
- **Flooring Material:** Hard flooring, such as concrete, can cause discomfort and fatigue. Using softer flooring materials or anti-fatigue mats can help alleviate this issue.
- **Footwear:** Ensure that shoe sole materials provide sufficient friction to prevent slipping and improve efficiency.

8.5.6.1.3 Environmental and Task Factors

When assessing task characteristics, consider the following factors:

- **Workplace Geometry:** Evaluate the spatial properties of the task during the pickup, handling, and placement of loads.
- **Task Frequency, Duration, and Pace:** Understand how often tasks are performed, how long they last, and the speed at which they are completed.
- **Task Complexity and Precision:** Assess the complexity or precision required in load handling and placement.
- **Environmental Conditions:** Take into account noise, temperature, vibration, and other environmental factors that could affect the worker's comfort and safety.
- **Flooring Type:** Consider the impact of different flooring types on the worker during load-handling tasks.

8.5.6.2 Material/Container Characteristics

The following are useful guidelines and considerations for optimizing manual material handling.

8.5.6.2.1 Key Considerations for Load Handling

- **Size and Weight:** Smaller and lighter loads are easier to handle. Large or awkward loads can obstruct vision or make contact with the legs, increasing the risk of injury.
- **Proximity to Spine:** Loads should be carried close to the spine to reduce compressive forces on the lower back.
- **Load Stability:** Containers should have stable loads to prevent contents from shifting, which can cause sudden changes in the center of gravity and increase the risk of dropping the load or straining the back.
- **Edge Design:** Container edges should be rounded to minimize contact stress between the load and the handler's hand, arm, and body.

8.5.6.2.2 Container Design for Safe Handling

- **Handles and Handholds:** Good coupling is essential for safe handling. Containers should have handles or hand cutouts for a better grip. According to the Revised NIOSH Lifting Equation:
  - **Handles:** Should be cylindrical, smooth, and nonslip, with a diameter of 1.9–3.8 cm and a minimum length of 11.5 cm. They should provide at least 5.0 cm of hand clearance.
  - **Handhold Cutouts:** Optimal dimensions include a height of 3.8 cm or greater, a minimum length of 11.5 cm, and 5.0 cm hand clearance. Cutouts should be semi-oval, smooth, and non-slip with a minimum wall thickness of 0.635 cm.
  - **Positioning:** Handholds near the bottom of the container allow carrying at knuckle height, reducing static muscle loading on the upper extremities.

8.5.6.2.3 Summary of Material-Handling Factors

Assessing the impact of MMH tasks requires the consideration of multiple factors. A comprehensive evaluation of a load should consider the following categories:

- **Load Mass:** Assess the weight or force required for carrying, pushing, or pulling.
- **Container Dimensions:** Ensure appropriate size and shape to avoid obstructing vision or causing contact with the legs.
- **Load Distribution:** Maintain stability to prevent shifting during handling.
- **Coupling Quality:** Design handles or handholds for secure and ergonomic gripping.

- **Load Stability:** Ensure the contents do not shift, which could affect the handler's balance and control.

8.5.6.3 Work Practice Characteristics

When designing work practices for manual material handling, it is essential to apply ergonomic principles across three broad categories: engineering controls, administrative controls, and personal protective equipment (PPE). Addressing each area helps create a safer and more efficient working environment.

8.5.6.3.1 Engineering Controls

Engineering controls are preferred in ergonomic interventions because they address the root causes of hazards, aiming to reduce physical exertion and stamina requirements. By implementing effective work systems—including equipment, tools, furniture, processes, methods, workflow, and environment—workers can perform tasks safely and efficiently. Engineering controls often incorporate assistive technologies or mechanical aids to minimize the risk factors associated with musculoskeletal injuries, particularly low-back pain.

8.5.6.3.2 Key Considerations for Engineering Controls in MMH

- **Workplace Design:** Ensure the workspace is adequately sized to accommodate task performance, equipment, and worker movement without causing strain or awkward postures.
- **Lighting:** Provide appropriate lighting quantity and quality to ensure visibility and reduce eye strain.
- **Optimal Load Placement:** Avoid placing materials on the floor. Instead, position loads at heights that minimize the need for bending and reaching.
- **Use of Momentum and Gravity:** Design workflows that conserve momentum and utilize gravity to reduce the effort required in material-handling tasks.
- **Flooring and Environment:** Consider the type of flooring and floor finishes to enhance safety and reduce the risk of slips, trips, and falls.

8.5.6.3.3 Implementation of Engineering Controls

1. **Ergonomic Equipment:** Use tools and machinery such as conveyors, hoists, and lift tables to reduce manual handling.
2. **Redesign Tasks:** Modify tasks to eliminate unnecessary bending, twisting, reaching, and prolonged static postures.
3. **Mechanical Aids:** Implement mechanical aids like carts and dollies to assist in the movement of heavy loads.
4. **Adjustable Workstations:** Design workstations that can be adjusted to fit the worker's height and reach, promoting neutral postures.

8.5.6.3.4 Administrative Controls

While engineering controls are the preferred method for mitigating ergonomic hazards, administrative controls, personal protective equipment, and personnel selection are alternative strategies. These methods are generally less effective as they do not eliminate the root cause of the hazard. The primary goal of administrative controls is to limit the duration of personal exposure to the risk factors associated with manual material-handlingtasks.

8.5.6.3.5 Key Administrative Controls

1. **Job Rotation:** Rotate workers between physically demanding and less demanding tasks to prevent the overuse of specific muscle groups.
2. **Job Enlargement or Enrichment:** Add variety to tasks by incorporating less taxing activities and distributing the workload among different muscle groups.
3. **Self-Paced Work:** Allow workers to set their own pace to reduce the risk of overexertion and fatigue.
4. **Teamwork:** Increase the number of people performing a task to distribute the physical load and reduce individual exposure duration.
5. **Training:** Provide comprehensive training on safe handling techniques to improve workers' ergonomic practices.
6. **Worker Selection and Placement:** Match workers to tasks that fit their physical capabilities to minimize the risk of injury.

8.5.6.3.6 Personal Protective Equipment

When engineering and administrative controls are insufficient to mitigate risks, personal protective equipment becomes vital for workplace safety. PPE is designed to be worn or attached to the body to enhance safety, ease task performance, reduce exposure, and provide ergonomic benefits.

8.5.6.3.7 Desirable Qualities of PPE

- **Functionality:** PPE must effectively perform its intended safety function.
- **Proper Fit:** It should fit well without causing discomfort or requiring the user to exert extra force.
- **Ergonomic Design:** PPE should not force users into extreme postures or create additional strain.

8.5.6.3.8 Common Types of PPE in MMH

- **Safety Shoes:** Protect feet from heavy objects and provide stability.
- **Gloves:** Protect hands from cuts, abrasions, and exposure to harmful substances.
- **Eye Protection:** Shields eyes from debris, chemicals, and other hazardous materials.
- **Hard Hats:** Protect the head from impact and falling objects.

PPE should be available in a variety of sizes to accommodate the diverse anthropometric measurements of the workforce. Proper training in the use, limitations, care, useful life, and disposal of PPE is crucial. Workers must understand how to use PPE to ensure maximum protection.

8.5.6.3.9 Regulatory Guidance

The Occupational Safety and Health Administration provides comprehensive guidelines on PPE through the Occupational Safety and Health Standards for General Industry, specifically in sections 1910.132 to 1910.137.

8.5.6.4 Worker Characteristics

The primary goal of ergonomics is to adapt tasks to fit the individual, rather than selecting individuals who can perform a job. Therefore, evaluating worker characteristics should not be used to exclude potential workers. Instead, understanding the risk levels associated with a worker population can guide task design decisions and support economic justification for ergonomic interventions that benefit a significant portion of the workforce.

8.5.6.4.1 Risk Factors in Material-Handling Tasks

Various individual characteristics have been identified as potential risk factors for developing low-back pain and other injuries in material-handling tasks. These include:

- **Previous Back Injury:** Previous injuries can increase susceptibility to further problems.
- **Fitness Levels:** Poor physical fitness can contribute to a higher risk of injury.
- **Second Jobs and Hobbies:** Additional physical activities outside of primary work can impact overall risk.
- **Smoking:** Known to affect overall health and recovery.
- **Age and Gender:** Different age groups and genders may have varying susceptibilities.
- **Obesity and Physical Stature:** Excess weight and body dimensions can influence ergonomic risk.
- **Psychosocial Issues:** Factors such as family and financial difficulties, job dissatisfaction, lack of job control, and work-related stress can affect injury risk.

8.5.6.4.2 Categories for Evaluating Worker Characteristics

To assess the likelihood of injury, worker characteristics should be considered in the following categories:

1. **Physical Attributes:** Age, sex, and anthropometry (body measurements).
2. **Sensory Capabilities:** Visual, auditory, and tactile senses.
3. **Psychomotor Skills:** Measures of worker capability and coordination.
4. **Personality Traits:** For example, a tendency toward risk-taking behavior.
5. **Health Status:** Previous injuries and overall fitness levels.
6. **Leisure Activities:** Recreational activities that could influence physical health.

8.5.6.4.3 *Importance of Holistic Evaluation*

Each of these factors should be taken into account to assess the overall risk of injury in MMH tasks. By understanding these characteristics, employers can design tasks that minimize risk and improve safety for the entire workforce.

8.5.7 Risks of Manual Material-Handling Tasks

Overexertion injuries occur when the physical demands of a task exceed the capabilities of the worker, typically when handling loads that are either too heavy or positioned improperly. These injuries place significant strain on the musculoskeletal system, particularly the lower back, leading to acute and chronic conditions. In the United States alone, overexertion injuries impact approximately 500,000 workers annually, affecting roughly 1 in 20 workers (OSHA, 2021). These injuries are often severe, with more than 60% of individuals who suffer from lower back pain attributing it to overexertion. Furthermore, only about one-third of workers who experience substantial lost time due to low-back pain are able to return to their previous job roles. Of these injuries, approximately two-thirds involve lifting, while around 20% result from pushing and pulling tasks (OSHA, 2021).

8.5.8 The Mechanisms of Overexertion Injuries in MMH

Handling loads, even relatively light ones, can lead to musculoskeletal strains due to dynamic and static forces that affect the fingers, arms, trunk, and legs. Strain occurs when workers engage in stretching, bending, twisting, or straightening movements, often under suboptimal conditions. These injuries can happen both during rapid or slow movements and can result from a single instance of overexertion or from repetitive tasks that, over time, cause wear and tear on the musculoskeletal system.

One of the primary concerns with MMH tasks is the wear and tear on the lumbar spine discs. This repetitive stress on the lower back often leads to severe back pain, prolonged absences from work, and, in many cases, permanent disability. Improper lifting techniques, especially those performed outside the sagittal plane (the vertical plane dividing the body into left and right), increase the risk of injury. Additionally, fatigue from inadequate recovery time between physically demanding tasks contributes to cumulative strain on the musculoskeletal structures (McCauley & Bartram, 2016).

The musculoskeletal structures of the trunk, including the vertebral column, experience significant stress during MMH tasks. Injuries can arise from linear elongation, bending, or twisting strains, which not only affect the lower back but also extend to the shoulders and arms, leading to widespread musculoskeletal issues (Salvendy, 2014b).

8.5.9 Risk Evaluation in MMH Tasks

To address the risks associated with MMH, it is essential to evaluate the physiological and biomechanical stresses imposed by tasks such as lifting, pushing, pulling, and carrying. These assessments help identify key risk factors and inform the development of ergonomic interventions designed to prevent overexertion injuries. While much of the focus remains on protecting the lower back, the entire vertebral column must be considered to ensure comprehensive injury prevention. Strains on the cervical and thoracic regions, as well as on the shoulders and arms, must be addressed as part of a holistic approach to reducing the risks of MMH.

8.5.10 Preventing Overexertion Injuries: Ergonomic Solutions

Implementing effective ergonomic designs and preventive measures can dramatically reduce the incidence of overexertion injuries and improve workplace safety and productivity. Key strategies include:

- **Ergonomic Training:** Workers should receive training on proper lifting techniques that minimize strain on the spine. This includes maintaining a neutral spine posture, lifting with the legs rather than the back, and avoiding twisting movements during lifting tasks.
- **Assistive Devices:** The use of mechanical aids, such as hoists, conveyors, lift tables, and carts, can alleviate the physical burden on workers. These devices reduce the need for manual handling and help prevent musculoskeletal injuries (OSHA, 2021).
- **Workspace Design:** Optimizing the layout of workspaces can minimize the need for excessive lifting, bending, and reaching. For example, placing materials at waist height reduces the frequency of bending and stretching, which are common contributors to back strain.
- **Task Rotation and Recovery Time:** Rotating workers through different tasks that require varying levels of physical exertion allows muscles to recover between high-intensity tasks, reducing the cumulative strain that leads to injury (McCauley & Bartram, 2016).
- **Risk Assessment Tools:** Employers should conduct regular ergonomic assessments to evaluate the risks associated with specific tasks. These assessments can be supplemented with biomechanical tools that model the forces acting on the body during MMH tasks, allowing for a more accurate understanding of injury risks.

Through proper training, the use of assistive devices, and optimizing workspace design, employers can reduce the incidence of injuries and improve overall productivity. Addressing MMH-related risks holistically, with a focus on the entire vertebral column and body mechanics, is essential for creating a safer and more efficient work environment.

8.5.11 Assessment of Workload for Manual Material-Handling Tasks

Various assessment methods are commonly used to measure or predict the energy costs and human capabilities associated with different MMH activities. An effective model for assessing workload should consider several parameters, including:

- **Body Weight of Individuals:** The range of body weights of individuals performing the task.
- **Weight of the Load:** The mass of the objects being handled.
- **Vertical Start and End Positions:** The initial and final heights of the load during a lift.
- **Dimensions of the Load:** The size and shape of the objects.
- **Frequency of Handling:** How often the task is performed.
- **Gender:** As physiological differences can influence capabilities.

Each model has its own advantages and disadvantages, and it is important to exercise caution when using any model to predict workload. Understanding these parameters helps in designing safer and more efficient MMH tasks.

8.5.11.1 Biomechanical Models

Biomechanical methods have long been used to assess strains and human capabilities, primarily focusing on the spinal column and its responses to compression. However, calculating compression strain is complex due to the spine's non-linear structure. The human spine's elasticity, necessary for absorbing shocks during daily activities, is provided by its elongated "S" shape. This includes a backward curve at the chest level (kyphosis) and a forward curve in the lumbar region (lumbar lordosis), with intervertebral discs separating the vertebrae.

Aging and repetitive motions can cause disc degeneration, but severe injuries often result from overexertion. Sudden compressive forces on a degenerated disc can exert pressure on the spinal cord or the nerves extending from it, leading to herniated discs. The greater the weight sustained by the upper body, the larger the force exerted on the spinal column. The spine, as the only solid bony structure in the human rump, prevents the rib cage from collapsing into the pelvis. Already supporting the weight of the head, neck, arms, hands, and upper trunk, any additional load serves to further compress the spine; this can occur through activities such as walking, running, bending, or twisting. The increased pressure on nerves, narrowing of spaces between vertebrae, and the pulling and squeezing of spinal tissues, muscles, and ligaments cause various discomforts, aches, muscular cramps, and conditions such as lumbago and sciatica.

Bending the back while keeping the knees straight places significantly more stress on the lumbar discs than bending the knees and keeping the back straight. Initially, the measured, modeled, and calculated spinal column strains were treated as static. However, due to the complexity and risks involved, dynamic examinations have been integrated into these models.

8.5.11.2 Intra-Abdominal Pressure

Lifting tasks exert increased pressure between the inferior and superior diaphragms of the trunk cavity due to the contraction of the abdominal muscles. Initially, it was believed that intra-abdominal pressure helped minimize the compression strain on the spine when lifting loads; however, recent research indicates that intra-abdominal pressure does not actually reduce spinal compression loading.

8.5.11.3 Physiological Approach

The physiological approach is an effective method for assessing manual material-handling abilities, particularly for tasks performed frequently and over extended periods, such as an 8-hour workday. This approach focuses on energy consumption and the stresses exerted on the cardiovascular system. Elevated heart rates can indicate stress due to various conditions, including:

- Physical Effort
- Environmental Heat and/or Humidity
- Psychological Stress and/or Time Pressure
- Cognitive Load (Decision-Making and Perceptual Work)
- Environmental Factors
- Chemical Exposure
- Noise
- Combinations of the Above Factors

Stress disrupts the body's equilibrium, known as homeostasis. The human body strives to maintain homeostasis during heavy physical exertion. According to the *American Heritage® Medical Dictionary*, homeostasis is the "tendency of an organism or cell to regulate its internal conditions, such as the chemical composition of its body fluids, to maintain health and functioning, regardless of outside conditions" (American Heritage Dictionary, 2011). The body maintains homeostasis by monitoring internal conditions and responding appropriately when they deviate from their optimal state.

During heavy work, the body undergoes several adaptations to maintain homeostasis, including:

- Deeper and More Rapid Breathing
- Increased Heart Rate
- Increased Blood Flow to Muscles
- Vasomotor Adaptations
- Dilation of Blood Vessels in Active Organs
- Redirection of Blood from Less Critical Organs to Those Needing More Oxygen and Nutrients
- Rise in Blood Pressure to Speed Up Blood Flow
- Increased Supply of Sugar
- Rise in Body Temperature and Increased Metabolism to Convert More Chemical Energy into Mechanical Energy

8.5.11.4 Psychophysical Approach

One of the most effective methods for establishing criteria in manual material-handling tasks is the psychophysical approach. This methodology combines both biomechanical and physiological stresses with a subjective evaluation of perceived physical effort to determine safe load-handling limits. Unlike purely objective approaches, which rely solely on physical measurements, the psychophysical method takes into account how individuals perceive the physical strain associated with lifting, pushing, pulling, and carrying tasks.

According to Mital et al. (2014), the psychophysical approach allows individuals to adjust a load to the maximum amount they feel they can handle safely, without experiencing strain, discomfort, or unusual fatigue. This self-selected limit is referred to as the Maximum Acceptable Weight Limit (MAWL). By allowing workers to set their own limits based on their perception of stress and exertion, the MAWL becomes a practical benchmark for designing safe and sustainable MMH tasks.

The psychophysical method operates on the principle that individuals have an inherent ability to judge their physical capacity for work and will adjust load-handling tasks accordingly. In a typical assessment, participants are asked to perform a material-handling task (e.g., lifting or carrying) with progressively heavier loads. They are instructed to stop or adjust the weight when they reach a point where the task still feels manageable but would not lead to strain or excessive fatigue over time.

This process incorporates several key factors:

- **Perceived Exertion:** Participants monitor their physical responses, such as muscle fatigue or discomfort, while performing the task.
- **Biomechanical Loads:** The task is analyzed for its impact on muscles, joints, and the spine, particularly the lower back, which is often a focal point in MMH tasks.
- **Physiological Response:** Heart rate and energy expenditure are often measured to correlate subjective feelings of exertion with objective physiological data.

This approach ensures that material-handling guidelines are based not only on biomechanical data but also on how workers feel during and after performing tasks, which is critical for reducing injury risks in the workplace (Mital et al., 2014).

8.5.11.5 Maximum Acceptable Weight Limit (MAWL)

The Maximum Acceptable Weight Limit is a key outcome of the psychophysical approach. It reflects the heaviest load that a worker can handle without discomfort over an extended period and varies by worker based on factors such as physical fitness, experience, and task conditions. The MAWL provides a practical standard for determining load limits in MMH tasks and is widely used to develop ergonomic guidelines aimed at minimizing injury risks.

For example, a MAWL assessment for a lifting task might involve gradually increasing the weight of a load while monitoring the participant's feedback and physiological responses. The MAWL is the load at which the participant reports that the task is manageable but approaches their limit. By using MAWL data, employers can set safe weight limits for tasks that balance productivity with worker safety.

8.5.11.6 Applications and Benefits of the Psychophysical Approach

The psychophysical approach is particularly valuable because it accounts for the subjective experience of workers, offering several key advantages over purely objective measures:

- **Customization:** The approach can be tailored to individual workers, allowing for variability in physical strength, endurance, and experience.
- **Prevention of Overexertion:** By relying on workers' perceptions of strain, this method helps prevent overexertion injuries, which are common in MMH.
- **Practical Implementation:** It is relatively simple to implement in workplace settings, requiring only minimal equipment (e.g., weights and a feedback system).

By integrating physiological data with subjective feedback, the psychophysical approach enhances our understanding of how workers perceive and manage their physical capabilities, allowing for the development of ergonomic interventions that are more closely aligned with real-world conditions.

8.6 ASSESSMENT OF ENERGY EXPENDITURES: DIRECT AND INDIRECT MEASURES

Direct and indirect measures of energy expenditure are essential for evaluating the physical demands of manual material-handling tasks. Modern methods allow for precise and reliable assessments, contributing to safer and more effective task design.

8.6.1 Direct Measures of Energy Expenditure

Direct measures involve computerized measurement of gas exchange, such as using computers and mouthpieces to calculate oxygen uptake during tasks. One traditional method is the Douglas bag technique, where the subject inhales air from the atmosphere and exhales it through a mask connected to a large bag. The oxygen content of the expired air in the bag is compared to that of the atmosphere to determine the amount of oxygen metabolized by the subject. By knowing the time taken to fill the bag, the subject's rate of oxygen uptake can be calculated.

8.6.2 Indirect Measures of Energy Expenditure

Indirect measures include subjective assessments of physical effort, such as the Borg Rating of Perceived Exertion (RPE) scale. The RPE scale ranges from no exertion to maximal exertion. Heart rate monitoring is also a common indirect measure, capturing continuous or intermittent heart rate changes during tasks.

Key heart rate values used in assessments include:

- **Resting Pulse:** Average heart rate before work begins.
- **Working Pulse:** Average heart rate during work.
- **Work Pulse:** Difference between resting and working pulses.
- **Total Recovery Pulse (Recovery Cost):** Sum of heartbeats from the end of work until the pulse returns to its resting level.
- **Total Work Pulse (Cardiac Cost):** Sum of heartbeats from the start of work until the resting level is restored.

The energy required for physical work can be correlated to the percent heart rate change, estimated using the following formula:

$$\% \text{ heart rate change} = (HR_{-task} - HR_{-rest}) / (HR_{-max} - HR_{-rest})$$

Where:
HR_{rest}: Resting Heart Rate
HR_{task}: Task Heart Rate
HR_{max}: Maximum Heart Rate

8.7 EVALUATION OF PHYSICAL EFFORTS IN MANUAL MATERIAL HANDLING

Estimating energy consumption and task demands involves analyzing the intensity and duration of the work being performed. These task elements are categorized into primary and secondary physical effort requirements. Primary physical efforts include activities such as lifting, pulling, and climbing. The intensity of task performance is influenced by factors such as muscular requirements, task duration, and the mental stress associated with the task.

Most models measure task intensity by considering physical expenditure and activity duration. These models classify intensity levels into five categories:

- Level I: Light intensity/occasional
- Level II: Light intensity/frequent or moderate intensity/occasional
- Level III: Light intensity/constant, moderate intensity/frequent, or heavy intensity/occasional
- Level IV: Heavy intensity/frequent or moderate intensity/constant
- Level V: Heavy intensity/constant

8.8 NIOSH LIFTING GUIDELINE

In 1981, the US National Institute for Occupational Safety and Health (NIOSH) developed a work practice guide for manual lifting, introducing distinct recommendations for acceptable lifting weights. Unlike previous guidelines, which suggested a single safe weight for men, women, and children, the 1981 guide established two thresholds. The lower threshold, known as the action limit (AL), was deemed safe for 99% of working men and 75% of working women. The action limit values depended on factors such as the starting height of the load, the vertical distance of the lift, the horizontal distance from the body, and the frequency of lifting. If the existing weight exceeded the action limit, engineering or managerial controls were required to reduce the load to an acceptable level. The maximum permissible load (MPL) was set at three times the action limit, and under no circumstances was continuous lifting allowed if the load was three times the AL.

The guidelines stated:

- Under optimal conditions, a maximum of 40 kg can be lifted.
- The MPL corresponds to the capacity of a 75th percentile male.
- The AL corresponds to the 99th percentile woman.
- MPL = 3(AL)
- The AL should not exceed 3,400 N at the L5/S1 vertebral joint.
- Loads exceeding the AL are considered a normal risk.
- Workloads exceeding the MPL are unacceptable.
- Loads between the AL and MPL require training and/or workforce selection.
- Aim to keep loads at or below the AL whenever possible.

However, flaws in these assumptions led to a revision in 1991. NIOSH introduced a new technique for assessing overexertion hazards in manual lifting, replacing the dual weight limits with a single recommended weight limit (RWL). The RWL represents the maximum weight that approximately 90% of American industrial workers, male or female, physically fit and accustomed to physical labor, can safely lift or lower. This new formula includes multipliers for factors such as asymmetry and hand-load coupling quality, resembling the 1981 AL formula but with additional considerations.

The revised equation is as follows:

$$RWL = LC \times HM \times VM \times DM \times AM \times FM \times CM$$

Where:

- LC: Load constant of 23 kg (51 lb).
- HM: Horizontal multiplier (distance of hands from the midpoint of the ankles).
- VM: Vertical multiplier (height of the hands above the floor at the start and end points of the lift).

- DM: Distance multiplier (vertical travel distance from the start to the end points of the lift).
- AM: Asymmetric multiplier (angle of symmetry, such as the angular displacement of the load from the medial plane).
- FM: Frequency multiplier (rate of lifting, expressed in lifts per minute).
- CM: Coupling multiplier (quality of coupling between hand and load).

Each multiplier ranges from zero to one. Additionally, the Lifting Index (LI) is calculated by dividing the actual weight of the load by the RWL:

$$LI = \frac{Actual\,weight\,lifted}{RWL}$$

A Lifting Index greater than 1.0 indicates an increased risk of low-back pain for some workers, and the higher the LI, the greater the risk. Ideally, all lifting tasks should have an LI of less than 1.0.

Steps to Calculate the RWL Using the NIOSH Lifting Equation:

1. Determine the weight of the load (contents and container).
2. Assess the operator's position and lifting task by obtaining the following parameters:
   - Horizontal distance (H) of the lift.
   - Vertical distance (V) of the lift.
   - Carry distance (D).
   - Frequency of lift (F).
   - Angle of asymmetry (A).
   - Quality of coupling (C).
3. Apply NIOSH multiplier factors using standard tables or digital tools.
4. Calculate the Recommended Weight Limit (RWL) using:
   RWL = LC × HM × VM × DM × AM × FM × CM where LC = Load Constant (usually 23 kg)
5. Calculate the Lifting Index (LI):
   LI = Load Weight / RWL
6. Interpret the result:
   LI ≤ 1.0: Acceptable for most workers; LI > 1.0: Increased risk of low-back injury

A comprehensive manual for the application of the NIOSH Lifting Guideline is available on the NIOSH website (Waters et al,, 1994) and can provide additional details for applying this method. A few general design/redesign suggestions given the outcome of the parameters are provided in Table 8.4 (NIOSH, 1994).

8.8.1 Other Lifting Guidelines

The American Conference of Governmental Industrial Hygienists (ACGIH) provides guidelines for safe lifting practices through its Threshold Limit Values (TLVs).

TABLE 8.4
General Design/Redesign Suggestions for NIOSH Lifting Guidelines

If HM is less than 1.0	Bring the load closer to the worker by removing any horizontal barriers or reducing the size of the object. Lifts near the floor should be avoided; if unavoidable, the object should fit easily between the legs
If VM is less than 1.0	Raise or lower the origin or destination of the lift Avoid lifting near the floor or above the shoulders
If DM is less than 1.0	Reduce the vertical distance between the origin and the destination of the lift
If AM is less than 1.0	Move the origin and destination of the lift closer together to reduce the angle of twist, or move the origin and destination further apart to force the worker to turn the feet and step, rather than twist the body
If FM is less than 1.0	Reduce the lifting frequency rate, reduce the lifting duration, or provide longer recovery periods (i.e., light work period)
If CM is less than 1.0	Improve the hand-to-object coupling by providing optimal containers with handles or handhold cutouts, or improve the handholds for irregular objects
If the RWL at the destination is less than at the origin	Eliminate the need for significant control of the object at the destination by redesigning the job or modifying the container or object characteristics

Source: NIOSH, 1991.

These TLVs establish upper and lower limits for lifting based on factors such as frequency, duration, and other associated risk factors. These guidelines aim to minimize the risk of injury and ensure safer manual material handling.

8.8.2 University of Michigan 3D Static Strength Prediction Program

The 3D Static Strength Prediction Program software is designed to predict the static strength requirements for various tasks such as lifting, pressing, pushing, and pulling. This program simulates job scenarios by incorporating posture data, force parameters, and male/female anthropometry. The results provide valuable information, including the percentage of men and women who possess the necessary strength to perform the specified job, spinal compression forces, and data comparisons to NIOSH Guideline.

Key features of the software include the ability to analyze torso twists and bends, make detailed hand force entries, and utilize an automatic posture generation feature. The program also offers three-dimensional human graphic illustrations to enhance the analysis.

8.8.3 Ohio State Lumbar Motion Monitor

Traditionally, workplace ergonomic assessments have primarily focused on joint loading in static postures. However, recent epidemiological studies have indicated that three-dimensional dynamic motion is associated with a higher risk of occupational injury and illness. To address this, the Biodynamics Lab has developed a specialized research program that studies occupational joint loading under realistic dynamic motion conditions. The goal of this program is to gain a better understanding of the thresholds for exposure to realistic risk factors and determine how much is too much.

8.8.4 Snook's Psychophysical Tables

In 1978, Liberty Mutual published extensive tables of acceptable loads and forces for continuous manual material-handling jobs, based on research by Snook and Cirello. These tables were first updated in 1983, revised in 1991, and again updated in 1993 by Snook, Cirello, and Hughes. The guidelines assume several conditions for safe lifting:

- Two-handed symmetrical material handling in the medial (mid-sagittal) plane, directly in front of the body, allows for slight body twists during lifting or lowering.
- Moderate load width, typically 75 cm (28.5 inches) or less.
- Good coupling, such as hands with handles and shoes with the floor.
- Unrestricted working postures.
- Favorable physical environment, approximately 21°C with 45% relative humidity.
- Minimal other physical work activities.
- Physically fit workers accustomed to labor.

These guidelines are captured in the Liberty Mutual Lifting Tables, which differ from the 1991 NIOSH Guidelines by providing additional details regarding lifting capability. The Liberty Mutual Tables categorize recommendations based on the percentage of the worker population for whom the values are acceptable and provide gender-specific data. Lifts are subdivided into three height areas: floor to knuckle height, between knuckle and shoulder heights, and shoulder to overhead reach.

Both the new NIOSH Guidelines and the revised Liberty Mutual data indicate that the absence of handles reduces the loads people are willing to lift by an average of 15%. The NIOSH Lifting Guideline is widely regarded as a primary method for assessing lifting loads, but other guidelines exist as well. Maximum recommended weights range from 51 pounds using the NIOSH Guideline to 70 pounds using the American Conference of Governmental Industrial Hygienists (ACGIH) Threshold Limit Values for lifting.

A study by Russell et al. (2007) compared five lifting analysis tools, assessing and comparing maximum lifting capabilities using the following approaches:

- NIOSH Lifting Guideline
- ACGIH Threshold Limit Values (TLVs) for Lifting
- Liberty Mutual Manual Material-Handling Tables
- University of Michigan 3-D Static Strength Prediction Program (3DSSPP)

- Washington State Department of Labor and Industries (WA L&I) Lifting Calculation

To enable comparisons between these five lifting assessment instruments, the output of each approach was converted to an exposure index similar to the NIOSH Lifting Index.

The study results indicate that the NIOSH Guideline, ACGIH TLV, and Liberty Mutual instruments provided similar outcomes when evaluating musculoskeletal exposures associated with lifting tasks. Furthermore, the study found that the ACGIH TLV, Liberty Mutual, and WA L&I methods were easier to use because they required fewer inputs compared to the six inputs needed for the NIOSH Lifting Equation. However, the NIOSH Guideline offered more comprehensive interpretive capabilities, allowing for a better understanding of which aspects of the lift would benefit most from changes (Russell et al., 2007).

8.8.5 General Ergonomic Rules for Lifting of Loads

Implementing the following guidelines can significantly reduce the risk of injury associated with industrial manual handling tasks, especially when lifting loads:

- **Reduce Sizes and Weights:** Minimize the size, weight, and force required to handle objects.
- **Provide Good Handholds:** Ensure that objects have secure handholds to improve grip and control.
- **Keep Objects Close:** Hold objects close to the body, directly in front, and avoid twisting movements.
- **Use Proper Lifting Technique:** Keep the trunk upright and bend the knees when lifting.
- **Minimize Movement Distance:** Reduce the distance that objects need to be moved to lessen physical strain.
- **Prefer Horizontal Movement:** Move objects horizontally rather than vertically. Whenever possible, convert lifting and lowering tasks to pushing, pulling, or carrying by providing materials at the appropriate working height.
- **Plan and Smooth Movements:** Plan your movements carefully and execute them smoothly to avoid sudden strain.
- **Avoid Unnecessary Lifting:** Do not lift or lower objects more than necessary.

8.9 FATIGUE

In the industrial environment, fatigue can be broadly categorized into three primary types: muscular fatigue, mental fatigue, and visual fatigue. Understanding these categories and their impact is crucial for designing better work practices to reduce the risks associated with each type and improve overall worker health and productivity.

8.9.1 Muscular Fatigue

Muscular fatigue is commonly characterized by a reduction in muscle power and slower movements, often accompanied by significant pain in overstressed muscles. During muscular contraction, chemical reactions occur to generate the energy needed for mechanical work. After a muscle contracts, its energy reserves need to be replenished. However, if the demand for energy exceeds approximately half of a person's maximal oxygen uptake, lactic acid and potassium ions accumulate within the muscle tissue, leading to muscle fatigue.

In the context of sustained static muscle contraction, muscular effort exceeding about 15% of maximal voluntary contraction can diminish or even halt blood flow through the muscle. This lack of blood flow results in the accumulation of potassium ions and a depletion of sodium in the extracellular fluid, disrupting the biochemical balance necessary for proper muscle function. Additionally, the intracellular build-up of phosphate, due to ATP breakdown, interferes with the coupling between nervous excitation and muscle fiber contraction, causing muscle fatigue.

To minimize muscular fatigue, it is essential to avoid overexertion and adhere to established guidelines for muscular duration in task performance. Static loading should be minimized as much as possible. Effective fatigue management also involves incorporating several short rest periods rather than fewer long rest periods, as the former has a higher "recovery value" for alleviating fatigue.

8.9.2 Mental Fatigue

Mental fatigue is the result of overloading sensory organs and mental capacities, impacting both physically intensive and cognitive tasks. It is caused by prolonged mental or intellectual work, which can affect individuals over short or long periods. Symptoms of mental fatigue typically become evident toward the end of the workday, when concentration becomes more challenging, tasks seem more complicated, and mistakes are more frequent. Late-night work or studying can also lead to mental fatigue, making it harder to retain information and increasing the likelihood of errors and clumsiness.

While there is no direct cure for mental fatigue, maintaining a well-balanced and nutritious diet can help alleviate some of its effects. Regular physical exercise is another effective way to combat mental fatigue, as it increases oxygen levels in the bloodstream, which in turn helps improve cognitive processes and overall mental performance.

8.9.3 Visual Fatigue

Visual fatigue, also known as eyestrain (asthenopia), includes subjective visual symptoms or discomfort resulting from prolonged use of the eyes. Tasks requiring extended near work, such as reading or working with a microscope, are common causes of visual fatigue. Temporary shifts in the near point of accommodation and the resting position of convergence can also contribute to visual fatigue, indicating a loss of refractive power, sometimes referred to as accommodation strain.

Additional factors that lead to visual fatigue include tasks requiring vigilant attention, using viewing equipment like microscopes or binoculars, and frequent accommodation due to changes in viewing distance. Occupational factors such as the rigidity of work routines and the duration of work periods exacerbate visual fatigue. Symptoms of visual fatigue can include eye discomfort, blurred vision, headaches, and increased sensitivity to light.

Minimizing visual fatigue involves assessing tasks, processes, and equipment to ensure they align with human visual capabilities. This may include optimizing lighting conditions, ensuring regular breaks to rest the eyes, and using equipment designed to reduce visual strain.

8.9.4 General Fatigue

General fatigue is the degree of fatigue a person experiences depending on the various stresses endured throughout the day. To recuperate, rest periods are crucial, especially during nighttime. However, taking rest breaks throughout the day and during work shifts is also highly beneficial. General symptoms of fatigue can be both subjective and objective, with common indicators including:

- **Subjective Symptoms:**
  - Feelings of weariness, drowsiness, and faintness
  - Aversion to work
  - Sluggish thinking
  - Reduced alertness
  - Poor and slow perception
  - Unwillingness to work
- **Objective Symptoms:**
  - Decline in physical performance
  - Decline in mental performance

8.9.5 Measuring Fatigue

For ergonomists, obtaining quantitative measurements of fatigue is crucial to understanding the relationship between fatigue, work output, and stress levels. This understanding helps in developing methods to improve work efficiency and reduce strain. Although no direct method exists for measuring fatigue, various studies have identified indicators of fatigue. Some of the commonly used methods include:

8.9.5.1 Methods for Measuring Fatigue

1. **Quality and Quantity of Work Performed:**
   - **Indirect Measurement:** Fatigue is often assessed by examining the quantity and quality of work output.

- **Quantity:** Metrics such as the time taken to produce one unit, the number of items processed, or the number of operations performed per unit time are used.
- **Quality:** The number of mistakes and faulty items produced can also indicate fatigue levels.

2. **Subjective Perceptions:**
   - **Questionnaires:** Used to gather subjective feelings of fatigue from workers.
3. **Electroencephalography (EEG):**
   - **Indicators:** Increases in alpha and theta rhythms and reductions in beta waves signify weariness and sleepiness.
4. **Frequency of Flicker–Fusion of Eyes:**
   - **Method:** Subjects are exposed to a flickering lamp, and the flickering rate is increased until it appears as continuous light. A lower flicker–fusion rate indicates fatigue.
   - **Usage:** This technique has become less common due to controversial results and weak correlations with other fatigue symptoms.
5. **Psychomotor Tests:**
   - **Functions Measured:** These tests assess perception, motor reactions, and interpretation.
   - **Examples:** Simple and selective reaction times, tests involving touching or pricking squares in a grid, skill tests, driving simulations, and typing.
   - **Interpretation:** Decreased performance is seen as an increase in fatigue, although results can be skewed by worker excitement or interest in the test.
6. **Mental Tests:**
   - **Types:** Include arithmetic problems, estimation, concentration, and memory tests.
   - **Interpretation:** Decreased performance is interpreted as increased fatigue, with similar potential inaccuracies as psychomotor tests.

8.9.5.2 Considerations in Measuring Fatigue

- Measurements are typically taken before, during, and after task execution.
- Results are relatively significant and need to be compared with a baseline of a non-stressed individual.

8.9.5.3 Additional Types of Fatigue

- **General Bodily Fatigue:** Results from physical overloading of the entire body.
- **Nervous Fatigue:** Caused by overstressing a part of the psychomotor system, often seen in skilled and repetitive work.
- **Chronic Fatigue:** Accumulation of long-term physical and psychological stress effects, potentially requiring medical intervention.

TABLE 8.5
Comparison of Five Analysis Tools

Power grip

This handgrip provides maximum hand power for high force tasks. All the fingers wrap around the handle

Contact pressure

Pressure from a hand surface, point, or edge on any part of the body

Pinch grip

This hand grip provides control for precision and accuracy. The tool is gripped between the thumb and fingertips

Single-handle tools

Tube-like tools measured by handle length and diameter. Diameter—the length of a straight line through the center of the handle

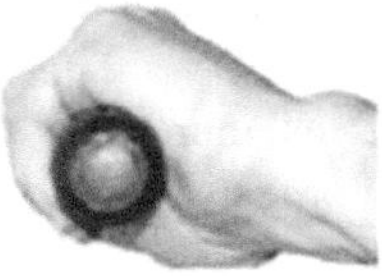

Double-handle tools

Plier-like tools measured by handle length and grip span. Grip span—the distance between the thumb and fingers when the tool jaws are open or closed

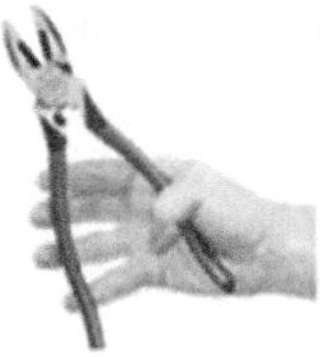

Source: NIOSH website, accessed February 16, 2011.

8.9.6 Shift Work

Research on rotating shift work and night work has shown that such schedules impose physiological strain on individuals (Rodahl, 1989). A typical work shift is defined as a period of no more than eight consecutive hours during the day, five days a week, with a minimum of an hour rest period before returning to work. Shifts that exceed these hours, require more consecutive workdays, or take place during the evening are considered extended shifts. These extended shifts are often utilized to optimize limited resources.

8.9.6.1 Impact of Shift Work

Shift work can disrupt normal circadian rhythms, leading to various health issues, including sleep disorders, gastrointestinal problems, and an increased risk of cardiovascular diseases. The physiological strain from shift work arises from the body's need to adjust to irregular sleep–wake cycles and the challenges of maintaining alertness and performance during non-standard hours.

TABLE 8.6
Work Systems Definitions

Shift	The hours of a given day that an individual or a group of individuals is scheduled to be at the workplace
Off time	The hours of a given day that an individual or a group of individuals is not normally required to be at the workplace
Schedule	The sequence of consecutive shifts and off time assigned to a particular individual or group of individuals as their usual work assignment
Permanent hours	A schedule for an individual or a group of individuals that does not normally require them to work more than one type of shift. That is, the time of day one works is constant
Rotating hours	A schedule that normally requires an individual or a group of individuals to work more than one type of shift That is, the time of day one works changes
Basic cycle	The minimum number of days required to complete the specific sequence of shifts and off time constituting a given schedule The number of days until a schedule begins to repeat
Major cycle	The minimum number of days required to arrive at a point where the basic sequence of a schedule begins to repeat on the same days of the week The number of days until the basic sequence falls on the same days when it repeats
Work system	All of the schedule(s) implemented in a given workplace to meet the real or perceived requirements of a given plant, process, or service

Source: Gavriel, S. (Ed.): *Handbook of Human Factors and Ergonomics*. 1997. Copyright Wiley-VCH Verlag GmbH & Co. KGaA.

TABLE 8.7
Work Shift Definitions

	Three-Shift Systems
First shift	A work period of about 8 hours in duration that generally falls between the hours of 0600 and 1700. Also known as morning or day shift.
Second shift	A work period of about 8 hours in duration that generally falls between the hours of 1500 and 0100. Also known as afternoon–evening or swing shift.
Third shift	A work period of about 8 hours in duration that generally falls between the hours of 2200 and 0700 Also known as night or graveyard shift
	Two-Shift Systems
Day shift	A work period of about 10 or more hours in duration that generally falls between the hours of 0800 and 2200
Night shift	A work period of about 10 or more hours in duration that generally falls between the hours of 2200 and 0800
	Other Work Shift Classifications
Split shift	Any work period that is regularly scheduled to include two or more work periods less than 7 hours, separated by more than 1 hour away from work, on the same day.
Irregular shifts non-workday	Work periods that vary their shift starting time and duration in an erratic way. Any calendar day in which only off time is scheduled.

Source: Gavriel, S. (Ed.): *Handbook of Human Factors and Ergonomics*. 1997. Copyright Wiley-VCH Verlag GmbH & Co. KGaA.

8.9.6.2 Recommendations for Managing Shift Work

To mitigate the negative effects of shift work, several strategies can be employed:

- **Rotate Shifts Forward:** Implementing forward rotation (morning to afternoon to night) is generally easier for the body to adapt to than backward rotation.
- **Limit Consecutive Night Shifts:** Reducing the number of consecutive night shifts can help prevent cumulative fatigue.
- **Ensure Adequate Rest:** Providing longer rest periods between shifts can improve recovery and reduce fatigue.
- **Promote Healthy Habits:** Encouraging proper nutrition, regular exercise, and good sleep hygiene can help workers manage the demands of shift work.

Shift work is not a recent phenomenon. Night watches have been documented since the earliest written history, and many historical battles were initiated in the early hours to gain an advantage over resting enemies. The industrial revolution, especially around the 1920s, led to the adoption of three-shift rotations to maximize the use of mechanical equipment around the clock. Additionally, many service industries have always required continuous presence for critical tasks.

- Emergency telephone operators
- Police officers
- Firefighters

TABLE 8.8
Five-Day Shift Schedules—12 Hour Shifts

	A Shift											
	1	2	3	4	5	6	7	8	9	10	11	12
Employee	W	T	F	S	S	M	T	W	T	F	S	S
A1	D	D	—	—	—	D	D	—	—	D	D	D
A1	D	D	—	—	—	D	D	—	—	D	D	D
A2	N	N	—	—	—	N	N	—	—	N	N	N
A2	N	N	—	—	—	N	N	—	—	N	N	N
Supervisor	12P	12P	—	—	—	12P	12P	—	—	12P	12P	12P
	B Shift											
	1	2	3	4	5	6	7	8	9	10	11	12
Employee	W	T	F	S	S	M	T	W	T	F	S	S
B1	—	—	D	D	D	—	—	D	D	—	—	—
B1	—	—	D	D	D	—	—	D	D	—	—	—
B2	—	—	N	N	N	—	—	N	N	—	—	—
B2	—	—	N	N	N	—	—	N	N	—	—	—
Supervisor	—	—	12P	12P	12P	—	—	12P	12P	—	—	—

Source: Eastman Kodak, 1983.

Notes: D, day shift; N, night shift; day's work week: Saturday 1200 to Saturday 1200; night's work week: Sunday 0000 to Sunday 0000; supervisor works noon to midnight; work week is Saturday to Saturday 1800

- Hospital staff
- Military personnel
- Restaurant workers

8.9.6.3 Current Regulations and Risks

Although OSHA does not have specific regulations addressing shift work schedules, design, or safety, it is governed under the OSH Act's General Duty Clause, Section 5(a)(1). This clause requires employers to provide a workplace free from recognized hazards that can cause death or serious physical harm.

Statistical data indicate that accident rates are higher during night shifts, and employees working long or irregular hours are at a greater risk of injury. Furthermore, shift workers face a higher risk of stress-related health problems compared to employees who work regular hours. This increased risk includes the potential for sleep disorders, gastrointestinal issues, and cardiovascular diseases.

8.9.7 Organizational Approaches to Shift Work

When designing shift schedules, it is crucial to consider multiple factors to optimize job performance and minimize negative impacts on workers. Key considerations include the job performance capabilities of people at different hours, the physiological effects of night work, and the psychosocial effects on workers and their families. The Ergonomics Group of Eastman Kodak (1983) outlined a process for designing shift schedules that accommodate five-day and seven-day schedules, including eight-hour, 12-hour, and other work-hour combinations.

8.9.8 Effective Shift Schedule Design

Effective shift schedule design can significantly reduce ergonomic, health, and safety problems while addressing the social challenges faced by shift workers. Satisfaction with a shift system results from balancing personal, psychological, social, and medical concerns. The Canadian Centre for Occupational Health and Safety (CCOHS) lists the following factors to consider in shift work design.

8.9.9 Length of the Rotation Period

Optimal Rotation Length: The ideal rotation length has been a subject of debate. A common system uses a one-week rotation with five to seven consecutive night shifts. This duration is often considered because it provides a consistent routine within a typical work week. However, one drawback is that it may not allow sufficient time for circadian rhythm adjustment, as it generally takes at least seven days for the body's internal clock to adapt to a new sleep–wake cycle. Just as adjustment starts to occur, workers may have to rotate to a different shift, disrupting this adaptation process.

Longer Rotation Periods: Some researchers advocate for longer rotation periods, ranging from two weeks to one month, on the same shift. This longer duration can help circadian rhythms adjust more fully to a particular schedule, reducing the physiological strain of constantly changing sleep patterns. However, a significant challenge with longer rotations is that workers often revert to a "normal" day-night schedule on their days off, potentially negating any adaptation achieved during the work period. Additionally, longer rotations can result in prolonged periods of social isolation, as workers on night shifts, for example, may find it difficult to engage in regular social activities.

Rapid Shift Rotation: Another approach is rapid shift rotation, where shifts change every two to three days. This system may reduce disruption to body rhythms because the readjustment of circadian rhythms is minimized. Workers do not spend enough time on any one shift to fully adapt, which might help in maintaining a more stable overall rhythm. Moreover, rapid rotations provide time for some social interaction each week, as workers have more frequent changes in their schedules, allowing for varied social and family activities. Individual differences and preferences also play a significant role in determining the best rotation period. Some workers may prefer longer periods on a single shift, while others may adapt better to more frequent changes.

8.9.10 Direction of Rotation

Forward Rotation: It is generally recommended that shifts should rotate forward (from day to afternoon to night) rather than backward. This recommendation is based on the fact that circadian rhythms adjust more easily when moving forward. The body's natural tendency is to extend the day, making it easier to adapt to a schedule that progressively delays the start time. Forward rotation aligns better with the natural circadian rhythm and can help reduce sleep disruption and fatigue.

8.9.11 Start and Finish Times

Early Shifts: Early morning shifts, especially those starting around 5:00 a.m., are associated with shorter sleep durations and greater fatigue. These shifts can disrupt the natural sleep cycle, leading to sleep deprivation and decreased alertness. It is advisable to avoid shift start times as early as 5:00 a.m. when possible to mitigate these issues.

Late Shifts: The finish times of shifts should be consistent with standard shift endings to allow workers to obtain a reasonable amount of nighttime rest. For example, finishing a shift at 3:00 a.m. can significantly disrupt sleep patterns and should be avoided if possible. Shifts should end at times that enable workers to have sufficient rest before their next shift, contributing to better overall health and performance.

8.9.12 Length of Rest Between Shifts

Adequate Rest Periods: Allowing adequate rest periods between shifts is crucial for recovery. A rest period of at least 24 hours should be provided after each set of night shifts. The more consecutive nights worked, the longer the rest period should be before the next rotation occurs. This practice helps reduce cumulative fatigue and supports better overall health.

8.9.13 Alternative Work Schedules

Extended Work Days: Some workplaces adopt extended workdays of 10 or 12 hours. This approach has the advantage of reducing the number of consecutive night shifts and providing longer blocks of time off. However, extended work hours can also lead to additional fatigue, particularly if the tasks are physically or mentally demanding. The physical and mental load of the task should be carefully considered when selecting the length of a work shift. Ergonomic hazards and exposure to chemical or physical agents should also be taken into account when choosing a shift system.

8.9.14 Additional Considerations

When designing shift schedules, it's important to consider the well-being and social needs of workers. The following guidelines can help optimize shift schedules to reduce stress and improve overall satisfaction:

1. **Provide Time Off at Socially Advantageous Times:** Whenever possible, schedule time off during weekends or other socially beneficial periods. This allows workers to engage in family activities and maintain a social life, which is essential for their overall well-being.
2. **Implement Special Shift Systems for Extended Overtime:** If production demands lead to prolonged periods of overtime, consider starting a special shift system to manage the increased workload. This approach can help distribute work more evenly and prevent burnout.
3. **Communicate Work Schedules in Advance:** Inform shift workers of their work schedules well ahead of time. Early notification enables employees and their families to plan their activities and manage their time effectively.
4. **Allow Flexibility for Shift Changes:** Offer as much flexibility as possible for shift changes. Allowing workers to swap shifts or adjust their schedules can help them manage personal commitments and reduce stress.
5. **Maintain Simple and Predictable Schedules:** Keep shift schedules as simple and predictable as possible. Complex and constantly changing schedules can lead to confusion and increased stress. Consistent and straightforward schedules help workers establish routines and maintain a better work-life balance.

8.9.14.1 Fatigue in Shift Work

Fatigue from shift work often arises from an individual's inability to rest adequately during off-hours, coupled with the body's natural tendency to slow down in the evening and night hours. This slowdown is regulated by the circadian rhythm, which affects bodily functions by signaling them to reduce activity at night and increase during the day. Circadian rhythms are physical, mental, and behavioral changes that follow a roughly 24-hour cycle, primarily responding to light and darkness in an organism's environment. These rhythms are found in most living things, including animals, plants, and many microbes. The study of circadian rhythms is known as chronobiology.

8.9.14.2 Impact of Circadian Rhythms on Shift Work

Circadian rhythms play a crucial role in shift work by influencing alertness and the ability to rest during off-hours. Daytime sleep generally offers less quantity and quality compared to nighttime sleep. Shift work is often less productive, and workers typically have difficulty adjusting. While the circadian clock can be reversed after several days of shift work, this adjustment requires significant time and is not easily maintained.

8.9.14.3 Challenges of Shift Work

Shift workers face numerous challenges, including:

- Social interactions that do not align with nighttime work schedules.
- Difficulty in obtaining sufficient quality sleep.
- Maintaining awareness of clock time.

According to OSHA (2010), fatigue is a common issue in shift work, and employers should be mindful of its symptoms. These symptoms vary among individuals and depend on the degree of overexertion. Common symptoms of fatigue include:

- Weariness
- Sleepiness
- Irritability
- Reduced alertness, lack of concentration, and memory issues
- Lack of motivation
- Increased susceptibility to illness
- Depression
- Headaches
- Giddiness
- Loss of appetite and digestive problems

8.10 SUMMARY

Despite advances in mechanization that have reduced the need for workers to perform strength-intensive tasks, many industries still require substantial physical effort. Consequently, the human body must generate sufficient energy to perform these tasks. The respiratory, circulatory, and metabolic systems are crucial in producing the energy needed for work performance. Individual muscle strength varies based on factors such as age, gender, health, training, and other variables. However, the biomechanical layout of the human body allows us to use muscular strength efficiently with minimal effort. Manual material handling, such as lifting, often leads to severe lower back injuries, primarily due to overexertion. Preventing unnecessary strains, overexertions, and other injuries related to manual material handling is essential for both ethical and economic reasons. Proper lifting guidelines and detailed recommendations for handling loads have been established to make material handling easier and reduce the risk of injuries.

Key considerations for safe manual material handling include understanding the role of mechanization and its impact on reducing physical strain, recognizing the importance of energy production systems (respiratory, circulatory, and metabolic) in optimizing work performance, and acknowledging the factors that affect individual muscle strength. The biomechanical efficiency of the human body allows for the effective use of muscular strength, which can help reduce effort and prevent injuries. Preventing injuries not only protects workers' health but also has significant economic benefits by reducing healthcare costs and improving productivity.

To enhance safety in manual material handling, it is important to follow proper lifting guidelines, use mechanical aids when possible, provide comprehensive training on safe handling practices, and regularly assess and adapt work environments to identify potential hazards. By integrating these considerations and recommendations, employers can create a safer work environment that reduces the risk of injury and enhances overall productivity.

8.11 CASE STUDIES

Problem Overview

Manual material-handling tasks involving heavy loads present significant challenges in industrial environments. The physical demands placed on workers during these tasks can lead to a variety of ergonomic issues, including musculoskeletal disorders. Factors influencing the safety and efficiency of these tasks include physical strength, fatigue, temperature, slopes, and load dimensions. Improper lifting techniques or overexertion can result in immediate injuries or long-term health risks, primarily MSDs, affecting the back, shoulders, and arms.

Ergonomic Design Process

To mitigate these risks, several ergonomic principles are employed in task design:

1. **Assessment of Physical Strength and Fatigue:** Workers' physical capabilities must be considered to prevent overexertion. Implementing proper task rotation and rest periods reduces fatigue and the likelihood of injuries. Strength-testing methods, such as isometric testing, can be used to match workers with tasks suited to their physical capacity.
2. **Temperature and Environmental Factors:** Temperature extremes can affect worker performance. High heat can increase fatigue, while cold temperatures can reduce dexterity, both of which compromise safety.

Modifying work schedules and ensuring proper clothing and hydration are critical for minimizing these risks.

3. **Load Considerations:** The weight, size, and shape of the load are important factors. Large or awkwardly shaped objects increase the strain on the body, making it difficult to maintain safe lifting postures. Employers should train workers to use mechanical aids, such as forklifts or dollies, to assist with heavy loads.
4. **Slope and Ground Conditions:** Lifting or carrying heavy loads on slopes can increase the risk of injury, as it requires additional effort to maintain balance. Identifying and minimizing these environmental hazards is key to maintaining safety.
5. **NIOSH Lifting Equation:** One of the primary tools for designing safe lifting tasks is the NIOSH Lifting Equation. This equation helps determine the recommended weight limit for specific tasks, considering factors such as horizontal and vertical distance from the load, asymmetry, and frequency of lifting. By calculating the overall load dimensions and determining the RWL, employers can redesign tasks to minimize physical strain and prevent injury.

Outcome

Through proper application of ergonomic principles, including task design and the use of the NIOSH Lifting Equation, the risks associated with manual material-handling tasks can be significantly reduced. Workers experience fewer cases of musculoskeletal disorders, and overall productivity improves as tasks are designed to be more efficient and safer.

Conclusion

The manual handling of heavy loads remains a high-risk task in industrial settings, contributing to long-term health issues for workers. By employing ergonomic principles, such as understanding physical limits, environmental considerations, and utilizing tools like the NIOSH Lifting Equation, employers can create safer and more effective work environments, reducing the risk of injury and improving overall job satisfaction.

Exercises: Heavy Work, Evaluating Physical Workloads and Fatigue

1. **Defining Heavy Work:**
   - Define heavy work and describe industry and task characteristics that contribute to high physical demands.
   - **Extended Activity:** Use AI-powered tools (e.g., ChatGPT or PubMed) to identify three recent studies or reports on heavy work in different industries. Summarize the findings and compare them to textbook definitions.

2. **NIOSH Load-Lifting Analysis Data Requirements:**
    - List the specific data necessary to perform a NIOSH load-lifting analysis (e.g., load weight, lifting frequency, posture).
    - **Extended Activity:** Search the NIOSH Lifting Equation calculator online or use AI to generate an example dataset. Apply it to calculate the recommended weight limit for a task of your choice.
3. **Job or Task Conditions Not Included in NIOSH Analysis:**
    - Identify job or task conditions that NIOSH does not account for in its lifting analysis.
    - **Extended Activity:** Research real-world case studies of material-handling injuries or ergonomic failures where NIOSH limitations played a role. Use OSHA, CDC, or industry-specific reports. Summarize key takeaways.
4. **Lifting Analysis Using Liberty Mutual Tables:**
    - Conduct a lifting analysis using the Liberty Mutual Tables to determine acceptable weight limits for different lifting tasks.
    - **Extended Activity:** Find an online ergonomic lifting calculator (such as one provided by an industry organization or university). Compare its output with results derived from the Liberty Mutual Tables.
5. **Primary Risk Factors in Manual Material Handling:**
    - List the primary risk factors associated with manual material-handling (MMH) tasks.
    - **Extended Activity:** Use AI to generate an example scenario of a workplace with multiple MMH risks. Propose an ergonomic intervention plan based on research from online OSHA and NIOSH resources.
6. **Types of Occupational Fatigue and Ergonomic Solutions:**
    - Describe the primary types of occupational fatigue (e.g., physical, mental, and psychosocial). Provide an ergonomic strategy to mitigate each type.
    - **Extended Activity:** Use AI tools or an academic database to find a recent scientific article on fatigue in the workplace. Summarize its findings and compare them to the textbook recommendations.
7. **Ergonomic Guidelines for Material-Handling Tasks:**
    - Identify a federal guideline, industry standard, or ISO standard that provides ergonomic recommendations for material-handling tasks.
    - **Extended Activity:** Research how companies apply these guidelines in real-world settings. Use industry reports, OSHA case studies, or corporate sustainability reports to find examples of best practices.
8. **Occupational Ergonomics Concerns in Shift Work:**
    - Explain the ergonomic challenges associated with shift work, including circadian rhythm disruption and fatigue.
    - **Extended Activity:** Use AI to conduct a **meta-analysis** of online resources, blogs, and research papers to identify modern solutions companies implement to reduce shift work fatigue.

9. **NIOSH Guide for Manual Lifting—Literature Review:**
   - Identify the four different types of literature and data sources that NIOSH experts consulted while developing the NIOSH Guide for Manual Lifting.
   - **Extended Activity:** Use Google Scholar or AI tools to find one of these original sources. Summarize its contribution to the NIOSH Guidelines.

REFERENCES

American Heritage Medical Dictionary. (2011). Homeostasis. Houghton Mifflin Harcourt Publishing Company.

Bassett, D. R., & Howley, E. T. (2014). Maximal oxygen uptake: "Classical" versus "contemporary" viewpoints. *Medicine and Science in Sports and Exercise*, *42*(5), 456–466.

Bouchard, C., Rankinen, T., & Timmons, J. A. (2015). Genetics and genomics of VO2 max: State of the art and future directions. *Medicine and Science in Sports and Exercise*, *44*(1), 133–139.

Bridger, R. S. (1995). *Introduction to ergonomics*. London: McGraw-Hill.

Deloitte Insights. (2024). *Ergonomic strategies in the workplace: Balancing productivity and well-being*. Deloitte Insights.

Dul, J., & Weerdmeester, B. (2008). *Ergonomics for beginners: A quick reference g uide* (3rd ed.). Taylor & Francis. Available at: https://www.taylorfrancis.com/books/mono/10.1201/9781420077520/e

Dul, J., & Weerdmeester, B. (2014). *Ergonomics for beginners: A quick reference guide*. CRC Press.

Eastman Kodak Company. (1983). Kodak's Ergonomic Design for People at Work. Van Nostrand Reinhold.

Fleg, J. L., & Lakatta, E. G. (2014). Role of muscle loss and cardiovascular function in VO2 max declines with aging. *Journal of Applied Physiology*, *97*(1), 3–12.

Jones, A. M., & Carter, H. (2014). The effect of endurance training on parameters of aerobic fitness. *Sports Medicine*, *29*(6), 373–386.

Karwowski, W. (2015). *Handbook of human factors and ergonomics*. Wiley.

Lehmann, G. (2014). Energy efficiency in staircase design: A biomechanical approach. *Journal of Occupational Ergonomics* , *13*(2), 45–53.

McCauley, M., & Bartram, S. (2016). *Occupational ergonomics: Design and management of work systems*. Springer.

Mital, A., Nicholson, A. S., & Ayoub, M. M. (2014). *Manual materials handling: Psychophysical and physiological considerations*. CRC Press.

Nature. (2023). The role of sleep in occupational performance and safety. *Nature*, *586*(8), 1254–1260.

NIOSH. (1994). Applications manual for the revised niosh lifting equation (DHHS (NIOSH) Publication No. 94-110). Cincinnati, OH: U.S. Department of Health and Human Services, Public Health Service, Centers for Disease Control and Prevention, National Institute for Occupational Safety and Health. https://www.cdc.gov/niosh/docs/94-110/

NIOSH (2015). *Elements of ergonomics programs: A primer based on workplace evaluations of musculoskeletal disorders*. National Institute for Occupational Safety and Health.

NIOSH. (2021). *Ergonomic guidelines for manual material handling*. National Institute for Occupational Safety and Health.

Orwell, G. (1961). *Down and out in Paris and London*. Penguin Books.

OSHA. (2010). Long Work Hours, Extended or Irregular Shifts, and Worker Fatigue. U.S. Department of Labor. Retrieved from https://www.osha.gov/worker-fatigue

OSHA. (2015). *Guidelines for stair and ladder safety.* Occupational Safety and Health Administration.

OSHA. (2016). *Ergonomics and manual material handling.* U.S. Department of Labor. https://www.osha.gov/ergonomics

OSHA (2021). *Preventing work-related musculoskeletal disorders in material handling.* U.S. Department of Labor, Occupational Safety and Health Administration.

OSHA. (2022). *Respiratory protection in construction.* Occupational Safety and Health Administration.

Plowman, S. A., & Smith, D. L. (2014). *Exercise physiology for health, fitness, and performance.* Lippincott Williams & Wilkins.

Rodahl, K. (1989). *The physiology of work* (4th ed.). London: Taylor & Francis.

Russell, P. J., Gregory, D. E., Gaul, C. A., & Braun, W. A. (2007). Evaluation of lifting equation methodologies: A comparative study of five lifting analysis tools. *International Journal of Industrial Ergonomics, 37*(7), 571–577

Saltin, B., & Calbet, J. A. L. (2014). Point: In health and in a normoxic environment, VO2 max is limited primarily by cardiac output and locomotor muscle blood flow. *Journal of Applied Physiology, 97*(2), 458–459.

Salvendy, G. (2014a). *Handbook of human factors and ergonomics.* Wiley.

Salvendy, G. (2014b). *Handbook of industrial engineering: Technology and operations management.* Wiley.

Santos, A., Sousa, I., & Sousa, P. (2018). Exoskeletons in construction: Addressing ergonomic and productivity challenges. *International Journal of Industrial Ergonomics, 66*, 1–10.

Sleep Doctor. (2021). *Fatigue and safety in the workplace: NIOSH lifting equation.* Sleep Doctor.

Sleep Foundation. (2024). *Circadian rhythm disruptions in shift workers.* Sleep Foundation.

SleepScore. (2024). *Enhancing productivity through ergonomic sleep interventions.* SleepScore.

Waters, T. R., Putz-Anderson, V., Garg, A., & Fine, L. J. (1994). Revised NIOSH equation for the design and evaluation of manual lifting tasks. *Ergonomics, 37*(7), 1025–1043.

Westgaard, R. (2014). *Musculoskeletal disorders: Prevention and management in the workplace.* Elsevier.

9 Information Ergonomics, Cognition, Controls, and Displays

9.1 LEARNING GOALS

The learning goals of this chapter are focused on the cognitive aspects of ergonomics, including information processing and measurement of mental workload, as well as an introduction to ergonomic design of controls and displays. The student will be able to perform a subjective analysis of mental workload upon understanding the material contained within this chapter.

9.2 KEY TOPICS

- Information processing
- Information processing models
- Cognition and ergonomics
- Information workload measurement
- Controls and displays

9.3 INTRODUCTION AND BACKGROUND

Mental workload, in the context of occupational ergonomics, refers to the cognitive demands placed on individuals when engaged in work-related tasks. It encompasses both the objective requirements of the task and the subjective perceptions of the individual's effort. Mental workload is influenced by multiple factors, including task complexity, time constraints, environmental conditions, and the worker's cognitive abilities (Longo, 2014). From an ergonomics perspective, understanding these factors is critical to designing jobs that optimize worker performance, enhance task productivity, and minimize mental fatigue and errors.

In ergonomics, mental workload is increasingly viewed as a dynamic interaction between task demands and the operator's cognitive resources. These resources include memory, attention, and the capacity to allocate mental effort, all of which vary among individuals and fluctuate during task execution (Liang et al., 2014). An imbalance can lead to increased mental fatigue, degraded performance, and higher risk of accidents or errors (Charbonnier et al., 2016).

DOI: 10.1201/9780429342806-9

While earlier models emphasized the demands of the task and subjective assessments of difficulty (Cain, 2007), contemporary research in ergonomics considers a broader spectrum of influencing factors. These include the individual's mental state, experience, and external environmental conditions, such as noise, lighting, and workspace design, which can exacerbate cognitive load. The concept of mental overload is particularly important in occupational ergonomics. Mental overload occurs when task demands exceed the operator's available cognitive resources, leading to increased stress, reduced task efficiency, and heightened fatigue (Liang et al., 2014). For instance, identical tasks performed in a noisy environment versus a quiet one may generate different levels of mental workload, demonstrating the impact of situational factors on cognitive demands and task performance (Longo, 2014) .

Understanding and mitigating mental overload is essential for improving safety and productivity in work environments. Ergonomic interventions, such as task redesign, adequate rest breaks, and environmental modifications, are often employed to manage mental workload effectively and reduce the risk of mental overload (Charbonnier et al., 2016).

9.4 INFORMATION PROCESSING

When information is received, it must be compared to and combined with data already stored within the brain. This information is then encoded into memory in order to initiate the decision-making process. This process is influenced by factors such as prior experience, knowledge, cognitive agility, and the ability to process and generate new ideas. Thus, the effective exchange of information between the user and the system is a key component of human-machine interaction, as designers provide precise models of machines to improve user experience and ensure a smooth interaction.

Information processing is a key aspect of both human cognition and machine intelligence. Defined as the interpretation of data through human sensory systems, information is often measured in bits (binary digits), with each bit representing the amount of information needed to choose between two equally probable alternatives. This is represented in the following formula:

$$H = \log 2n$$

where n is the number of equally probable alternatives. The more alternatives involved, the more information is required to make a decision.

Recent advances in human-machine interaction emphasize the importance of emotional and cognitive processing in improving the efficiency and adaptability of these systems. As machine intelligence progresses, it is still largely specialized, whereas human intelligence encompasses broader capabilities, such as social and emotional understanding. While human intelligence remains more versatile, improvements in AI and robotics can enhance technical systems that focus on specialized tasks like logical reasoning and symbol manipulation.

9.4.1 Perception

Perception is the process of becoming aware of or understanding sensory information as it is interpreted. Perceptual processing has two key characteristics: first, it typically occurs automatically and quickly, requiring minimal conscious effort; and second, it is influenced by both sensory input and cognitive factors. Sensory input is analyzed by the brain through what is known as bottom-up processing. This type of processing involves the analysis of environmental stimuli, which is relayed from the sensory receptors via "lower" neural pathways (Goldstein, 2014).

In contrast, when sensory input is unclear or ambiguous, perception may rely more heavily on expectations shaped by prior experiences, leading to top-down processing. Both bottom-up and top-down processes usually operate in tandem to enable rapid and accurate perception (Goldstein, 2014). The level of perceptual processing depends on the complexity of the stimulus and the task being performed.

The simplest form of perception is detection, which involves determining whether a signal or target is present. Signal detection theory is a widely used method to assess perception, especially in scenarios with binary outcomes—such as whether a sound is detectable or not. This theory is particularly useful in evaluating a person's ability to detect stimuli in emergency situations. The possible outcomes of signal detection theory are as follows:

- **Hit:** Correctly detecting a signal when it is present.
- **False Alarm:** Incorrectly identifying a signal when none is present.
- **Miss:** Failing to detect a signal when it is present.
- **Correct Rejection:** Correctly identifying that no signal is present.

9.4.2 Humans as Information Processors

Various theories of information processing aim to describe how information flows through the human mind. Key processes typically included in information processing theories are perception, encoding and decoding, learning, memory and recall, reasoning, decision-making, transmitting information, and executing physical responses. These theories are generally categorized into two broad types: serial and parallel processing.

Serial models conceptualize information processing as a series of discrete, sequential stages where information is processed one step at a time. In contrast, parallel (or distributed/connectionist) models suggest that information is processed simultaneously across multiple pathways, with various processes occurring at different locations within a network of connections.

There are several models used to describe information processing:

- Mathematical models quantify the amount of information processed at various stages.

TABLE 9.1
Reference for Some of the More Common Workload Assessment Techniques

Types of Measures/Techniques	References
Subjective	
Bedford	Roscoe (1987), Roscoe and Ellis (1990)
Modified Harper-Cooper	Wierwille and Gasali (1983)
NASA-TLX	Hart and Staveland (1988)
Psychophysical scaling	Gopher and Braune (1984)
SWAT	Reid and Nygren (1988)
WORD	Vidulich et al. (1991)
Workload profile	Tsang and Velazquez (1996)
Performance	
AGARD STRESS battery	AGARD (1989)
Choice reaction time	Kalsbeek and Sykes (1967), Krol (1971)
Criterion task set	Shingledecker (1984)
Multi-attribute task battery	Comstock and Arnegard (1992)
Time estimation/interval production	Wierwille and Connor (1983)
Mental arithmetic	Brown and Poulton (1961), Harms (1986)
Sternberg memory	Wickens et al. (2004)
Tracking task	Levison (1979), Wickens (1986)
Psychophysiological EEG/ERPS	Biological Psychology (1995), 40 (1 & 2)
Eye blinks	Stern and Dunham (1990)
Heart rate variability	Biological Psychology (1992), 34 (2 & 3)
Respiration	Biological Psychology (1992), 34 (2 & 3)
Analytic	
PROCRU	Baron and Corker (1988)
Queuing theory	Moray et al. (1991)
Timeline analysis	Kirwin and Ainsworth (1992)
TAWL	Hamilton et al. (1991)
TLAP	Parks and Boucek (1989)
VCAP	McCracken and Aldrich (1984)
W/INDEX	North and Riley (1989)

Source: Gavriel, S. (Ed.): *Handbook of Human Factors and Ergonomics*. 1997. Copyright Wiley-VCH Verlag GmbH & Co. KGaA.

- Physical models aim to explain the actual physical processes involved in information processing.
- Structural models outline the major components and systems involved in processing information.

9.4.2.1 Serial Models

In the field of occupational ergonomics, understanding information processing models is essential for optimizing task performance and reducing cognitive load in the workplace. The "stage theory" of information processing, originally proposed by Atkinson and Shiffrin (1968), outlines three key stages in how individuals process and store information. This theory, commonly applied to cognitive ergonomics, helps to explain how workers interact with and respond to stimuli in their environment.

The first stage in Atkinson and Shiffrin's model is *sensory memory*, where sensory input from the environment is captured and stored for a brief duration—usually just milliseconds. This stage is critical in occupational settings where workers must rapidly detect and respond to external stimuli, such as visual signals on a control panel or auditory warnings (Huitt, 2003). A proper ergonomic design ensures that stimuli are presented clearly and effectively, allowing workers to capture and process the necessary information without overloading sensory memory.

The second stage, *short-term memory* (also referred to as working memory), serves as a temporary holding area where information is maintained just long enough to be used. In the workplace, this stage supports real-time decision-making and task execution, such as when an operator must momentarily hold information to adjust equipment settings or solve a problem. However, short-term memory has limited capacity and can be easily overwhelmed, making it critical to reduce unnecessary complexity in task design to prevent cognitive overload (Huitt, 2003).

The final stage, *long-term memory*, is responsible for storing information more permanently. Occupational tasks that require the recall of procedural knowledge or previously learned skills rely heavily on long-term memory. Ergonomic interventions can aid in the development of effective training programs that reinforce long-term retention and recall of essential job-related information (Huitt, 2003).

Another prominent model is *Welford's three-stage information processing model*, which parallels the operations of a computer. In this model, the first stage, *perception*, involves receiving inputs from the environment, similar to sensory memory. In ergonomic contexts, this stage highlights the importance of how information is presented in the work environment to maximize accuracy and efficiency in perception (Jarvis, 1999).

The second stage, *decision-making*, occurs when the perceived information is processed and analyzed. For example, a worker in a control room may analyze incoming data and decide on the appropriate course of action. This stage is critical in time-sensitive tasks, such as emergency response, where decision-making speed and accuracy are paramount (Welford, 1968).

In the third stage, *response*, the worker executes an appropriate action based on the decision-making process. Occupational ergonomics emphasizes the design of tools and systems that allow for quick and accurate responses. Tools should be designed to minimize physical strain while ensuring that workers can perform tasks effectively, even under high-stress conditions (Welford, 1968).

Whiting's three-stage model of information processing is another approach commonly referenced in ergonomic design. This model begins with *sensory inputs* received by perceptual mechanisms, which then pass through *translator mechanisms* where decisions are made. Finally, *effector mechanisms* carry out the necessary actions (Mac, 2010). This model closely aligns with task design in physical work environments, where the interplay between perception, decision-making, and action is vital for efficient performance and safety.

Understanding these models helps occupational ergonomists design systems and work environments that align with human cognitive capabilities, ensuring both worker well-being and productivity. By reducing unnecessary cognitive load and supporting effective information processing, ergonomic interventions can enhance task performance, minimize errors, and prevent mental fatigue in high-demand occupational settings.

9.4.2.2 Parallel Models

In occupational ergonomics, understanding how individuals process information is essential for designing systems that optimize performance, minimize cognitive load, and reduce errors. While traditional *serial models* conceptualize information processing as a step-by-step, linear sequence, *parallel-distributed processing (PDP) models* propose a more complex and efficient approach. PDP models view information processing as occurring simultaneously across multiple interactions within the brain's neural network, a concept first described by Rumelhart and McClelland (1986).

In PDP models, information is processed "through the interaction of a large number of simple processing elements," often referred to as nodes, which work in parallel rather than in sequence. This perspective aligns with the brain's physiological structure, where neural networks process diverse stimuli concurrently. From an ergonomic standpoint, this simultaneous processing offers significant advantages, particularly in complex occupational environments that demand the integration of multiple sources of information (Rumelhart & McClelland, 1986).

One key advantage of the parallel-distributed processing model is its ability to handle complex and simultaneous tasks more efficiently than serial models. In occupational settings, workers often encounter tasks requiring the simultaneous processing of visual, auditory, and cognitive inputs. For example, a control room operator may need to monitor multiple screens, listen to alarms, and make critical decisions under time pressure. A serial information processing model would struggle with these concurrent demands, leading to bottlenecks and delays. PDP models, by contrast, better mirror the brain's natural multitasking capabilities, allowing for the efficient handling of simultaneous and sometimes conflicting information (Rumelhart & McClelland, 1986).

Ergonomic applications of PDP models could include the strategic placement of visual and auditory signals to reduce cognitive overload, or the integration of user-friendly interfaces that allow workers to switch seamlessly between tasks. For example, well-designed dashboards in complex work environments can help operators simultaneously monitor various inputs without becoming overwhelmed, leveraging the brain's ability to engage in parallel processing.

However, it is important to note that individuals do not rely exclusively on either serial or parallel processing. In practice, workers may switch between these strategies based on the task's demands. For simple, well-defined tasks requiring high precision, a serial approach may be more effective. In contrast, when tasks involve high levels of complexity or require the integration of multiple inputs, a parallel processing strategy is often more appropriate (Rumelhart & McClelland, 1986).

This alternation between serial and parallel strategies can be attributed to the level of focus and attention required for different components of task performance. For instance, in a manufacturing setting, a worker may switch from parallel processing while monitoring several machines to serial processing when performing a detailed quality check on a single component. Understanding when and how these processing strategies are deployed can inform ergonomic interventions aimed at improving task performance and reducing mental fatigue.

9.4.3 Channel Capacity Theory

The channel capacity theory explains that a channel can only handle a certain amount of information at any given time. Initially, when the input is small, all information is transmitted efficiently. However, as the input increases, it reaches a threshold where the channel can no longer process everything, and some information gets lost. This concept also applies to human cognitive processing; humans are often bombarded with more information than they can process, leading the brain to filter out less relevant data and focus on what is deemed most important.

9.4.4 Memory

Memory encompasses the brain's processes dedicated to encoding, storing, and retrieving information. Typically, only a small portion of incoming information is retained in the brain. Human memory can be separated into three main categories:

- sensory memory, which briefly holds incoming sensory information;
- working memory, also known as short-term memory, which processes and manipulates information currently in use; and
- long-term memory, where information is stored indefinitely and can be recalled later. Each type plays an essential role in our ability to understand, react to, and remember our experiences.

9.4.4.1 Sensory Memory

Sensory memory acts as a brief reservoir for incoming sensory information, allowing it to persist momentarily after the original stimulus has ended. This system is vital for processing sensory input from different modalities such as vision and hearing. Iconic memory, which pertains to the visual sensory channel, maintains images for a very brief period, generally under a second, aiding in the perception of continuous visual experience despite rapid eye movements and blinking. Echoic memory, related to the auditory channel, can retain sounds for several seconds, which helps in understanding speech and processing auditory sequences.

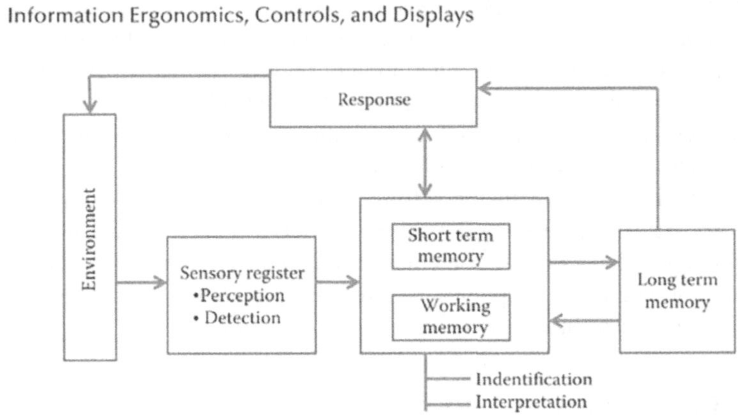

FIGURE 9.1 Components of human memory. (Adapted from Gagne and Driscoll, 1988)

These temporary storage systems are essential for integrating sensory information into coherent perceptions, supporting activities like reading and conversational listening. Iconic and echoic memories play an essential role in the stabilization and transition of sensory inputs into more sustained forms of memory, such as short-term and eventually long-term memory. This process involves not only the simple retention of sensory information but also the selective attention and processing that determines what information is transferred further into the cognitive system.

9.4.4.2 Short Term (Working) Memory

Working memory is a primary element of the cognitive system responsible for the temporary storage and manipulation of information necessary for complex tasks such as language comprehension, learning, and reasoning. It consists of three core components:

1. **Phonological Loop:** This component handles verbal and auditory information. It is subdivided into two subsystems: the phonological store, which retains words and sounds in linguistic form, and the articulatory control process, also known as the articulatory loop, which rehearses this information vocally or sub-vocally to maintain it in an active state.
2. **Visuospatial Sketchpad:** This subsystem stores and manipulates visual and spatial information, representing details like the location, shape, and color of objects in an analog spatial format.
3. **Central Executive:** This is the supervisory component that directs attention and coordinates the activities of the phonological loop and visuospatial sketchpad. It plays an essential role in task management and decision-making by focusing on relevant information while suppressing distractions.

Research has shown that the phonological short-term memory (STM) capacity is a central component aspect of learning new phonological forms such as vocabulary, suggesting a direct link between STM and language acquisition. Additionally, semantic short-term memory may support the learning of new semantic content, such as understanding complex concepts or instructions.

Short-term memory is inherently limited in both duration and capacity. This limitation is famously encapsulated in Miller's Chunking Theory, which suggests that short-term memory can hold about seven (plus or minus two) chunks of information at a time.

According to Miller's Chunking Theory (1956), individuals can retain 7 ± 2 chunks of information in their working memory, depending on how the information is organized.

A chunk refers to any meaningful unit, such as numbers, words, or patterns, which the brain organizes to optimize memory retention. By grouping information into larger, meaningful chunks, individuals can more efficiently process and recall data. This principle has influenced instructional design and memory research, forming the basis for strategies that enhance learning and cognitive load management.

To enhance the retention of information in working memory, rehearsal is essential. Rehearsal can be passive, such as simple repetition, or active, involving deeper cognitive processes like forming associations or creating mental images.

In the context of errors and memory failures, short-term memory is prone to quick decay and interference from similar information, which can lead to forgetting even before the information is consolidated into long-term memory. This is particularly evident in tasks that require maintaining or manipulating several pieces of information simultaneously.

Recent studies continue to explore the intricacies of these components, emphasizing the dynamic and flexible nature of working memory and its critical role in human cognition. This understanding aids in developing strategies for education and cognitive enhancement, as well as interventions for memory-related disorders.

9.4.4.3 Long-Term Memory

Long-term memory (LTM) stores information transferred from working memory. This process involves semantic encoding, where information is associated with existing knowledge to create meaningful connections. LTM is diverse, containing procedural memory (knowledge of how to perform tasks) and declarative memory (knowledge of facts). Declarative memory can further be divided into:

- **Semantic Memory:** General knowledge, like vocabulary and facts.
- **Episodic Memory:** Personal experiences and specific events.

Cognitive research has shown that LTM's stability is remarkable and resilient; it can resist disruptions from brain disturbances or external shocks, maintaining integrity over extensive periods.

9.4.4.4 Deletion

The fading of memories from LTM, often referred to as forgetting, can occur due to decay (the natural loss of a memory over time) and interference (the disruption of a memory by other information). Emotional factors, physical trauma, and intense emotional experiences also play significant roles. Some theories suggest that memories are never truly "deleted" but become difficult to retrieve due to disuse, interference, or other physiological factors.

9.4.4.5 Retrieval

Retrieval from LTM can occur through two primary methods:

- **Recall:** Actively retrieving a memory with minimal external cues, like remembering a home address.
- **Recognition:** Identifying whether presented information matches stored memory, such as recognizing a face in a photograph.

Failures in these processes can lead to performance errors. Forgetting often arises from the lack of external cues or prolonged disuse, exemplifying the "out of sight, out of mind" principle. Additionally, the "recency effect" underscores that more recent events are typically easier to recall due to the fresher memory traces.

9.4.5 Attention

Attention is a cognitive function that facilitates the processing of specific information within our environment. According to Pashler (1997), there are four primary types of attention:

1. **Sustained Attention (Vigilance):** This involves maintaining focus over extended periods, in tasks such as driving or operating machinery. To manage fatigue and maintain vigilance consider the following actions in task design:
    - Implement appropriate work–rest schedules.
    - Enhance the salience of signals (e.g., louder or larger signals).
    - Reduce uncertainty about the timing and location of signals.
    - Provide feedback and artificial signals to keep operators engaged.
    - Offer training and motivation, while monitoring environmental factors.
2. **Selective Attention:** This requires monitoring multiple information sources to focus on a single task, such as distinguishing one instrument in an orchestra. Guidelines include the following activities:
    - Minimize the number of channels to monitor.
    - Clarify the importance of different information channels.
    - Reduce user stress and provide previews of where signals might occur.
    - Train users in effective scanning techniques.

3. **Focused Attention:** This involves concentrating on a single channel or a few channels of information without getting distracted. To enhance focused attention the following strategies in task design are beneficial:
   - Make target channels distinctly different from competing channels.
   - Spatially separate target and competing channels.
   - Reduce the number of competing channels.
   - Emphasize the target channel through size, brightness, volume, or location.
4. **Divided Attention:** This pertains to managing multiple tasks simultaneously, using different sensory modalities, such as reading while watching TV. Recommendations for improving divided attention in task design include the following:
   - Limit the number of information sources.
   - Provide clear priorities for tasks to optimize attention division.
   - Ensure tasks are dissimilar to minimize cognitive overload.
   - Train users in manual tasks to reduce disruptions to mental tasks.

These types of attention and the associated guidelines help optimize performance and reduce errors in cognitive tasks, ensuring that attentional resources are effectively allocated.

9.4.6 Stimuli Response

The study of reaction time is important for both psychologists and ergonomists, offering insights into cognitive processes and workplace design. Reaction time is defined as the duration between the onset of a stimulus and the initiation of a response. This can be a simple physical response to a sensory stimulus like a light flash or a more complex decision-making process involving several potential responses.

9.4.6.1 Types of Reaction Times

1. **Simple Reaction Time:** This occurs when the stimulus is known and only one response is required. Typically, the simple reaction time for visual, auditory, and tactile stimuli is around 0.2 seconds under optimal conditions.
2. **Choice Reaction Time:** This involves a scenario where multiple stimuli could be presented, requiring the selection of the correct response or requiring the individual to make a choice in response selection. The time required increases logarithmically with the number of choices, described by Hicks' Law.

$$\mathrm{RT} = a + b \log_2 N$$

Where:

RT is the reaction time,
a and b are constants dependent on the situation, and
N is the number of choices, with
$p = 1/N$ representing the probability of any particular alternative.

9.4.6.2 Motion Time

Following reaction time is motion time, which refers to the time it takes to initiate and complete the physical movement required in response to a stimulus. Fitts' Law provides a model to calculate motion time based on the distance and precision required for the movement. Fitts Law has a long history of use in control and display design and is mathematically described in the equation below:

Fitts' Law: $MT = a + b \times \log_2(2A/W)$

Where:
MT represents motion time,
A is the distance of the movement, and
W is the width or precision of the target.

The constants *a* and *b* vary based on factors like the body parts involved and the equipment used.

9.4.6.3 Factors Affecting Reaction and Motion Time

Several factors influence the rate at which individuals will react, respond and move. The primary factors are the following:

- **Auditory Factors:** Reaction times can be slower when tones similar to the stimulus are used.
- **Visual Factors:** Variables such as the intensity, duration, and size of the visual stimulus impact reaction times.
- **Tactile Factors:** Tactile response times are generally consistent across different body parts but vary based on the required movement time.
- **Uncertainty and Number of Choices:** Increased uncertainty or a higher number of choices both lead to longer decision times.
- **Aging:** Reaction times tend to be slower in individuals younger than 15 or older than 60.

9.5 MENTAL WORKLOAD MEASUREMENT

Cognitive science has shown mental capacity to be a limited resource, and thus workload is defined as the portion of these resources expended in performing a specific task. Various models exist to measure workload in the occupational ergonomics field. Young et al. (2015) explored drivers' mental workload while performing secondary tasks alongside the primary task of driving. Cognitive ergonomics or information processing research has shown that mental workload can be determined using the following approaches:

- Measures of primary and secondary task performance.
- Objective physiological measurements.
- Psychophysical assessment.

- Subjective assessment approaches.

Additional commonly used methods for workload assessment and their relative references are listed in the references section. According to Hart (2006) and updated by Young et al. (2015), these methods provide a comprehensive framework for understanding and measuring mental workload in various contexts (Young et al., 2015).

9.5.1 Measures of Primary and Secondary Task Performance

Primary task performance assesses an individual's ability to perform the main task, assuming known minimum and maximum capacity levels. As workload increases, notable changes in performance occur, monitored through secondary tasks. Secondary task performance evaluates the residual capacity left after primary task resource allocation. Misallocation of these resources can lead to diminished primary task performance. Recent research by Matthews, et al. emphasizes the dynamic interaction between primary and secondary task performances and their impact on overall workload management.

9.5.2 Physiological Measures

Physiological measures, including heart rate, eye movement, pupil diameter, muscle tension, electroencephalogram (EEG), electrooculogram (EOG), and electrocardiogram (ECG), are useful in assessing mental workload. These measures assume that an individual's arousal levels vary in response to changes in workload or stress. While these physiological indicators offer critical insights, they present challenges in collection and interpretation. When properly analyzed, however, these measures provide valuable information, such as physiological responses to increasing workload indicative of stress. Recent studies by Charles and Nixon (2019) further delve into the complexities and utility of these physiological measures in assessing mental stress under varying workloads.

9.5.3 Psychophysical Assessment

Psychophysics explores the connection between the physical characteristics of stimuli and their subjective perception by an individual. Recent advances emphasize that individuals can discern the strain their bodies endure during physical or mental exertion and can make comparative assessments of the perceived effort (Mehta & Agnew, 2015). The use of psychophysical scaling methods enables individuals to quantify the perceived intensity of exerted effort, both mental and physical, thereby enhancing workload assessments and ergonomic planning to improve safety and efficiency in various environments (Parasuraman & Rizzo, 2017).

9.5.4 Subjective Workload Assessment

Subjective assessments utilize questionnaires and employee feedback to gauge perceived workload. While these approaches to workload assessment have been useful in some cases, they are often criticized. The criticism is largely due to the fact that they do not offer real-time evaluations when conducted post-task, and may disrupt performance or inadvertently increase workload if administered during the task. A summary of these subjective approaches is detailed in Table 9.2 (Hart, 2018a).

Subjective assessment approaches can offer distinct advantages, and the specifics of the evaluation should be carefully considered when selecting an assessment approach. Three widely recognized subjective workload assessment methods in human factors are the NASA Task Load Index (TLX), the Subjective Workload Assessment Technique (SWAT), and the Modified Cooper-Harper Scale (Hancock & Warm, 2019).

9.5.4.1 NASA Task Load Index (TLX)

The NASA Task Load Index is a multidimensional subjective workload rating tool that calculates an overall workload score from a weighted average of ratings across six subscales: mental demands, physical demands, temporal demands, performance, effort, and frustration (Hart, 2018b).

This overall workload score is derived from the weighted averages of these six dimensions (Byers et al., 2019). The TLX is adaptable for assessing workload across various human-machine environments, including aircraft cockpits, command, control, and communication (C3) stations, supervisory and process control settings, as

TABLE 9.2
Subjective Assessment Approaches

	Approaches		
Subjective Rating Instruments	**Unidimensional vs. Multidimensional**	**Absolute vs. Relative**	**Immediate vs. Retrospective**
Bedford	U	A	I
MKH	U	A	I or R
Psychophysical	U	RS	R
SWORD	U	RR	R
NASA-TLX	M	A	I
SWAT	M	A	I
Workload profile	M	A	R

Source: Gavriel, S. (Ed.): *Handbook of Human Factors and Ergonomics*,1997. Copyright Wiley-VCH Verlag GmbH & Co. KGaA.

Notes: MCH, Modified Cooper-Harper; U, unidimensional; M, multidimensional; A, absolute rating; RS, relative rating to a single standard; RR, redundant relative rating; I, immediate; R, retrospective.

TABLE 9.3
NASA-TLX Rating Scale Definitions

Workload Component	Endpoints	Description
Mental demand	Low/high	How much mental and perceptual activity is required (e.g., thinking, deciding, calculating, remembering, looking, searching, etc.)? Was the task easy or demanding, simple or complex, exacting or forgiving?
Physical demand	Low/high	How much physical activity is required (e.g., pulling, pushing, turning, controlling, activating, etc)? Was the task easy or demanding, slow or brisk, slack or strenuous, restful or laborious?
Temporal demand	Low/high	Did you feel the time pressure due to the rate or pace at which the task or task elements occurred? Was the pace slow and leisurely or rapid and frantic or somewhere in between?
Effort	Low/high	How hard did you have to work mentally and physically to accomplish your level of task performance?
Frustration level	Low/high	How insecure, discouraged, irritated, stressed, and annoyed vs. secure, gratified, content, relaxed, and complacent did you feel during the task?
Performance	Good to poor	How successful do you think you were in accomplishing the goals of the task set by the experimenter (or yourself)? How satisfied were you with your performance in accomplishing these goals?

Source: Taken directly from Hart, NASA AMES Research Center, 2007

well as simulations and laboratory tests. Current software developments have further refined this approach for measuring mental workload and are available for use without charge (Matthews et al., 2015).

9.5.4.2 Subjective Workload Assessment Technique (SWAT)

The Subjective Workload Assessment Technique remains a prominent tool for measuring mental workload, incorporating dimensions such as time load, mental effort, and psychological stress. The SWAT is based on the additive multidimensional model of mental workload (Reid & Nygren, 1988). SWAT has been used in a number of studies and measures workload in three dimensions including time load, mental effort load, and psychological stress load, where each of these factors is measured on a scale with threelevels. Subjects are asked to rank the 27 possible combinations of levels on the threescales before providing ratings for particular tasks or events. SWAT has been used in diverse environments, for example: high-G centrifuge; command, control, and com-munications centers; nuclear power plants; domed flight simulators; task simulators;real flight; and laboratory settings (Potter & Bressler, 1994)

This multidimensional assessment approach has been detailed comprehensively in various user guides, outlining the development phases and scoring processes used in diverse settings like flight simulators and laboratory environments.

9.5.4.3 Modified Cooper and Harper Scale

The Modified Cooper and Harper (MCH) Scale, primarily developed for assessing workload in cognitive-intensive systems, continues to be a relevant tool in various fields beyond its initial aviation focus. It features a decision tree and a ten-point scale, where a rating of 1 indicates minimal effort and 10 signifies an impossible task. The scale's flexibility and adaptability have allowed for its integration with other assessment tools like the NASA-TLX, enhancing its utility in comparing subjective and physiological indicators of workload across different tasks and environments.

Recent adaptations have modified the MCH scale for specific applications such as unmanned vehicle operations, tailoring it to better measure the cognitive demands associated with modern automated systems. These studies highlight the scale's consistent sensitivity and reliability, underscoring its continued evolution to meet contemporary ergonomic and human factor research needs

9.5.4.4 Selection of a Subjective Assessment Tool

Subjective measures have gained recognition as critical tools in system evaluations, particularly for assessing operator workload. These methods are lauded for their reliability in predicting mental workload and are favored for their practical benefits such as ease of implementation and non-intrusiveness. Modern studies continue to underscore the significance of these measures in providing precise and sensitive assessments of operator load. Building upon the foundational work of Eggemeier et al. (1991), more recent guidelines emphasize several essential criteria for evaluating the effectiveness of procedures designed to assess mental workload:

1. **Sensitivity:** The ability of the measure to discern variations in task difficulty or demands.
2. **Diagnosticity:** The capacity of the tool to pinpoint changes in workload and offer insights into the causes of these changes.
3. **Selectivity/Validity:** The requirement that the measure should respond solely to variations in cognitive demands, and not be affected by other factors such as physical workload or emotional stress that are unrelated to the mental workload of the specific task.
4. **Intrusiveness:** The importance of ensuring the measure does not disrupt or interfere with the primary task or activity being evaluated.
5. **Reliability:** The necessity for the measure to consistently and reliably assess mental workload.
6. **Implementation Requirements:** The ease of implementation concerning time, instruments, and software necessary for data collection and analysis.
7. **Subject Acceptability:** The perception by subjects of the validity and usefulness of the procedure.

More recent research studies underscore the continued relevance and refinement of these subjective measures in system evaluations, highlighting their adaptability to various operational settings and advancements in cognitive assessment technologies. Notably, Matthews et al. (2015) provide a comprehensive review of subjective workload assessment techniques, affirming their critical role in ergonomic and human factors research (Matthews, et al., 2015) .

9.6 CONTROLS AND DISPLAYS

During World War II's initial years, the US faced over 2,000 significant multi-engine aircraft accidents. In response, US Air Force human factors researchers conducted a laboratory study to identify the optimal configuration of control shapes in aircraft crew stations. This research aimed to enhance the identifiability and distinctiveness of controls and to ensure consistency across different aircraft models. The findings from this study influenced the standardization of control arrangements in the late 1940s and 1950s, pinpointing the instruments most critical to safe flight. The standardized control design implemented across aircraft not only led to a substantial decrease in system-induced pilot errors but also bolstered overall aviation safety. Additionally, this seminal study laid the groundwork for ongoing research and development in aircraft control design, shaping modern aviation practices that continue to prioritize pilot ease-of-use and safety.

9.6.1 Controls

Controls, which are termed as activators in ISO standards, are essential components that transmit inputs to machinery. These controls can be manually operated by hand or foot, and the outcomes of these inputs are conveyed to the operator through displays, indicators, or the resulting actions of the equipment. The roles of these controls encompass a range of functions such as starting or stopping a machine, discrete settings like channel selection on a television or data entry via a keyboard, quantitative adjustments such as temperature regulation on a thermostat, and continuous operations like steering a vehicle. For optimal task performance, it is important that the chosen type of control aligns with common or population-based expectations. For instance, a button, rather than a lever, should be used for activating a horn due to conventional usage patterns. Additionally, the size and motion of the control should reflect common experiences and past practices; for example, a steering wheel is generally preferred over dual levers for driving a car. The direction of operation should also meet standard expectations, as seen in throttle controls where pushing forward typically increases speed, and pulling back decreases it. Fine manipulations and operations requiring minimal force are best handled by hand, whereas tasks needing greater force are more suitably performed with foot controls. Safety considerations must be integral to both the design and maintenance of these controls (Helander, 2014).

9.6.2 Guidelines for Control Layout and Design

The election of control layout and design should consider cognitive factors, human expectations, and occupational ergonomic norms. The following are widely researched guidelines used in the arrangement and grouping of controls to enhance usability and ergonomics:

1. **Locate for Ease of Operation:** Controls should be placed conveniently for easy access and operation.
2. **Orientation Relative to Operator:** Controls should be oriented based on the operator's typical position. If the operator's position varies, the controls should adjust accordingly.
3. **Consistent Arrangement Across Operators:** In each position, the layout of controls should be uniform across all operators to prevent confusion.
4. **Prioritize Primary Controls:** The most important and frequently used controls should be positioned optimally for ease of operation and reach.
5. **Group Related Controls:** Controls should be grouped logically:
   - Controls that are used sequentially.
   - Controls related to a specific function.
   - Functions that are typically operated simultaneously.
6. **Arrange for Sequential Operation:** Controls used in a specific sequence should be arranged to facilitate this order. A common arrangement follows the western reading pattern: left to right and top to bottom.
7. **Ensure Consistency:** The arrangement of functionally identical or similar controls should be consistent across different panels.
8. **Dead-Man Control:** Systems should be designed to shut down automatically if the operator becomes incapacitated or loses contact with the controls.
9. **Guard Against Accidental Activation:** Measures such as shields or specific force requirements should be used to prevent inadvertent activation, although these can decrease operation speed.
10. **Balance Density and Accessibility:** Controls should be closely packed yet not crowded, balancing ease-of-use with feedback to prevent unintentional activations caused by vibrations or sudden movements.
11. **Appropriate Resistance and Feedback:** The force required to activate a control should be minimal yet sufficient to provide feedback, depending on the control's design and the frequency of use.
12. **Choice of Control Type:** Continuous controls are ideal for operations needing adjustment within a range (like a stereo volume control), whereas detent controls (like a car's gear shift) are better for discrete settings where the control rests at specific points.

These principles ensure that the interface is intuitive, reducing the cognitive load, improving performance, and enhancing safety. This approach is supported by modern ergonomic research which suggests that the design of control interfaces should cater to user expectations and habitual usage to optimize functionality and safety (Stanton et al., 2013).

9.6.3 Types of Controls

The following section provides an overview of the terminology used in the occupational ergonomics community to describe or classify different types of controls.

Discrete Controls (Detent):

1. **Key Lock or Key Operated Switches:** Prevent unauthorized use.
2. **Bar Knob or Rotary Selector Switches:** Suitable for operations requiring multiple set positions.
3. **Detent Thumbwheel:** Compact device for stepwise input.
4. **Push Button:** Switches conditions, enters commands, or releases locks; offers momentary or sustained contact.
5. **Push–Pull Switch:** Typically used for on/off settings.
6. **Legend Switch:** Displays qualitative status information needing operator attention.
7. **Toggle Switch:** Ideal for operations requiring two distinct positions.
8. **Rocker Switch:** Similar to toggle switches but less protrusive from the panel.

Continuous Controls:

1. **Knob:** For precise, low-force adjustments.
2. **Crank:** Suitable for multiple rotations and moderate force applications.
3. **Hand Wheel:** Two-handed operation for tasks requiring significant force.
4. **Lever:** For large force or displacement in continuous adjustments or multi-directional movements.
5. **Continuous Thumbwheel:** Compact alternative to knobs.
6. **Slide:** Facilitates continuous settings.

Foot-Operated Controls:

- Best used for binary (on/off) conditions requiring high force or minimal precision. While sometimes used for fine adjustments, this is generally not recommended due to control challenges.

9.6.4 Coding of Controls

Control coding is a vital aspect element in human factors design, enhancing the differentiation and recognition of controls in systems with multiple mechanisms. There are six primary types of control coding.

1. **Location:** The positioning of controls is highly influential. Controls that serve similar functions should be consistently located, as seen in the standardized placement of many vehicle controls across different countries, aligning with driver expectations for consistency.

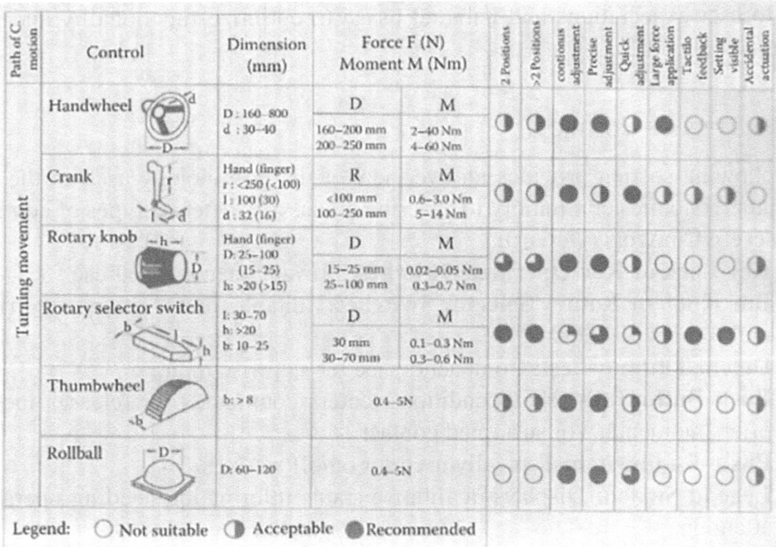

Path of C. motion	Control	Dimension (mm)	Force F (N) Moment M (Nm)		2 Positions	>2 Positions	contionus adjustment	Precise adjustment	Quick adjustment	Large force application	Tactile feedback	Setting visible	Accidental actuation
Turning movement	Handwheel	D: 160–800 d: 30–40	D 160–200 mm 200–250 mm	M 2–40 Nm 4–60 Nm	◑	◑	●	●	◑	●	○	○	◑
	Crank	Hand (finger) r: <250 (<100) l: 100 (30) d: 32 (16)	R <100 mm 100–250 mm	M 0.6–3.0 Nm 5–14 Nm	◑	◑	●	◑	●	◑	◑	◑	○
	Rotary knob	Hand (finger) D: 25–100 (15–25) h: >20 (>15)	D 15–25 mm 25–100 mm	M 0.02–0.05 Nm 0.3–0.7 Nm	◕	◕	●	●	◑	○	○	○	◑
	Rotary selector switch	l: 30–70 h: >20 b: 10–25	D 30 mm 30–70 mm	M 0.1–0.3 Nm 0.3–0.6 Nm	●	●	◔	◕	◔	◑	●	●	◑
	Thumbwheel	b: >8	0.4–5N		◑	◑	●	●	◑	○	○	○	◑
	Rollball	D: 60–120	0.4–5N		○	○	●	●	◕	○	○	○	◑

Legend: ○ Not suitable ◑ Acceptable ● Recommended

FIGURE 9.2 Hand- and foot-operated control devices and their operational characteristics and control functions. Source: Sanders, M. S., & McCormick, E. J. (1993). *Human Factors in Engineering and Design* (7th ed.). McGraw-Hill.

2. **Color:** Color coding is effective in well-lit conditions but can require longer reaction times as operators must interpret color meanings before acting. Common associations, such as using red for emergency functions, should be maintained to meet stereotypical expectations.
3. **Size:** Different sizes can help distinguish controls. To avoid cognitive overload, it is advised to use no more than three different sizes. Controls with identical functions across various devices should be consistently sized.
4. **Shape:** The shape of a control can indicate its function, making shape an effective coding method.
5. **Labeling:** Labels provide direct information about controls but require adequate lighting to be useful. Labels should be positioned for easy visibility and readability, avoiding vertical orientations which are harder to read. Symbols often offer clearer communication than text alone.
6. **Mode of Operation:** Controls should have distinct operational feels that match their functions, consistent with user expectations based on common stereotypes.

Coding methods can be mixed, but designers must be cautious as combinations can create new interpretations. Coding also serves as a redundancy measure; for example, traffic lights use both color and positional codes to convey information clearly and effectively.

9.6.5 Emergency Controls

In emergency situations where stress can lead to forgetfulness or panic, the ergonomic design of emergency controls is paramount. Here are refined design guidelines for emergency controls that enhance safety and efficacy:

1. **Separation from Regular Controls:** Emergency controls should be distinctly positioned away from frequently used controls to prevent accidental activation.
2. **Accessibility:** Emergency controls must be easily reachable for all operators, regardless of their physical position or situation.
3. **Visibility and Size:** Controls should be large enough to be quickly identified and operated, even under stressful conditions.
4. **Activation Force:** These controls should require moderate force to activate—enough to prevent accidental usage but not so much that they're difficult to use in emergencies.
5. **Color Coding:** Color them red, a universally recognized color for caution and emergencies, to enhance quick recognition.
6. **Special Features for Industrial Settings:**
    - **Dead-Man's Switch:** Incorporate systems like the dead-man's switch, which turns machinery off if the operator stops actively engaging with the control, such as when incapacitated.
    - **Automatic Activation:** Include sensors that trigger the emergency controls automatically under specific conditions, such as trip cords or emergency signals, to ensure rapid response to potential hazards.

These recommendations aim to optimize emergency control designs to be foolproof and efficient, significantly mitigating risks during high-pressure scenarios. Implementing these principles helps ensure that emergency systems are both intuitive and reliable, safeguarding both operators and equipment.

9.6.6 Displays

Displays play a vital role in providing operators with essential information about the status of the work environment, vehicles, or equipment. They facilitate interaction by presenting data in various sensory forms—visual, auditory, and tactile. Each type of display is chosen based on the specific needs of the task, the operator, and established expectations. Here are some key principles for effective display design:

1. **Clarity and Perception:** Information should be presented in a manner that is straightforward and easy for operators to perceive and interpret quickly.
2. **Relevance of Information:** Only display the information that is necessary for the performance of the task. Overloading displays with irrelevant data can lead to confusion and slow down decision-making.

3. **Accuracy:** Provide information with the level of accuracy required for the operators' decisions and actions. Overly precise information can be as detrimental as imprecise data if it complicates the user's task.
3. **Simplicity and Usability:** Information should be straightforward, simple to understand, and easy to use. Complex displays can hinder quick comprehension and responsiveness.
4. **Fail-Safe Design:** It must be immediately obvious if a display fails or malfunctions. This ensures that operators are quickly aware of any issues and can take appropriate action without delay.

9.6.7 Types of Displays

The choice of display type depends heavily on the environmental conditions and the requirements of the operator. The following provides an overview of how auditory and visual displays can be optimized based on their operational context.

9.6.7.1 Auditory Displays

Auditory displays are ideal in scenarios where visual attention is compromised, such as in poorly lit environments or when the operator needs mobility. The design of auditory displays should adhere to the following principles:

- **Simplicity and Urgency:** Sounds should be simple, brief, and convey urgency to capture immediate attention.
- **Differentiation:** Auditory signals must be easily distinguishable from each other and from background noise to prevent confusion.
- **Volume and Frequency:**
    - The volume should exceed ambient noise by at least 10 dB to ensure audibility.
    - Optimal signal frequency ranges from 200 to 5,000 Hz, with the most effective range being 500 to 1,500 Hz for clarity and distinction.
    - Lower frequencies (below 500 Hz) are preferable for long-distance transmission and when signals need to circumnavigate sound barriers.
    - Enhancing the conspicuity of tones can be achieved by varying intensity, altering the pattern of repetition, or changing the frequency.

9.6.7.2 Visual Displays

Visual displays are more suited to environments where noise levels are high and the operator remains stationary relative to the display. Visual information can be complex and relate to spatial details, categorized into four types:

1. **Check Display:**
    - Indicates the presence or absence of a condition.
    - Generally provides limited detail.
2. **Qualitative Display:**

- Shows an approximate value or the trend of a changing variable.

3. **Quantitative Display:**
   - Provides precise numerical values or detailed information that requires careful reading.
4. **Electronic or Video Display:**
   - Highly versatile, capable of presenting static or dynamic content.
   - Can incorporate elements of check, qualitative, quantitative, and video displays in various combinations.

Each type of display has its specific applications and must be tailored to fit the environmental conditions and the informational needs of the operator.

9.6.7.3 Guidance on Color Coding in Displays

When designing displays that incorporate color, it's important to use colors effectively to ensure that information is easily recognizable and interpretable by users. Some refined guidelines for using colors in displays are as follows:

1. **Limit the Number of Colors:**
   - For inexperienced users or infrequently used displays, limit the palette to no more than four colors.
   - In general, do not exceed seven colors in a single display to avoid overwhelming the user.
2. **Color Wavelength Separation:**
   - Choose colors that are distinctly different in wavelength to ensure each is clearly distinguishable from the others.
   - This separation helps prevent confusion, especially in critical information displays.
3. **Avoid Similar Shades:**
   - Do not use multiple shades of the same color as this can lead to confusion and difficulty in distinguishing information quickly.
4. **Suggested Color Combinations:**
   - To maximize clarity and visibility, consider using these recommended combinations:
     - **Green, Yellow, Orange, Red, and White:** This combination covers a broad spectrum and is useful for indicating various levels of urgency or status.
     - **Blue, Cyan, Green, Yellow, and White:** Ideal for displays requiring a calm aesthetic but with necessary differentiation, such as in medical or aquatic environments.
     - **Cyan, Green, Yellow, Orange, and White:** Effective for environments that need a clear distinction between normal operations and alert states.

9.6.7.4 Emergency Displays

For emergency displays and signals to be effective, they must capture attention immediately and convey urgency effectively. Useful guidelines to enhance the visibility and impact of such emergency displays are provided below:

1. **Multimodal Alerts:**
   - Pair visual signals with auditory alerts or flashing lights to ensure they catch the operator's attention even in noisy or visually cluttered environments.
2. **Size and Visibility:**
   - Emergency displays should be larger than general status indicators to stand out during critical situations.
   - Ensure a luminance contrast ratio of at least 3:1 against the immediate background to make the displays easily discernible.
3. **Flashing Specifications:**
   - Set the flash rate of emergency lights between 3 to 5 pulses per second. This rate is optimal for drawing attention without being misinterpreted as a less urgent signal.
   - In the event of a flasher device failure, the light should remain on steadily to continue signaling the presence of an emergency condition.
4. **Clear Messaging:**
   - Accompany visual and auditory signals with explicit warning messages such as "Danger" or "System Failure." These messages should be simple, direct, and unmistakable to communicate the severity of the situation immediately.

9.6.7.5 Electronic Displays

Electronic displays revolutionized information presentation in the 1980s, introducing capabilities far beyond those of mechanical displays. However, their complexity demands careful consideration of human factors to ensure they effectively communicate information without overwhelming the operator. Here's a breakdown of essential aspects to consider in the design of electronic displays:

1. **Geometry:**
   - **Size:** The display size should be appropriate for the viewing distance and the type of information presented, ensuring all text and graphics are easily readable.
   - **Viewing Angle:** Design the display to be easily viewable from different angles, accommodating various operator positions to prevent posture-related strain and enhance visibility.
2. **Spatial Qualities:**
   - **Imaging Capacity:** The display should have high resolution to render detailed images clearly.
   - **Raster Modulation:** Manage the raster (the pattern of image data on the screen) to minimize distortion and enhance clarity.

- **Pixel Faults:** Minimize pixel errors as these can lead to misinterpretation of information.
- **Distortion:** Ensure the display geometry does not distort images, which could mislead the operator.

3. **Temporal Qualities:**
   - **Image Formation Time:** The time it takes for an image to form on the display should be quick enough to not delay information delivery.
   - **Display Time:** Information should remain on the screen for a sufficient duration to ensure it can be comfortably read and understood.
   - **Flicker:** Minimize flicker to reduce eye strain and potential headaches, particularly in environments where operators must focus on screens for extended periods.
4. **Photometry and Colorimetry:**
   - **Illuminance:** The display should be well-lit, matching the ambient light conditions to reduce eye strain and enhance readability.
   - **Contrast:** High contrast between the foreground and background is useful for quick and easy reading of the display.
   - **Color:** Use colors judiciously to aid in information categorization and prioritization but ensure they are distinguishable by all users, including those with color vision deficiencies.

9.7 TECHNOLOGY IMPACT ON CONTROL AND DISPLAY DESIGN

New technologies in control display design have significantly enhanced the usability and functionality of various devices across different environments. These advancements have introduced a new generation of displays that are more efficient, adaptable, and user-friendly. Here's a look at some key products benefiting from these technologies:

1. **Flat Screen Monitors:** Utilizing LED backlighting, these monitors offer sharper images and more vibrant colors while being energy efficient and thinner than traditional CRT monitors.
2. **Handheld Displays:** Personal Digital Assistants (PDAs) and similar devices have been transformed with touch screens, providing more interactive and accessible user interfaces.
3. **Electronic Books:** E-readers use e-ink and other advanced display technologies to simulate the appearance of ink on paper, which is easier on the eyes and readable in various lighting conditions.
4. **Tactile Displays:** These provide feedback through touch, often used in devices for visually impaired users or in applications where direct physical interaction with the device is preferable.
5. **GPS Navigational Displays:** Advanced LEDs and touch screen technologies have greatly enhanced the clarity and responsiveness of GPS displays, making them integral in vehicles and portable electronics.
6. **Virtual Reality Displays/Head Mounted Displays (HMD):** These incorporate sophisticated display technologies to provide immersive visual

experiences, significantly impacting gaming, training simulations, and even therapeutic settings.

9.7.1 Touch Screens

Touch screen technology has revolutionized how users interact with a wide range of devices, becoming a fundamental component in modern digital interactions. From smartphones, laptops, and iPods to a vast array of other consumer electronics, touch screens have expanded their influence significantly. Below is an overview of how touch sensor technology has been integrated into various aspects of daily life and potential areas for additional research and development:

9.7.2 Integration Across Sectors

1. **Assistive Technology:** Touch screens offer customizable interfaces that greatly enhance accessibility for users with disabilities. Features like screen magnification, voice-over text, and adjustable touch sensitivity make digital devices more accessible.
2. **Retail Environments:** In retail, touch screens facilitate interactive customer service experiences, including self-service kiosks that streamline transactions and inventory checks, enhancing customer independence and operational efficiency.
3. **Entertainment:** The entertainment industry has been transformed by touch technology, with devices offering intuitive gaming experiences, interactive displays in museums, and digital panels in theaters.
4. **Healthcare:** Touch screens in healthcare settings help manage patient data more efficiently, allowing for quick updates and access to health records, thus improving workflow and patient care.
5. **Education:** In educational settings, touch screens are used in devices that support interactive learning, enabling features like digital textbooks and collaborative projects.

9.7.3 Need for Ergonomic Research

Despite their widespread adoption, the long-term ergonomic implications of touch screens are not fully understood. Research areas that could be explored include:

- **Posture and Physical Strain:** Investigating how prolonged use of touch screens affects posture and whether it contributes to strain injuries, particularly in the neck, shoulders, and wrists.
- **Cognitive Load:** Examining how the interface design of touch screens impacts cognitive load and how this can be optimized to reduce mental fatigue.
- **Visual Strain:** Assessing the impact of long-term use of touch screens on eye health, given the proximity and duration of screen exposure.

- **Hygiene Factors:** Considering the health implications of frequent contact with touch screens, especially in public and healthcare settings, and identifying best practices for hygiene.
- **Accessibility Improvements:** Continuously improving touch screen technology to enhance accessibility for users with various disabilities, ensuring that technological advancements do not inadvertently exclude users.

The ongoing integration of touch screen technology across multiple domains underscores the need for comprehensive research to understand its long-term impacts better and to develop guidelines that maximize benefits while minimizing potential adverse effects.

9.8 SUMMARY

Understanding the principles of information processing, mental workload, and control/display design guidelines is essential for the effective development of human-machine systems in ergonomics. By applying these guidelines, developers can create systems that are well-suited to operator capabilities, thereby reducing the likelihood of errors, enhancing safety in task performance, and promoting greater efficiency. This involves managing cognitive load by designing systems that provide the right amount of information without overwhelming the user, ensuring that attention distribution is optimized. Interfaces should be intuitive, allowing for easy navigation and quick access to necessary information to help minimize cognitive strain. Additionally, consideration of the mental workload is crucial; systems should be designed to simplify tasks and reduce the mental effort required to operate them. This not only improves user satisfaction but also boosts productivity by enabling more efficient interactions with the system. Ultimately, the integration of these ergonomic principles into the design and operation of human-machine systems serves to foster a safer and more productive environment.

9.9 CASE STUDY

Problem Overview

Air traffic control (ATC) is a high-stakes environment where controllers manage multiple aircrafts at once, ensuring safety, efficiency, and adherence to flight schedules. The role of an air traffic controller is heavily dependent on information processing, requiring the individual to process a high volume of visual and auditory information while making critical decisions in real-time. The challenge lies in ensuring that the mental workload imposed on air traffic controllers is neither too high, leading to errors and fatigue, nor too low, which could result in reduced vigilance.

Ergonomic Design Process

The design of air traffic control systems involves applying principles of cognitive ergonomics, particularly focusing on information processing, measurement of mental workload, and ergonomic design of controls and displays. These elements are critical for maintaining an optimal level of performance.

1. **Information Processing:** Air traffic controllers must process data from various sources, such as radar screens, flight plan information, and communications with pilots. Cognitive ergonomists designed the system to minimize unnecessary information and focus the controller's attention on the most relevant data. Visual displays were optimized for clarity, ensuring that all important data is easily accessible and reducing the time spent interpreting information.
2. **Measurement of Mental Workload:** To evaluate the mental workload of controllers, subjective methods such as the NASA-TLX (Task Load Index) were used. Controllers rated their perceived mental workload across dimensions like mental demand, physical demand, and frustration. This information helped ergonomists identify points of excessive workload and provided insights into how system design could be adjusted to reduce cognitive load. For example, simplifying data presentation or automating routine tasks helped in balancing workload.
3. **Ergonomic Design of Controls and Displays:** The ergonomic design of controls and displays is critical to reducing mental strain. The ATC system's control panels were designed for intuitive use, minimizing the need for controllers to switch between devices or reach for controls. Displays were color-coded and used contrast and size effectively to guide the user's attention toward urgent information. Additionally, auditory alerts were implemented to quickly draw attention to critical issues without overwhelming the controller.

Outcome

Through these ergonomic interventions, the air traffic control system saw a significant reduction in controller errors and improved response times. Controllers reported lower levels of perceived mental workload, particularly during peak traffic periods, due to improved information clarity and workload distribution. The integration of automated systems further reduced the need for constant manual intervention, allowing controllers to focus on higher-priority tasks. Overall, the changes led to a safer, more efficient, and less stressful working environment.

Conclusion

This case study demonstrates how ergonomic principles, particularly those related to cognitive ergonomics, can significantly improve performance in high-stress environments such as air traffic control. By focusing on information processing, mental workload measurement, and the ergonomic design of controls and displays, the system becomes not only more efficient but also safer for users. Understanding these concepts enables the conceptualization of the process of subjective analyses of mental workload in various settings, contributing to better system designs across industries.

Exercises

1. Describe the elements of the human information processing model presented in this chapter.
    - Conduct an internet search for variations of human information processing models in cognitive psychology and ergonomics.
    - Use AI tools to summarize how different models compare and evolve.
    - Interview a cognitive scientist, psychologist, or ergonomist to gain practical insights.
    - Create a visual representation of the model and discuss real-world applications in different industries.
2. Explain signal detection theory and the potential outcomes.
    - Use AI tools and online literature databases (e.g., Google Scholar, PubMed) to explore research articles on signal detection theory (SDT).
    - Search for real-world applications of SDT in areas such as aviation, healthcare, or cybersecurity.
    - Conduct a short experiment (e.g., testing how quickly people detect signals in a noisy environment) and analyze the results.
    - Prepare a case study on SDT's importance in decision-making.
3. Describe the different types of approaches to study or measure mental workload.
    - Research different methods of assessing mental workload using academic databases and AI-generated summaries.
    - Compare subjective (e.g., NASA-TLX) and objective (e.g., heart rate variability, EEG) measures.
    - Interview professionals in high-workload environments (e.g., air traffic controllers, medical professionals) to understand how workload is managed.
    - Create a chart comparing various workload measurement approaches and their advantages/disadvantages.
4. Explain Miller's Chunking Theory and provide an example.
    - Search the literature for studies on Miller's Chunking Theory and its applications in memory and user interface design.
    - Use AI tools to generate modern examples of chunking in digital interfaces, education, and cognitive load management.
    - Conduct an informal experiment where participants recall information in chunked vs. non-chunked formats.
    - Present findings in a short report or infographic.

5. Explain the NASA-TLX approach to subjective workload assessment.
    - Research NASA-TLX applications in different fields such as healthcare, aviation, and software usability.
    - Use AI tools to simulate a workload analysis based on NASA-TLX.
    - Conduct an interview with a human factors professional who has used NASA-TLX in practice.
    - Design and implement a mini-study where participants rate their workload using NASA-TLX while performing a task.
6. Describe the different types of attention and how tasks should be designed to optimally use human attention capacities.
    - Conduct an internet search on selective, divided, sustained, and alternating attention.
    - Use AI tools to summarize best practices for designing tasks that optimize attention.
    - Interview professionals in high-attention-demand occupations (e.g., emergency responders, pilots) about strategies they use to manage attention.
    - Design an infographic illustrating the types of attention and recommendations for task design.
7. Provide a detailed explanation of Fitts' Law.
    - Research how Fitts' Law applies to human-computer interaction, sports, and industrial tool design.
    - Use AI tools to analyze the relationship between movement time, target size, and distance.
    - Interview a UX designer or ergonomist about how Fitts' Law is applied in product design.
    - Conduct a simple experiment to test Fitts' Law using a pointing task and document the results.
8. Explain the difference in perception, reaction time, and response time.
    - Search for cognitive psychology studies explaining these differences.
    - Use AI tools to generate real-world examples across different fields (e.g., sports, driving, workplace safety).
    - Interview an expert in cognitive psychology, neuroscience, or human factors engineering.
    - Design a table or infographic summarizing key differences and examples.
9. Describe how reaction time is measured.
    - Research different experimental methods for measuring reaction time, including computerized and manual tests.
    - Use AI tools to find historical and modern approaches to reaction time measurement.
    - Conduct an experiment using online reaction time measurement tools and analyze the data.
    - Interview a cognitive scientist or sports coach to understand applications in their field.

10. Explain why applying ergonomic guidelines to task design in cognitive tasks contributes to worker safety.
    - Use AI tools to generate key insights on cognitive ergonomics and task design.
    - Research case studies where poor cognitive task design led to errors or accidents.
    - Interview a safety specialist or human factors expert about best practices in cognitive task design.
    - Develop a short report discussing the impact of ergonomic principles on worker safety, supported by real-world examples.

REFERENCES

AGARD. (1989). AGARD-STRESS battery: A tool for measuring pilot workload. NATO Advisory Group for Aerospace Research and Development, AGARDograph No. 289.

Atkinson, R. C., & Shiffrin, R. M. (1968). Human memory: A proposed system and its control processes. In K. W. Spence & J. T. Spence (Eds.), *The psychology of learning and motivation* (Vol. 2, pp. 89–195). Academic Press.

Baron, S., & Corker, K. (1988). PROCRU: A process resource model for operator task performance. NASA Contractor Report 177611.

Biological Psychology. (1992). Special issue on psychophysiological measures of workload, 34(2–3).

Biological Psychology. (1995). Special issue on workload assessment, 40(1–2).

Brown, I. D., & Poulton, E. C. (1961). Measuring the spare "mental capacity" of car drivers by a subsidiary task. *Ergonomics*, *4*(1), 35–40.

Byers, J. C., Bittner, A. C., & Hill, S. G. (2019). Toward a comprehensive model of workload: A comparative study of workload measures and their relationships. *Ergonomics*, *62*(3), 397–420.

Cain, B. (2007). *A review of the mental workload literature.* Defence Research and Development Toronto.

Charbonnier, S., Roy, R. N., Bonnet, S., & Campagne, A. (2016). EEG index for control operators' mental fatigue monitoring using interactions between brain networks. *Computers in Biology and Medicine, 64*, 1–9.

Charles, L. E., & Nixon, A. E. (2019). Evaluating stress and workload in the workplace: An overview of physiological measurement techniques. *International Journal of Psychophysiology, 143*, 79–89.

Comstock, J. R., & Arnegard, R. J. (1992). The multi-attribute task battery for human operator workload and strategic behavior research. NASA Technical Memorandum 104174.

Eggemeier, F. T., Shingledecker, C. A., & Biers, D. W. (1991). Workload analysis in human-machine systems. In G. R. S. Hockey (Ed.), *Stress and human performance* (pp. 207–248). John Wiley & Sons.

Gagné, R. M., & Driscoll, M. P. (1988). *Essentials of learning for instruction* (2nd ed.). Englewood Cliffs, NJ: Prentice Hall.

Goldstein, E. B. (2014). *Cognitive psychology: Connecting mind, research, and everyday experience* (4th ed.). Cengage Learning.

Gopher, D., & Braune, R. (1984). On the psychophysics of workload: Why bother with subjective measures? *Human Factors*, 26(5), 519–532. https://doi.org/10.1177/001872088402600502

Hamilton, W. I., et al. (1991). Task Analysis and Workload Measurement in Aviation Systems. NASA Contractor Report 177649.

Hancock, P. A., & Warm, J. S. (2019). A dynamic model of stress and sustained attention. *Human Factors, 31*(5), 519–537.

Harms, L. R. (1986). The utility of mental arithmetic as a performance measure of cognitive workload. *Human Factors, 28*(2), 147–160.

Hart, S. G., & Staveland, L. E. (1988). Development of NASA-TLX (Task Load Index): Results of empirical and theoretical research. In P. A. Hancock & N. Meshkati (Eds.), *Human mental workload* (pp. 139–183). North-Holland.

Hart, S. G. (2006). NASE-TLX: 20 years later. *Proceedings of the Human Factors and Ergonomics Society Annual Meeting, 50*(9), 904–908.

Hart, S. G. (2018a). Development of NASA-TLX (Task Load Index): Results of empirical and theoretical research. In P. A. Hancock & N. Meshkati (Eds.), *Human mental workload.* North-Holland.

Hart, S. G. (2018b). *NASA-Task Load Index (NASA-TLX); 20 years later.* Human Factors and Ergonomics Society Annual Meeting Proceedings.

Helander, M. G. (2014). *Handbook of human-computer interaction.* Elsevier. This reference provides updated insights into the design and ergonomic considerations of controls in machinery.

Huitt, W. (2003). The information processing approach to cognition. In *Educational psychology interactive.* Valdosta State University. Retrieved from http://www.edpsycinteractive.org/topics/cognition/infoproc.html

Jarvis, M. (1999). *Theoretical approaches in psychology.* Routledge.

Kalsbeek, J. W. H., & Sykes, R. N. (1967). Choice reaction time and attention. *Ergonomics, 10*(1), 35–48.

Kirwan, B., & Ainsworth, L. K. (1992). A guide to task analysis: The task analysis working group. Taylor & Francis.

Krol, J. R. (1971). A simplified criterion task set for evaluating workload. NASA Technical Memorandum X-62045.

Levison, W. H. (1979). Tracking performance as a measure of workload. In *Proceedings of the Human Factors Society Annual Meeting, 23*(1), 701–705.

Liang, Y., Cabeza, R., & Anderson, M. (2014). A cognitive neuroscience perspective on the role of attention in mental workload. *Neuroscience & Biobehavioral Reviews, 37*(6), 133–142.

Longo, L. (2014). Formalising human mental workload as non-monotonic concept for adaptive and personalised web-design. *International Journal of Human-Computer Studies, 72*(5), 415–435.

Mac, L. (2010). *Motor learning and control: Concepts and applications.* McGraw-Hill Education.

Matthews, G., Reinerman-Jones, L., & Barber, D. J. (2016). *Workload assessment and prediction.* Manuscript Submitted for Publication.

Matthews, G., Reinerman-Jones, L. E., Barber, D. J., & Abich, J. (2015). The psychometrics of mental workload: Multiple measures are sensitive but divergent. *Human Factors, 57*(1), 125–143.

McCracken, J. H., & Aldrich, T. B. (1984). Analysis of selected LHX mission functions: Implications for operator workload and system automation goals. U.S. Army Research Institute for the Behavioral and Social Sciences. Technical Report 694.

Mehta, R. K., & Agnew, M. J. (2015). Influence of mental workload on muscle endurance, fatigue, and recovery during intermittent static work. *European Journal of Applied Physiology, 115*(8), 1741–1751.

Miller, G. A. (1956). The magical number seven, plus or minus two: Some limits on our capacity for processing information. *Psychological Review, 63*(2), 81–97. https://doi.org/10.1037/h0043158

Moray, N., Dessouky, Y., Kijowski, S., & Adapathya, S. (1991). Queueing models of workload in human-machine systems. *Ergonomics, 34*(6), 833–853.

North, R., & Riley, M. (1989). W/INDEX: A Workload Index for Air Traffic Control Systems. Presented at the 33rd Annual Meeting of the Human Factors Society.

Parasuraman, R., & Rizzo, M. (2017). *Neuroergonomics: The brain at work.* Oxford University Press.

Parks, D. K., & Boucek, G. P. (1989). Task Load Assessment Procedure (TLAP): A Computer-Based Rating System.

Pashler, H. (1997). *The psychology of attention.* MIT Press.

Potter, S. and Bressler, J. (1994)Subjective Workload Assessment Technique (SWAT): A User's Guide, Systems Research Lab: Dayton, OH.

Reid, G. B., & Nygren, T. E. (1988). The subjective workload assessment technique: A scaling procedure for measuring mental workload. In P. A. Hancock & N. Meshkati (Eds.), *Human Mental Workload* (pp. 185–218). North-Holland.

Roscoe, S. N. (1987). *The airplane and the pilot.* Iowa State University Press.

Roscoe, S. N., & Ellis, D. R. (1990). A subjective rating scale for assessing pilot workload in flight. In *AGARD Conference Proceedings* No. 478.

Rumelhart, D. E., & McClelland, J. L. (1986). *Parallel distributed processing: Explorations in the microstructure of cognition* (Vol. 1). MIT Press.

Sanders, M. S., & McCormick, E. J. (1993). *Human Factors in Engineering and Design* (7th ed.). McGraw-Hill.

Shingledecker, C. A. (1984). A battery of performance measures for mental workload. In *Proceedings of the Human Factors Society Annual Meeting* (Vol. 28, No. 7, pp. 660–664).

Stanton, N., Salmon, P., Rafferty, L., Walker, G., Baber, C., & Jenkins, D. P. (2013). *Human factors in the design and evaluation of central control room operations.* CRC Press. This reference offers a modern perspective on the design and ergonomic use of control systems in technical environments.

Stern, J. A., & Dunham, D. N. (1990). The eyeblink and workload assessment. *Human Factors, 32*(2), 195–208.

Tsang, P. S., & Velazquez, V. L. (1996). Diagnosticity and multidimensional subjective workload ratings. *Ergonomics, 39*(3), 358–381.

Vidulich, M. A., Ward, K., & Schueren, J. (1991). NASA Task Load Index: A sensitive tool for evaluating training effectiveness. *Proceedings of the Human Factors Society Annual Meeting, 35*(2), 1291–1295.

Welford, A. T. (1968). *Fundamentals of skill.* Methuen.

Wickens, C. D. (1986). The effects of control dynamics on performance. In *Proceedings of the Human Factors Society Annual Meeting, 30*(2), 1138–1142.

Wickens, C. D., Lee, J., Liu, Y., & Gordon Becker, S. E. (2004). *An introduction to human factors engineering.* Pearson Education. This text offers updated insights into the role of coding in control design, emphasizing the importance of ergonomic principles in enhancing usability and safety.

Wierwille, W. W., & Casali, J. G. (1983). A validated rating scale for global workload assessment: A NASA TLX pilot study. *Proceedings of the Human Factors Society, 27*(1), 129–133.

Young, M. S., Brookhuis, K. A., Wickens, C. D., & Hancock, P. A. (2015). State of science: Mental workload in ergonomics. *Ergonomics, 58*(1), 1–17.

10 Product Liability, Warning Labels, and Instruction Design

10.1 LEARNING GOALS

The learning goals for this chapter include preparing the student to have the following knowledge:

1. Understand the key principles of product liability and the role of ergonomics in minimizing legal risks related to consumer product safety.
2. Recognize the importance of clear, effective warning labels and instructions in preventing accidents and injuries associated with product misuse.
3. Apply ergonomic principles to the design of warnings, symbols, and instructions to ensure they are accessible, legible, and understandable to diverse users.

10.2 KEY TOPICS

- Types of product liability
- Hazard classification
- Principles of warning label design

10.3 INTRODUCTION AND BACKGROUND

Ergonomics plays a crucial role in consumer product safety through the principles of safe design, clear communication, and human-centered engineering. Over the past several decades, the focus on consumer product safety has seen significant progress driven by regulatory agencies, litigation trends, and technological advancements.

The Consumer Product Safety Commission (CPSC)—an independent federal regulatory agency established by the Consumer Product Safety Act (CPSA) of 1972 (15 U.S.C. 2051 et seq.)—is tasked with minimizing risks associated with consumer products. The commission's core responsibilities include:

- Protecting the public from unreasonable risks of injury associated with consumer products.

DOI: 10.1201/9780429342806-10

- Evaluating consumer products for potential hazards and establishing safety standards.
- Harmonizing federal, state, and local safety regulations.
- Conducting recalls of hazardous products.
- Leading research efforts to identify and mitigate risks.

In addition to the CPSC, workplace safety is governed by the Occupational Safety and Health Administration (OSHA), which enforces hazard communication standards under 29 CFR 1910.1200, requiring clear and effective warning labels and instructions for hazardous materials. The National Institute for Occupational Safety and Health (NIOSH) also provides guidance on hazard assessment and risk mitigation, particularly in industrial settings. Furthermore, the American National Standards Institute (ANSI Z535 series) outlines best practices for warning labels, including signal words, color coding, and pictograms to enhance comprehension and reduce user error.

10.3.1 The Role of Warnings and Instructions in Safety

Warnings and instructions serve as vital components of risk communication, particularly when hazards cannot be entirely eliminated through design modifications. According to Laughery and Wogalter (2014), there is a key distinction between warnings—which explicitly address safety concerns—and instructions, which may or may not include safety-related information. Effective warnings must be:

1. **Noticeable:** Designed to capture attention through size, color, and placement.
2. **Comprehensible:** Easily understood by users of varying literacy levels and linguistic backgrounds.
3. **Specific:** Clearly stating the nature of the hazard and necessary precautions.
4. **Credible:** Aligning with user expectations and real-world risks to encourage compliance.

Warnings, however, should never be the sole method of risk mitigation. The hierarchy of controls, a fundamental principle in ergonomics and safety engineering, prioritizes hazard elimination and substitution over administrative controls (such as training and warnings) and personal protective equipment (PPE).

10.3.2 Product Liability and Human Factors in Design

With the rise of litigation related to product safety, ergonomics and human factors have become increasingly important in product development. Poorly designed products that fail to consider human capabilities and limitations can lead to injuries, consumer dissatisfaction, and costly legal claims. Recent research emphasizes the role of usability testing and human-centered design in minimizing risk. Measurement-based technologies, such as motion capture and biomechanical modeling, allow designers to evaluate how users interact with products, leading to safer and more intuitive designs.

Product testing is also essential for compliance with OSHA, CPSC, and ANSI standards, ensuring that products meet both regulatory and ergonomic criteria. In some cases, manufacturers may choose to delay or discontinue product releases to avoid potential liability, balancing innovation with safety and legal considerations. While some critics argue that excessive litigation stifles innovation, product liability laws ultimately promote higher-quality products and responsible manufacturing practices.

As technology evolves, new challenges emerge. For example, the proliferation of smart devices introduces risks related to cybersecurity and user data privacy, in addition to physical safety hazards. The rapid advancement of automation and artificial intelligence also raises questions about accountability in human-machine interactions. These issues highlight the growing intersection of ergonomics, law, and ethics in modern product development.

Understanding the principles of warning labels, instructions, and product liability is essential for engineers, designers, and safety professionals. As regulatory frameworks evolve and consumer expectations shift, integrating human factors into product design remains a cornerstone of injury prevention and legal compliance. Through careful application of ergonomic principles, the likelihood of product-related injuries can be minimized, ensuring that safety remains a priority in both workplace and consumer environments.

10.4 DEFINITIONS

This section defines key terms—hazard, danger, and risk perception—to establish clarity, particularly in the context of occupational safety and ergonomics. Understanding these concepts is essential for identifying and mitigating risks in workplace environments.

A *hazard* is any source of potential harm or adverse health effects to individuals or potential damage to property or the environment (OSHA, 2023). Hazards can be physical (e.g., noise, vibration, radiation), chemical (e.g., toxic substances, flammable materials), biological (e.g., viruses, bacteria), ergonomic (e.g., repetitive strain, poor posture), or psychosocial (e.g., stress, fatigue). A comprehensive hazard analysis considers both intrinsic product hazards and hazards related to human interaction with systems (NIOSH, 2023). The recognition and mitigation of these hazards are central to safety engineering and ergonomic intervention strategies (NIOSH, 2023; OSHA, 2022).

Danger refers to the combination of a hazard and the likelihood of its occurrence, leading to adverse consequences (ANSI Z535.4, 2022). While hazard represents potential harm, danger accounts for exposure levels, human error, and environmental conditions that amplify the risk of injury. Effective safety protocols, such as lockout/tagout procedures and hazard communication programs, are designed to reduce dangers associated with workplace hazards (OSHA, 2023).

Risk perception is an individual's assessment of the severity and likelihood of a hazard's consequences, influenced by personal experience, training, cognitive biases, and cultural factors (ISO 31010, 2021). Studies show that individuals often underestimate common hazards and overestimate rare but dramatic hazards, which has implications for safety training and communication strategies (Slovic, 2020). Poor risk perception has been linked to increased workplace accidents and non-compliance with safety measures. Therefore, enhancing risk perception through ergonomic training, safety culture development, and clear warning systems is a priority in occupational ergonomics.

10.5 PRODUCT LIABILITY AND RISK PERCEPTION IN ERGONOMICS

10.5.1 Hazard

A hazard refers to any condition, substance, activity, or process that has the potential to cause injury, illness, or property damage. Hazards can arise from various sources, including environmental factors (e.g., extreme temperatures, poor lighting), mechanical equipment (e.g., unguarded machinery, defective tools), and task-related variables (e.g., repetitive motion, heavy lifting). The recognition and mitigation of hazards are central to safety engineering and ergonomic intervention strategies (NIOSH, 2023; OSHA, 2022).

10.5.2 Danger

Danger is the presence of a hazard combined with the likelihood of that hazard causing harm. For instance, an unguarded rotating blade is a hazard, but if the blade is actively in use in an unmonitored workspace, it represents an imminent danger. Effective safety protocols, such as lockout/tagout procedures and hazard communication programs, are designed to reduce dangers associated with workplace hazards (OSHA, 2023).

10.5.3 Risk Perception

Risk perception refers to an individual's cognitive evaluation of a hazard, the probability of injury or harm, and the severity of potential outcomes. Research indicates that risk perception is influenced by factors such as experience, training, cognitive biases, and situational awareness (Slovic, 2020; Kahneman, 2018). Poor risk perception has been linked to increased workplace accidents and non-compliance with safety measures. Therefore, enhancing risk perception through ergonomic training, safety culture development, and clear warning systems is a priority in occupational ergonomics.

10.5.4 Product Liability in Ergonomics

Product liability refers to the legal responsibility of manufacturers, suppliers, and distributors for injuries caused by defective or unsafe products. Theories of product liability fall into three primary categories: negligence, strict liability, and breach of warranty. Each of these legal frameworks has implications for ergonomic product design and workplace safety.

10.5.5 Negligence

Negligence claims require proof of four elements:

1. **Duty of Care:** The manufacturer owes a duty to produce a reasonably safe product.
2. **Breach of Duty:** The manufacturer failed to mitigate foreseeable risks.
3. **Causation:** The defect directly caused harm.
4. **Damages:**The plaintiff suffered quantifiable injury or loss.

For example, if an office chair lacks adequate lumbar support and leads to chronic musculoskeletal disorders among users, a manufacturer may be held liable for failing to account for ergonomic design principles (ANSI/HFES 100-2022; OSHA, 2022).

10.5.6 Strict Liability

Strict liability does not require proof of negligence but instead focuses on the condition of the product. A plaintiff must demonstrate:

- The product had an inherent defect (design, manufacturing, or labeling).
- The defect caused harm while the product was used as intended.
- The product was not substantially altered after leaving the manufacturer.

A case example involves power tools with inadequate vibration damping leading to Hand–Arm Vibration Syndrome (HAVS). If alternative designs with superior vibration reduction existed, the manufacturer could be held strictly liable for failing to adopt safer designs (NIOSH, 2023).

10.5.7 Breach of Warranty

A breach of warranty occurs when a product fails to meet express or implied promises regarding safety and performance. This includes:

- **Express Warranties:** Explicit guarantees about product performance.
- **Implied Warranties:** Assurances that a product is fit for its intended purpose.

For example, a manufacturer promoting an ergonomic keyboard as reducing carpal tunnel syndrome without scientific backing may be liable for misleading claims if users develop repetitive strain injuries (RSI) (ANSI/HFES 100-2022).

10.5.8 Determining Defectiveness in Ergonomic Product Liability Cases

To assess whether a product is defective under strict liability, courts often evaluate:

1. **Utility vs. Risk:** Does the product's benefit outweigh its potential harm?
2. **Safer Alternatives:** Were feasible, safer designs available?
3. **Injury Likelihood and Severity:** How probable and severe is the potential injury?
4. **Obviousness of Danger:** Was the hazard clearly recognizable?
5. **Consumer Expectations:** Would an average user anticipate the risk?
6. **Avoidability:** Could users reasonably prevent harm through proper use?
7. **Cost and Feasibility of Safer Design:** Would safer alternatives be prohibitively expensive?

Recent ergonomic cases highlight the growing emphasis on human-centered design and compliance with industry standards such as ANSI/HFES 100-2022 (ergonomics for computer workstations) and ISO 9241 (ergonomic requirements for office environments).

Ergonomic principles play a crucial role in mitigating product liability risks. By designing for safety, ensuring clear hazard communication, and fostering accurate risk perception, manufacturers and employers can reduce liability exposure while enhancing user well-being. Regulatory compliance with OSHA, NIOSH, and ANSI standards is essential in both workplace and consumer product contexts to prevent injuries and improve long-term usability.

10.5.9 Steps in Hazard Analysis

The steps involved in an effective hazard analysis are as follows (ANSI/ASSP Z10.0-2019):

1. Identify hazards by level (see Table 10.1).
2. Identify the reasons and factors that can produce the hazard.
3. Evaluate and identify all potential effects of the hazard.
4. Categorize the identified hazards as catastrophic, critical, marginal, ornegligible.
5. Implement design changes to minimize the number and severity of hazards.
6. Determine if administrative controls (e.g., personal protective equipment [PPE]) should be applied.
7. Develop a plan to provide warnings and instructions to end users about any minimal or remaining hazards that could not be eliminated through design.

TABLE 10.1
DoD Hazard Classification

Severity	Category	Accident Definition
Catastrophic	I	Death or total system loss
Critical	II	Severe injury, severe occupational illness, or major system damage
Marginal	III	Minor injury, minor occupational illness, or minor system damage
Negligible	IV	Less than minor injury, occupational illness, or system damage

The US Department of Defense's Military Standard 882 (DoD MIL STD 882) lays out system safety program requirements and classifies hazard severity as shown in Table 10.1.

It is essential for manufacturers and employers to assess hazardous characteristics of products and workplace environments using methodologies such as preliminary hazard analysis, failure mode and effects analysis (FMEA), and job safety analysis (JSA) (NIOSH, 2022). These assessments ensure that hazards are identified and controlled before exposure to users or workers.

Common hazardous characteristics of products and processes include (ANSI/HFES 100-2022):

- Acceleration
- Chemical reactions
- Electrical risks
- Flammability and fire hazards
- Heat and temperature
- Pressure
- Vibration and noise

10.6 WARNINGS AND SAFETY COMMUNICATION

Warnings are designed to inform people of hazards and provide instructions on how to manage risks to avoid or minimize undesirable outcomes. OSHA's Hazard Communication Standard (HCS) mandates that hazards associated with chemical and industrial processes be clearly communicated through warning labels, signs, and Safety Data Sheets (SDS) (OSHA, 2023).

10.6.1 Objectives of Warnings

Warnings serve multiple functions, including:

1. **Reducing Accidents and Injuries:** Alerts users to potential dangers to prevent incidents resulting in harm.

2. **Modifying Behavior:** Encourages safer behaviors when interacting with hazardous products or environments.
3. **Providing Hazard Information:** Helps individuals understand risks and necessary mitigation actions.

10.6.2 Effective Warning Communication

Warnings function as a critical component of a communication model, which includes:

- **Sender:** Manufacturer, safety officer, or employer responsible for the warning.
- **Receiver:** User, worker, or any individual requiring safety information.
- **Channel/Medium:** Labels, manuals, signs, or digital alerts.
- **Message:** Information about hazards and safe use.

10.6.3 Safety Approaches in Product Design

Ensuring product safety involves three primary approaches:

1. **Designing Out the Hazard:** The most effective approach, eliminating risks through design modifications (ANSI Z590.3-2011).
2. **Guarding Against the Hazard:** When elimination is not possible, safeguards such as physical barriers and PPE should be implemented.
3. **Providing Warnings and Instructions:** When hazards remain, clear warnings and instructions must be provided to users to mitigate risks.

10.6.4 Warning Label Design Standards

The American National Standard for Product Safety Signs and Labels (ANSI Z535.4-2011) outlines five essential criteria:

TABLE 10.2
Factors Affecting the Written Communication Process

Factor	Description
Readability	Warnings should use clear and concise language
Visibility	Placement and font size must ensure easy reading
Interpretation	Symbols and pictograms should be easily understood
Compliance	Users must perceive the warning as credible and relevant

Source: Wogalter, M. S. (2006). *Handbook of warnings*. Lawrence Erlbaum Associates.

1. **Signal Words:** The appropriate hazard level should be communicated (e.g., Danger, Warning, Caution).
2. **Hazard Statement:** Clearly describes the nature of the hazard.
3. **Consequence Statement:** Specifies the risks of encountering the hazard.
4. **Avoidance Instructions:** Details safe practices to prevent injury.
5. **Design Elements:** Uses appropriate colors, symbols, and pictorials to enhance comprehension.

Signal Word Categorization

Signal Word	Description
Danger	Immediate hazard that may result in severe injury or death (highest level)
Warning	Serious hazard with potential for injury if not avoided
Caution	Minor hazard that could result in injury or property damage

10.6.5 Human Factors in Warning Effectiveness

Research has shown that effective warnings must be perceived, understood, and acted upon (Laughery & Wogalter, 2014). The effectiveness of warnings is influenced by:

- **Orientation and Location:** Proper positioning ensures users see the warning before exposure to hazards.
- **Size and Font:** Text should be legible under expected viewing conditions.
- **Shape and Color:** Should conform to user expectations for hazard communication.
- **Graphical Design and Contrast:** High-contrast enhances visibility.

In high-risk environments, warnings should undergo usability testing to ensure comprehension and compliance (ANSI/HFES 200, 2022). Testing should involve real-world scenarios to determine whether users notice, understand, and follow safety instructions.

Hazard identification and effective warning systems play a critical role in injury prevention and safety management. Integrating principles from human factors engineering, ANSI standards, and OSHA guidelines ensures that warnings are designed to be noticed, understood, and acted upon. By prioritizing hazard elimination through design, implementing protective measures, and providing clear warnings, industries can enhance safety outcomes for workers and consumers alike.

10.6.6 Effective Warning Label Design

Sensing a warning does not necessarily ensure that it will be received, understood, or acted upon. Several factors influence whether a warning is acknowledged, including familiarity with the product, warning complexity, and cognitive load.

FIGURE 10.1 Example of a warning label.

FIGURE 10.2 Samples of signal word usage. Source: American National Standards Institute (ANSI). (2007). ANSI Z535 Series – Safety Colors, Signs, and Labels. ANSI.

To enhance effectiveness, warning labels must adhere to principles outlined in ANSI Z535.4-2021, OSHA 1910.145, and NIOSH guidelines on hazard communication (CDC, 2021). These standards specify that effective warnings must include the following minimum information:

1. **Signal Word:** Conveys the severity or gravity of the risk (e.g., "Danger" for immediate severe hazards, "Warning" for moderate risks, and "Caution" for lower-level hazards).
2. **Hazard Description:** Clearly defines the nature of the danger.
3. **Consequences:** Outlines potential outcomes if the warning is ignored.
4. **Instructions:** Specifies actions needed to reduce or eliminate the risk.

Even when warnings are noticed and understood, compliance is not guaranteed. The "cost of compliance" plays a significant role, referring to the time, effort, and resources required to adhere to a warning. Studies indicate that individuals are less likely to comply when compliance imposes inconvenience or disrupts workflow (Wogalter, 2020). Additional factors influencing compliance include:

- **Risk Acceptance Level:** Individual perception of acceptable risk.
- **Memory:** The ability to recall and act on warnings when needed.
- **Trust in the Warning:** Users are more likely to comply with warnings from trusted sources (Laughery & Wogalter, 2014).

10.6.7 Prioritizing Warnings

Determining which hazards to warn about is critical, especially when multiple risks exist. According to Wogalter et al. (2020), prioritization should be based on:

1. **Likelihood:** The probability of an adverse event occurring.
2. **Severity:** The potential harm resulting from exposure.
3. **Practicality:**Constraints such as space limitations (e.g., small product labels) or time restrictions (e.g., brief television advertisements).

In compliance with OSHA's Hazard Communication Standard (HCS, 2012), priority should be given to new, severe, or likely hazards on primary warning labels. Less critical hazards may be communicated through supplementary materials such as instruction manuals (NIOSH, 2019).

10.6.8 Active vs. Passive Warnings

Warnings are classified as active or passive. Active warnings involve real-time monitoring (e.g., auditory alarms, flashing lights, sensor-based alerts) and provide immediate feedback, making them more effective in preventing incidents. Passive warnings, such as printed labels or instruction manuals, rely on users to notice and act on the information. Although passive warnings are more common, research suggests that active warnings significantly enhance hazard avoidance (Wogalter et al., 2020).

10.6.9 Design of Warnings

Effective warning labels should comply with ANSI Z535.2-2021 for environmental and facility safety signs and ANSI Z535.4-2021 for product safety labels. The fundamental hierarchy for product safety follows these steps:

1. **Eliminate the Hazard:** Design out potential risks where feasible.
2. **Limit Access:** Implement safeguards such as machine guards and barriers.
3. **Warn Users:** Provide clear labels, instruction manuals, and safety training.
4. **Train Users:** Ensure adequate understanding through structured education.

Warnings should never substitute for proper design but rather function as part of a broader risk mitigation strategy.

10.6.10 Steps for Designing Warning Signs or Labels

1. Define the risk, audience, and preventive measures.
2. Select the most effective communication medium (text, symbols, audio, etc.).
3. Draft clear and concise warning language.
4. Evaluate wording using communication experts and legal counsel.
5. Conduct user testing to assess:
   - Visibility and noticeability
   - Comprehension levels
   - Proper avoidance behavior
6. Revise and refine based on testing results.
7. Finalize, distribute, and periodically re-evaluate.

10.6.11 Criteria for Effective Warnings

- **Attention-Capturing:** Utilize contrast, colors (e.g., red for "Danger"), and icons.
- **Clear Hazard Information:** Define risks explicitly.
- **Explicit Consequences:** Describe the potential harm.
- **Action-Oriented Instructions:** Provide direct and simple avoidance guidance.
- **Comprehension:** Ensure accessibility for diverse audiences.
- **Motivation:** Encourage compliance through persuasive language.
- **Conciseness:** Maintain brevity while retaining key information.
- **Durability:** Ensure the warning remains intact and visible for the product's lifespan.

10.6.12 Receiver Characteristics in Warning Design

Warnings must be tailored to the target audience's cognitive abilities. Key considerations include:

1. **Technical Knowledge:** Ensure language and symbols align with users' expertise.
2. **Language Barriers:** Use multiple languages or pictograms when necessary.
3. **Reading Ability:** Warnings should target a reading level of 4th to 6th grade to maximize accessibility (NIOSH, 2019).

10.6.13 Principles for Designing Warnings Based on Receiver Characteristics

- **Understand the Audience:** Conduct usability research.
- **Design for the Least Knowledgeable Users:** Ensure accessibility for all.
- **Test with Target Subgroups:** Validate warnings with diverse groups before implementation.

By integrating regulatory standards, scientific research, and user-centered design principles, warnings can be optimized to reduce injury risk while improving compliance and comprehension.

10.7 SUMMARY

As product development evolves with continuous technological advances, the integration of ergonomics throughout every phase of the product lifecycle has become essential. Adoption of human-centered ergonomic principles at each stage of the product life cycle not only optimizes the usability and safety of products but also minimizes the risk of product liability.

Effective warning labels and instructions are vital to preventing user errors and accidents. Ensuring that critical information regarding product use, limitations, and potential hazards is effectively communicated throughout the product's life cycle helps mitigate potential legal and safety issues. Inadequate or unclear warnings are a major contributor to product misuse, injuries, and fatalities, resulting in liability lawsuits, which may be avoidable by adopting user-centered designs.

The incorporation of human factors principles helps designers understand the cognitive, physical, and emotional needs of users. When these needs are considered, products are more likely to be used correctly, resulting in safer, more reliable outcomes. Furthermore, following established safety standards and guidelines (e.g., OSHA regulations, NIOSH recommendations, ANSI standards) enhances the product's safety and reliability, aligning with regulatory requirements and industry best practices.

10.8 CASE STUDY: ERGONOMIC REDESIGN OF HAIR DRYER WARNINGS AND INSTRUCTIONS

Problem Overview

The role of clear, effective warnings and instructions cannot be overstated, especially in preventing product misuse and reducing injuries. Failure to design adequate user communication has led to incidents, injuries, and even deaths, resulting in product liability claims. Manufacturers are legally and ethically bound to ensure their products are safe for use, with appropriate precautions communicated to consumers. However, when risks cannot be fully eliminated, these risks must be clearly outlined to prevent harm and minimize liability.

This case study focuses on a hair dryer model that had been involved in multiple incidents of electrical shock and burns, primarily attributed to poorly designed and unclear warning labels and instruction materials. These shortcomings prompted a thorough ergonomic evaluation to determine whether the product's warnings and instructional materials were sufficient in addressing key safety concerns, including electrical shock, burns, and misuse.

Ergonomic Design Process

Several ergonomic principles were applied to enhance the usability and clarity of the hair dryer's warnings and instructions. The following considerations were incorporated into the redesign:

Readability and Font Size

The original warning labels used small, low-contrast fonts, which were difficult to read, especially under the dim lighting typical in bathroom environments. According to NIOSH (2021), poor visibility of warning information is a leading factor in preventing user adherence to safety protocols. The redesigned labels employed larger, high-contrast fonts, ensuring better legibility from a typical user distance, in alignment with ANSI Z535.4 guidelines for safety signs.

Color and Contrast

Effective color usage in warning labels is critical for conveying urgency and drawing attention to potential hazards (Wogalter, 2019). The original labels were printed using dull colors, failing to stand out in a cluttered or brightly lit environment. The redesign utilized high-contrast color schemes, incorporating red and yellow backgrounds for hazard warnings (consistent with ANSI Z535.1), and green for safe-use instructions. This ensured that users could quickly recognize key warnings about electrical shock risk when the hair dryer was used near water.

Symbol Usage

Pictograms and standardized symbols significantly enhance comprehension, especially for non-literate or non-native language speakers (ANSI Z535.6-2022). The original labels contained few symbols, relying mostly on text. The ergonomic redesign integrated internationally recognized symbols for electrical shock, heat hazards, and water exposure, ensuring that users across diverse linguistic and cultural backgrounds could understand the risks.

Instructional Clarity

The original instruction manual was dense, technical, and difficult for non-expert users to follow. Research shows that over-complicated instructions lead to misunderstandings and improper product use (Wogalter, 2019). The redesigned manual was simplified by breaking it into clear, actionable steps, using simple language and incorporating pictorial representations of safe operation, maintenance, and cleaning procedures. The use of visual aids is a well-documented ergonomic strategy to improve comprehension and reduce errors (OSHA 29 CFR 1910.1200).

Placement and Accessibility

Warning labels were initially placed in areas that were easily overlooked, such as the underside of the handle. Research by NIOSH (2021) emphasizes that the placement of warnings should ensure visibility during all stages of product use. In the redesign, critical warnings were relocated to more visible areas, such as near the power switch and on the cord, ensuring users encountered them during every use.

Outcome

Following the ergonomic redesign, user testing revealed significant improvements in the recognition and understanding of the hair dryer's hazards. Consumers were able to identify the warnings more quickly and follow instructions with greater ease. Subsequent user reports showed a noticeable decrease in electrical shock and burn incidents, as well as a reduction in product misuse. Additionally, the manufacturer avoided potential product liability issues by ensuring that the warnings and instructions met safety standards, aligned with guidelines from organizations like OSHA, NIOSH, and ANSI.

Conclusion

This case study highlights the importance of applying ergonomic principles in the design of warnings and instructions to enhance product safety. By addressing factors such as readability, color contrast, symbol usage, instructional clarity, and label placement, manufacturers can significantly reduce user errors and minimize liability risks. The integration of safety standards and ergonomic guidelines is essential not only for ensuring compliance but also for creating products that are safer, more reliable, and user-friendly.

Key References

- ANSI Z535.1-2022: American National Standard for Safety Colors

- ANSI Z535.6-2022: American National Standard for Product Safety Signs and Labels
- NIOSH (2021). *Guidelines for the Ergonomic Design of Hazardous Warnings and Instructions*
- Wogalter, M. S. (2019). *Handbook of Warnings* (2nd ed.). CRC Press.

Exercises

1. Explain the four types of product liability classifications from a legal perspective.
   Extended Activity: Research the four types of product liability classifications using online legal databases (such as LexisNexis or Google Scholar) or legal websites. Summarize how AI tools like legal chatbots or AI-powered document analysis can assist in identifying liability risks for manufacturers. Provide real-life case studies, including product recalls, to support your explanation.
2. What have been some of the historic benefits in manufacturing as a result of product liability cases?
   Extended Activity: Use online case law databases or websites like the National Consumer League (NCL) to research significant product liability cases and their impact on manufacturing practices. Create a visual timeline using an AI-based tool like Canva or Google Slides, highlighting key product liability cases and how they led to innovations in product design and safety standards.
3. Identify the branch of the government in your country (or the United States) responsible for consumer product safety and go to the website and obtain an overview of the process used to establish consumer product-safety guidelines.
   Extended Activity: Visit the website of your country's consumer product safety authority (e.g., US Consumer Product Safety Commission or the European Union's Consumer Safety Division). Use AI-powered summary tools like ChatGPT or summarization apps to condense the key points about the process for establishing safety guidelines. Share the link to the resource and summarize the steps.
4. Discuss the types of keywords that are used in warning design and give an example where each should be used in warning design.
   Extended Activity: Use an online resource, such as a product safety guide or an AI-driven keyword suggestion tool (e.g., Google Keyword Planner), to identify common keywords in warning design. Then, analyze and categorize these keywords in a blog post or article using an AI writing assistant. Discuss their applications in specific industries or real-world scenarios and create a mock-up of a product label using a design tool like Adobe Spark or Canva.
5. Design a warning label for a physically intensive computer-based game that is used for recreational purposes.

Extended Activity: Use AI tools to create a prototype of your warning label, incorporating symbols and text that reflect best practices in warning design. Leverage design platforms like Figma or Canva to integrate accessibility features, such as color-blind-friendly palettes. Research how AI-driven design tools can help ensure the label adheres to ergonomic and safety guidelines. Share your design and explain the rationale behind the choices of keywords, colors, and layout.

REFERENCES

American National Standards Institute (ANSI). (2007). ANSI Z535 Series – Safety Colors, Signs, and Labels. ANSI.

American National Standards Institute (ANSI) & Human Factors and Ergonomics Society (HFES). (2022). *ANSI/HFES 100-2022: Human factors engineering of computer workstations.*

ANSI Z535.6-2022: American National Standard for Product Safety Signs and Labels.

CDC. (2021). *National Institute for Occupational Safety and Health (NIOSH) Hazard communication guide.* Retrieved from https://www.cdc.gov/niosh

International Organization for Standardization (ISO). (2021). ISO 31010:2019 – Risk management – Risk assessment techniques. ISO. https://www.iso.org/standard/72140.html

Kahneman, D. (2018). *Thinking, fast and slow.* Farrar, Straus, and Giroux.

Laughery, K. R., & Wogalter, M. S. (2014). *Human factors perspectives on warnings.* CRC Press.

National Institute for Occupational Safety and Health (NIOSH). (2023). *Ergonomics and musculoskeletal disorders prevention.*

NIOSH. (2019). *Elements of effective warnings and labels: Guidance for workplace safety.* National Institute for Occupational Safety and Health.

NIOSH (National Institute for Occupational Safety and Health). (2021). *Publications on safety guidelines.*

NIOSH. (2022). Hierarchy of Controls. Centers for Disease Control and Prevention. https://www.cdc.gov/niosh/topics/hierarchy/

Occupational Safety and Health Administration (OSHA). (2022). *Hazard Communication Standard (HCS).*

Occupational Safety and Health Administration (OSHA). (2023). Occupational Safety and Health Act of 1970 (as amended). U.S. Department of Labor. https://www.osha.gov/laws-regs/oshact/completeoshact

OSHA. (2012). *Hazard Communication Standard (HCS 2012).* U.S. Department of Labor, Occupational Safety & Health Administration.

OSHA. (2023). Occupational Safety and Health Act of 1970 (as amended). U.S. Department of Labor. https://www.osha.gov/laws-regs/oshact/completeoshact

OSHA 29 CFR 1910.1200: Hazard Communication Standard.

Slovic, P. (2020). *The perception of risk.* Earthscan.

U.S. Consumer Product Safety Commission (CPSC). (2023). Fiscal Year 2023 Operating Plan.

Wogalter, M. S. (2019). *Handbook of warnings* (2nd ed.). CRC Press.

Wogalter, M. S., Conzola, V. C., & Smith-Jackson, T. L. (2020). *Warnings and risk communication.* Taylor & Francis.

Index

C

D

E

F

G

H

I

J

K

L

M

T

U

V

W